COMMERCIAL WINEMAKING

Processing and Controls

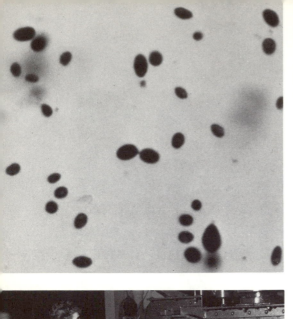

COMMERCIAL WINEMAKING

Processing and Controls

by Richard P. Vine

Cellarmaster
A. B. McKay Food and Enology Laboratory
Mississippi State University

AVI PUBLISHING COMPANY, INC.
Westport, Connecticut

© Copyright 1981 by
THE AVI PUBLISHING COMPANY, INC.
Westport, Connecticut

Library of Congress Cataloging in Publication Data

Vine, Richard P.
  Commercial wine making, processing and
controls.

  Bibliography: p.
  Includes index.
  1. Wine and wine making.   I.   Title.
TP548:V483   663;.22   81-10850
                                     AACR2

ISBN 87055-376-3

Printed in the United States of America by
Saybrook Press, Inc.

Preface

The very first winemaster may have been a cave man who discovered the magic of fermentation by tasting the result of some crushed grapes having been left inadvertently for a few days. Wine will, literally, make itself.

In simplest terms, yeast cells will collect on the outside of grape skins in the form of bloom and, when exposed to the natural sweetness inside the fruit, fermentation of the sugar into carbon dioxide gas and ethyl alcohol will commence.

During the millenia that have transpired since the cave man, the state of the art has evolved into five generally accepted categories of classification.

Table wines are usually dry (made with no appreciable amount of fermentable sugar remaining) or nearly so, and contain less than 14% alcohol by volume. They can be white, pink or red and are the result of uncomplicated processes of fermentation, clarification, stabilization, aging and bottling. The term table wine suggests the use for which these wines are intended—at the table with food. The overwhelming majority of the wine produced in the world is in this category. Table wines range from the obscure and ordinary to the most expensive classics known to man.

Sparkling wines are made from table wines called *cuvée*, that have been partially refermented in such a manner as to capture the carbon dioxide gas that is generated. The release of this gas causes the finished wine to effervesce or sparkle. Champagne is the most famous of sparkling wines and is traditionally made from white cuvée wines. Gaining rapidly in popularity is the very aromatic spumante from Italy. Today pink Champagne is produced as well as the red sparkling burgundy.

Dessert wines are normally sweet and are, as the name implies, usually consumed with dessert courses. Grape brandy (distilled wine) is added at some predetermined point during or after primary fermentation to a level totaling, usually, an alcohol content of from 18 to 20% by volume. This inhibits any further growth of yeast, leaving the desired portion of the natural sugars unfermented. Port types are primarily red, although some are "tawny" and others are white. Sherry ranges from a light straw gold to very dark amber.

Aperitif wines have been infused with different combinations of herbs, spices and essences, as well as other color and flavor components unnatural to the grape. Vermouth is the best known example and can be nearly colorless with a dry finish or dark amber with extraordinary sweetness. Aperitifs can be taken as appetizer wines or used as mixer ingredients in making cocktails.

A new category of wine has evolved during the past decade or so called pop wines—a term apparently derived from the soda-pop association given to wines of this type. These are somewhat related to aperitif wines in that flavor constituents from other fruits are added. Pop wines contain alcohol at about the same levels as described for table wines. Sangria is one example of pop wine, although many more exotic types exist.

Great wines are sometimes distinguished from ordinary ones because of grape and vineyard nobility. They can be a result of superb natural growth conditions during vintage years, but the trademark of a great wine is most often built upon consistently high quality production.

One cannot do much about the weather, and even less about nobility. No matter, as there can be no question that it is what transpires in the winery that determines the ultimate fate of any crop grown in the most prestigious vineyard during the best of years. There is no paradox: poor wines can easily be made from fine grapes, but it is impossible to make fine wines from poor grapes. The task of a reputable winemaster begins to emerge clearly.

There are hundreds of volumes already published that provide a wealth of information about wine. Some that come to mind are the works of Leon Adams, Maynard Amerine, Alexis Bespaloff, Ruth Ellen Church, Hugh Johnson, William Massee, Cyril Ray, Frank Schoonmaker, André Simon and Philip Wagner, to name only a few. These authors have made outstanding portrayals of classic and ordinary wine regions around the world, both historic and contemporary. A reader can gain a thorough knowledge of almost any aspect of enology, with abundant methodology, from the most intricate and technical chemistry to the more romantic aspects of wine cookery, service and tasting. There are even wine dictionaries and wine encyclopedias.

Why then, yet another book in concern for the subject?

During the past several decades a considerable interest has developed in America for the wines that are produced in small wineries in our country. This interest continues to intensify, especially for the truly good wines that are reasonably priced. There also seems to be a commensurate desire to learn more about just what functions take place in the small estate-type wine cellar and the controls that can be realistically exercised by the winemaster in the creation of superior products.

While wine can be a simple food to produce, it is a very vast topic. Perhaps much the same as with other art forms, it is the infinite variability of factors at the root of the subject that renders it so complex. There are thousands of different grape cultivars, a few hundred of which are grown commercially around the world. Combined with such factors as vineyard soils, climates, microclimates, cultivation techniques, harvesting methodology and overall operational philosophy, a great deal of variability surely exists. This diversity, along with more than 5000 years of enological development, generates a number of different wine possibilities that must be described exponentially. This text, however, is designed as a teaching aid which will provide a basic framework upon which the beginning winemaster may then apply an individual set of production variables.

In larger wineries one can usually find chemists, enologists, recordkeepers, viticulturists and other well trained professionals who respond and supervise within their respective fields of expertise. The comprehension and responsibility of estate-type winery production dynamics may weigh heavily, often entirely, upon the winemaster. It is, therefore, imperative that money and time inputs be carefully maximized when designing a small winery quality control program.

This book will also only be concerned with the most economical devices for generating and recording data, in a manner that fully supports the regulations and requirements of the U.S. Bureau of Alcohol, Tobacco and Firearms (ATF).

There seems to be a significant void between the highly technical information that is presently available and the practical comprehension of this material so as to apply it readily in the small winery laboratory, particularly in newly established winery premises. It is hoped that this work may serve to clarify academic technicality and simplify ATF regulations.

It is not necessary for the small winery vintner to have advanced degrees in the physical sciences, but a solid academic background is essential. The small winery winemaster should have completed significant studies of agronomy, botany, chemistry, biochemistry and horticulture in order to adequately interpret the basic profiles of analytical methodology provided and applied in this text. Agricultural economics and engineering, as well as accounting, business management and finance are also basic to small-scale winemaking.

Finally, the winemaster should be familiar with ATF regulations. *Wine, Part 240 of Title 27, Code of Federal Regulations*, hereafter referred to as Part 240, CFR, includes some approved laboratory equipment and procedures. Also contained are the essentials of wine production documentation and reporting, as well as charts, tables and other important items, all of which is designed to assist and guide the vintner in the maintenance of a legal and sound bonded winery operation. *Regulations under the Federal Alcohol Administration Act, Title 27, Code of Federal Regulations*, referred to as Title 27, CFR, is concerned with standards of wine identity, labeling requirements and certification, advertising and promotion of wine, standards of fill and other related provisions.

It is expected that, within this text, the reader can become familiar with the basic analytical functions that take place in a winery that is properly managed. Once confidence is established in the practical application of these principles, the winemaster may pursue a refinement of his or her specific quality control program. In similar fashion, this book has been outlined to introduce a broad association and involvement with ATF regulations. Acquaintance and communication with ATF personnel may result in the development of more meaningful data systems. While the purpose of this book is to bring forth an organized understanding of small winery quality control technology, it cannot be considered a total means to that end.

The author wishes to thank the following people for their time and effort in the review, criticism and suggestions for all or parts of the text: Dr. W. Lanny Bateman, Mississippi State University; Dr. Robert Bates, University of Florida; Dr. Warren C. Couvillion, Mississippi State University; Dr. James F. Gallander, Ohio State University; Mr. Sherman P. Haight, Jr., Haight Vineyards; Dr. C. P. Hegwood, Jr., Mississippi State University; Dr. R. E. Kunkee, Dept. of Viticulture and Enology, University of California, Davis California; Dr. Andrew C. Rice, Taylor Wine Co., Hammondsport, NY; Dr. W. B. Robinson, New York State Agricultural Experiment Station, Geneva, N.Y. Mr. Bertram E. Silk, Canandaigua Wine Co., Canandaigua, NY; Ms. Dody and Mr. Archie M. Smith, Jr., Meredyth Vineyards; Ms. Suzelle and Dr. Archie M. Smith, III, Meredyth Vineyards; Dr. Don F. Splittstoesser, New York State Agricultural Experiment Station Geneva, NY; Dr. Boris J. Stojanovic, Mississippi State University; Ms. Kay and Dr. James Truluck, Truluck Vineyards.

The efforts of Ms. Stella Phillips, who pleasantly endured all of the copy work necessary for compiling and preparing the manuscript is gratefully acknowledged.

Without the encouragement, help, sacrifice and understanding from my wife, Gaye, this effort would probably have never been completed.

RICHARD P. VINE

Starkville, Mississippi
January 8, 1981

Contents

Dedication

To
my family:
wife Gaye,
son Scott,
and daughters Sabrina and Stacia
who make my life loving, fun and worthwhile.

Wine and the History of Western Civilization

THE ANCIENT WORLD

There is plenty of evidence that the culture of the vine and the art of winemaking existed long before the chronicles found in Egyptian hieroglyphics. The very first chapter of the Old Testament describes how Noah landed his ark on Mount Ararat and promptly planted a vineyard in order to make wine.

The tombs of ancient Egypt have revealed a number of very detailed art treasures that clearly depict grape harvesting and wine production. One in particular, at the burial site of Phtah-Hotep, is judged to be about 6000 years old. This may mark the beginning of the wine industry in western history.

Numerous grape seeds have been found in the crypts of the Pyramids, apparently deposited by the Pharoahs so that grapes would be available in the hereafter. Queen Nefertiti is said to have used wine as a base for her perfume.

In Babylonia, the laws of King Hammurabi, written about eighteen centuries before the birth of Christ, promise harsh punishment for bad wines sold as good. This gives rise to the idea that there were probably many poor wines of the time as the natural alcohol content would not have been sufficient to serve as a reliable preservative against the action of vinegar bacteria.

Hebrew law provided for careful selection and cultivation procedures in remarkable detail. The famous Canaan grape was grown in Palestine to produce a broad selection of quality wines.

GREECE AND ROME

The great Phoenician traders are credited with taking vines and wines from the Nile delta to Greece. No doubt these items proved to be valuable tender as the works of Aeschylus, Plato and Socrates are laced with references to wine.

One of many Roman caves in the Province of Champagne now being used for the production of wine
Courtesy: Champagne News & Information Bureau

Homer accounts for some of the more notable Greek vineyards along with a few techniques used in the storage of good wines. He writes in Book IX of *The Odyssey:*

. . . But his gift most famed
Was twelve great vessels, filled with such rich wine
As was incorruptible and divine.
He kept it as his Iewell, which none knew
But he himselfe, his wife and he that drew.
It was so strong, that never any fill'd
A cup, where that was but by drops instill'd
And drunke it off; but 'twas before allaid
With twentie parts of water; yet so sward
The spirit of that litle, that the whole
A sacred odour breath'd about the boll.

A few museums are fortunate enough to possess the priceless remaining Greek *oinochoe*, which are small ceramic jugs that were used to carry and store wine during the sixth century B.C. The prefix, *oinos*, translates as wine from the classic Greek, but in earlier times was spelled *woinos*. In either case, the roots for the modern words wine and enology are obvious.

Two-handled versions of the *oinochoe*, called *amphorae*, were also used for wine, as well as for honey and oil. These containers were larger, of a more narrow and tall design and with a cover. Construction of the *amphora* was such as to allow it to be buried. This possibly indicates a technology which developed against the ill effects of high temperatures and air exposure to wine—matters of serious concern in small wineries even today. While burying the *amphorae* seems to indicate that wine cellars did not yet exist, there is some evidence in the Agora, just to the west of the majestic Acropolis in Athens, that wine was made, and certainly stored, in special rooms. In any case, if the measure of craftmanship of the Greek pottery, or Egyptian and Roman for that matter, is associated with the abilities of the ancient winemasters, then surely there were some wonderful wines at the foundations of our heritage.

On the beautiful island of Naxos the once great Temple of Dionysus endures today only as scant ruins. In the earliest accounts, this god of wine was held in the most divine reverence. The Dionysiac Festival, on the other hand, is among the most pagan selections of Greek mythology and illustrates the corruption leading to the fall of the Golden Age of Greece.

Wine was known in Rome long before Caesar could count the conquered Greeks as part of the great empire. Considerable amounts of wine were imported from Greece and the Greek colonists made notable wines, especially at Campania, several centuries prior to the fall of Carthage.

From *The Knights* of Aristophanes (circa 200 B.C.):

And dare you rail at wine's inventiveness?
I tell you nothing has such go as wine.

GREEK SKYPHOS DRINKING
CUP OF THE 5TH CENTURY,
B.C.

SYRIAN GLASS AM-
PHORA, 2ND—4TH
CENTURIES A.D.

GREEK KYATHOS OF
THE 5TH CENTURY B.C.
USED FOR LADLING
WINE FROM KRATERS

GREEK OINOCHOE OF
THE 5TH CENTURY B.C.
THE GOD OF WINE, DI-
ONYSUS, IS ILLUSTRAT-
ED BETWEEN THE TWO
SATYRS

GREEK KRATER OF THE 4TH
CENTURY B.C. USED FOR THE
MIXING OF WINE WITH WATER

CORINTHIAN AMPHORA
OF THE 7TH CENTURY
B.C.

*Courtesy: Lowie Museum of Anthropology, University
of California, Berkeley*

Courtesy of the German Wine Information Bureau

THE ROMAN WINE SHIP AT NEUMAGEN ON THE MOSEL

Why, look you now; 'tis when men drink they thrive,
Grow wealthy, speed their business, win their suits,
Make themselves happy, benefit their friends.
Go, fetch me out a stoup of wine, and let me
Moisten my wits, and utter something bright.

Accounting for the state of the art some 70 years before the birth of Christ,
Pliny offers some rather exacting data about Roman winegrowing. He declared that Rome had become the wine capital of the world in 154 B.C. and
wrote descriptions of no less than 91 grape varieties. It was also Pliny who
observed that *"in vino veritas."*

The poems of Virgil are steeped with observations, both academic and
romantic, about the growing technology of grape culture. It stands to reason

that enology had markedly advanced also, as some wines, principally Mamertine and Falernian, were often aged in excess of 50 years.

The pagan rites of Rome were much like those of the Greek Dionysiac Festival, but perhaps somewhat less violent. The spoils of war included the spirit of Dionysus, renamed Bacchus, and the drunken rages continued as the Roman Bacchanalia. Nero and Tiberius rewarded superior performances with favors of office and position.

Roman influence spread to Cadiz in Spain (very near the modern Spanish sherry region of Jerez) where, in the first century, Columella provided a remarkable thesis of viticulture—including procedures of propagation, both by cuttings and grafting, planting, cultivation, pruning, trellising, fertilization—and winemaking.

The armies of Rome marched up the Rhone valley in what is now France, westward to Bordeaux, north to Burgundy and beyond to Champagne, and easterly to the Rhine river in Gaul. Some seventeen centuries later, these places all remain important viticultural and vinicultural regions. In fact, many of the old Roman munitions storage vaults, carved out of the moist chalk subsoil in Champagne, are still used as storage and processing cellars by the French winemasters.

While the Roman Empire began to crumble, winegrowing carried on. At the Last Supper, Christ took wine as the symbol of his blood—a symbol still used at Communion, Eucharist and Mass. Thus, Christianity began with wine. Wine is mentioned in The Bible more than one hundred times. Winemaking had become a relatively sophisticated science as we read in Mark 2:22.

And no man putteth new wine into old bottles; else
the new wine doth burst the bottles, and wine is
spilled, and the bottles will be marred; but new wine
must be put into new bottles.

This is a rather perplexing statement. Certainly, the new wines were bottled as soon as possible in order to avoid vinegar spoilage and often before fermentation was totally completed. Perhaps new bottle resilience was such as to withstand a low pressure, suggesting that old bottles were too brittle and delicate for the gas yet to be released from the new wine. Good cases can be made for several other hypotheses.

THE POST-ROMAN WORLD

The Church grew to own practically the entire wine industry in post-Roman Europe. Abbeys became, in part, small wineries, the products of which funded their existence. Travelers, both noble and otherwise, appreciated good wine and would bequeath the abbots items of worldly value in exchange.

Wine had long since replaced polluted water and spoiled milk as the mealtime beverage. It was a common medicine, offering relief from the pains of disease and despair.

During the Dark Ages it was the self-sufficient Monks who were necessarily forced to develop strong economies for the scarce resources needed to continue vineyard and cellar operations. It may have been during this era, when the instabilities of social and political structures rendered private wine ventures as pursuits of exceptional risk, that the importance of winegrowing in our heritage was most in jeopardy.

The great Persian poet and astronomer of the eleventh century, Omar Khayyam, immortalized himself in these words from *The Rubaiyat*:

I often wonder what the Vintners buy
One half so precious as the stuff they sell.

And also:

And lately by the tavern door agape,
Came stealing through the dust an Angel shape
Bearing a vessel on his shoulder; and
He bid me taste of it: and 'twas the Grape!
The Grape that can, with logic absolute,
The two-and-seventy sects confute;
The subtle Alchemist that in a trice
Life's leaden metal into gold transmute.

Ancient relics of winegrowing are proudly displayed at the wine museum in Beaune, the capital city of Burgundy in France. Just to the north of Beaune is the famous Clos de Vougeot, established during the mid-1300s by the monks of Citeaux. While much of the winery remains intact as a prime example of monastic facility, the Clos is better known in modern times as the world headquarters of the *Confrerie des Chevaliers du Tastevin*, an international association of wine connoisseurs.

In Germany, a few miles west of Wiesbaden near the Rhine river, one can visit the magnificent Kloster Eberbach, at least the equal of the Clos de Vougeot in every respect, and about one century older. These two structures are perhaps the first such buildings designed and built for the express purpose of making wine. Both at Eberbach and Vougeot the use of different rooms by the winemaster for different winemaking purposes are readily apparent.

To the west of the Rhine, the famous Bernkastler Doktor cellars, on the Mosel river in Germany, have survived several centuries. Affectionately called the "Doktor" vineyard, it is believed that the legendary Dr. H. Thanish, the original owner, once prescribed ample wine for his patients, forming a dependable scheme of profit.

We owe to Geoffrey Chaucer much of what is known today about the wines most popular during the latter part of the fourteenth century. Beverage wines were known by color. Reds were very dark red—usually from Languedoc. Whites came from Anjou as a rule and claret, a nomer that has withstood more than six centuries, described the lighter red wines of Bordeaux.

Dessert wines were richer, higher in alcohol and more expensive than beverage wines, much the same as today. They were sipped and occupied the center of fellowship. Most notable dessert types were Malmsey from Cyprus, plus Grenache, Ribolle and Romany from Portugal and Spain, as well as Vernage from Italy. The chief medicinal wine was the aptly named Hippocras—usually, in one form or another, a base blend of honey and wine.

This was a very productive period and, by the end of the Middle Ages, German monasteries had developed woodworking to the state that it was widely used for wine casks. Some were of enormous size, such as the great Heidelburg Tun, built during the mid-seventeenth century, which could hold about 50,000 gallons of wine.

THE RENAISSANCE AND AFTER

The Renaissance, with a new appreciation for the arts and crafts, along with the monasteries, which were without real competition, set the stage for comparisons of wines from different regions. Having had the time to amass volumes of records, the monks were very instrumental in devising regional boundaries for the production of the best wine types known to be grown in specific areas. Classifications for wine quality were also developed, the framework for which is still in use in France.

Shakespeare, Rousseau, Voltaire, Bacon and Byron, as well as many other men classic to letters and science, all wrote appreciatively of wine. Although he loved wines of almost any type or variety, Shakespeare was most partial to Sack—a type of pale dry sherry from Spain that is still produced there. His quotes referring to wine are numerous. For example, from Act III of *Macbeth*:

. . . Come, love and health to all:
Then I'll sit down. - Give me some wine, fill full.
I drink to the general joy o' the whole table,
And to our dear friend Banquo, whom we miss;
Would he were here! To all and him we thirst,
And all to all.

And, of course, the very famous:

Give me a bowl of wine . . .
In this I bury all unkindness.

When Louis XIV came to power in the early 1640s, champagne, as we know it, had not yet been invented. Wine was made in the Champagne province, however, and had been since Roman times. Champagne was much the same as the lighter red burgundies to the south. The Sun King was one of the most powerful French rulers and his taxes were a heavy burden not only to vintners, but to all. He enjoyed wines from the Pinot Noir grape, and the wines of Burgundy and Champagne made from that culti-

THE *LIEBFRAUENKIRCHE* OF WORMS, WEST GERMANY—BIRTHPLACE OF THE FA-
MOUS *LIEBFRAUMILCH* WINES

var were in his favor—a situation that led to some fierce competition
between the two regions.

There was stiff competition among exporters for the international wine
trade, too. The English loved wine as much as the continental Europeans,
but could not grow it successfully. Consequently, the British market was a
prime target for both French and Spanish wines during the mid-1600s. At
the turn of the century, port overtook the Spanish sherry shipments to
England, and held that position for about 25 years until, once again, the
Portuguese wines fell behind the Spanish products.

Comparatively few German wines reached London, and all that did were
known as Rhenish, from the Rhineland regions where they were grown and
cellared. Prior to the year 1700, the Rhenish wine district had evolved into
two subdistricts known as Bacharach and Hochheim, the latter of which

gave birth to the generic nickname Hock, a term that still serves as a popular reference to German wines in general.

By the early part of the eighteenth century, America had been settled and colonized. However, most of the settlers in America were of British and Nordic descent, who knew little about the culture of the vine. Like England, the Colonies became a prime export market in which the wines of Madeira, among others, were popular.

Napoleon is not known to have preferred any particular wine type, except perhaps Chambertin from Burgundy. Nevertheless, wine must have been important in his lifestyle as, legend has it, he carried wine for his soldiers to every battle except one—Waterloo. (Supposedly, there was a logistical problem that delayed the wine wagons.) More important was the stability that Napoleon's wars brought to the French franc. Many of the wealthy saw fit to place large investments into Bordeaux estate wineries, many of which are still in operation.

But the magic era of winegrowing was about to end. Louis Pasteur discovered the mystique of fermentation to be a very small, egg-shaped, single-celled plant, the yeast cell at work. About 50 years earlier, Gay-Lussac, in 1810, formulated the now famous equation by which one molecule of sugar will ferment into two molecules each of ethyl alcohol and carbon dioxide.

PHYLLOXERA

Despite numerous trials by expert viticulturists, the European vines would not grow in the Colonies. This disappointing and perplexing turn of events will be discussed in more detail later in this chapter. However, in the early 1870s, some American vines were introduced in France for botanical studies. Little did anyone know that a root louse, Phylloxera, to which American stock was impervious, would be carried along. Nearly all of the vineyards of France and many in other European countries became infested with the invader. Grafting techniques had been studied for more than nineteen centuries in Europe and the Phylloxera plague provided a monumental chance for this technology to be useful. In dubious fashion America became an important factor in the history of European winegrowing as hundreds of thousands of American vinestocks were exported to Europe for rootstock propagation and grafting. Never before, or since, has a country utilized such massive resources in order to preserve an agricultural industry. The huge project proved successful and the great European vineyards live on in the twentieth century—on American roots.

If there was any saving grace from the Phylloxera disaster, it was that the chaos of the blight caused many European winegrowers to migrate, some to the United States, bringing an influx of diverse philosophies in mastery of the wine craft. This was the seed of American wine technology that has grown to rival Europe. Furthermore, many of the graft scions were from selected superior varietal clones, assigned to specific regions called appellations, reducing many of the vague boundaries and "succotash" vineyards in Europe that existed prior to the Phylloxera.

Another group of men, principally Messrs. Baco, Seibel, Seyve, Kuhl-mann and Villard, took a different tack in response to the Phylloxera blight. They took the American cultivars, especially those of hardy and bountiful character, and bred them to the quality-proven varieties of France. The results were not very encouraging, particularly in the beginning. It was often as long as 20 years before a new cultivar could be properly evaluated. It was found that the positive properties sought from each parent were usually recessive and the negative characteristics dominant. Seibel, whose work spans thousands of crosses, was by far the most successful, measuring about a dozen cultivars now grown in any relatively significant acreage. However, there continue to be purists who look upon even the best selections of the French-American hybrids as illegitimate. Cases are made for and against contemporary wines versus those of pre-Phylloxera, American wines compared to European, as well as the superiority of *Vitis vinifera* vines versus the hybrids.

WINEMAKING IN THE NEW WORLD

As with most other phases of New World history, that portion devoted to winegrowing in America is chronologically short compared to the longevity of the art throughout the whole of western civilization—some 400 years or so as compared to a total span of some 6000 vintages.

Nevertheless, during these four centuries, the United States has become an important viticultural nation, ranking fourteenth in vineyard acreage in 1971. It is even more competitive as a wine-producer and, more importantly, American wines have become recognized as some of the world's best quality. Even before the twentieth century American wines were winning medals of excellence in European tastings. At a recent comparison held in Paris, several Californian red wines made from the great Cabernet Sauvignon grape placed high among the most noble growths of Bordeaux, and one vintner, Stag's Leap, was awarded first place!

No other nation grows commercially as many cultivars of grapes as does America, primarily because of the wide diversities of climates and soil. There is also much more freedom to grow different selections of vines than there is in most important winegrowing countries in Europe. American viticultural and enological research ranks first, far and away, with major institutions conducting programs of varied intensity in many states.

But such positivity has not always existed. The development of wine-growing in America has endured severe biological, political and sociological resistance.

The Beginnings

The story can begin with a legend concerning Leif Erikson found in *The Discovery of America in the Tenth Century*, written several centuries ago by Charles C. Prasta:

Leif, son of Eric the Red, bought Byarnes' vessel, and manned it with thirty-five men, among whom was also a German, Tyrker by name ... And they left port at Iceland, in the year of our Lord 1000.

But, when they had been at sea several days, a tremendous storm arose, whose wild fury made the waves swell mountain high, and threatened to destroy the frail vessel. And the storm continued for several days, and increased in fury, so that the stoutest heart quaked with fear; they believed that their hour had come .... Only Leif, who had lately been converted to Christ our Lord, stood calmly at the helm and did not fear; ..... And, behold! while he spoke to them of the wonderful deeds of the Lord, the clouds cleared away, the storm lulled; and after a few hours the sea calmed down, and rocked the tired and exhausted men into a deep and calm sleep. And when they awoke, the next morning, they could hardly trust their eyes. A beautiful country lay before them ..... and they cast anchor, and thanked the Lord, who had delivered them from death.

A delightful country it seemed, full of game, and birds of beautiful plumage; and when they went ashore, they could not resist the temptation to explore it. When they returned, after several hours, Tyrker alone was missing. After waiting some time for his return, Leif, with twelve of his men, went in search of him. But they had not gone far, when they met him, laden down with grapes. Upon their enquiry, where he had stayed so long, he answered: "I did not go far, when I found the trees all covered with grapes; and as I was born in the country, whose hills are covered with vineyards, it seemed so much like home to me, that I stayed a while and gathered them." ... And Leif gave a name to the country, and called it Vinland, or Wineland.

This is still a fascinating tale, and wholly believable, as America continues to display a prolific exhibit of grapevines in the wild—more native species than can be found in any other nation.

From the fruit of wild vines the first American wine was made. During the mid-1500s the French Huguenots settled in Florida and made wines from the native grapes of *Vitis rotundifolia*, more commonly known today as the Muscadine varieties. *Vitis rotundifolia* is a much different grape than the *V. vinifera* grown in Europe. The Muscadines grow much like cherries with individual berries while the Old World grape grows in bunches or clusters. Also, the flavor values of the native Muscadines are very pronounced as compared to more subtle essences found in most cultivars of *V. vinifera*. No doubt this was a source of great discontent among the French settlers, as wine had long been a diet staple in their native culture.

Attempts to Grow European Cultivars

Captain John Smith was more explicit as he wrote in the early seventeenth century:

Of vines great abundance in many parts that climbe the toppes of highest trees in some places, but these beare but few grapes.

A ROMAN WINE-PRESS. BEAUNE WINE MUSEUM, JUNE, 1970

A BURGUNDIAN WINE PRESS OF THE MIDDLE AGES. BEAUNE WINE MUSEUM, JUNE, 1970

There is another sort of grape neere as great as a Cherry, they [probably Indian natives] call *Messamins*, they be fatte, and juyce thicke. Neither doth the taste so well please when they are made in wine.

This may have prompted the plantings of European *V. vinifera* cultivars in many of the northern Colonies during the 1600s. The first attempt for such an undertaking may have been by the London Company of Virginia circa 1620. Lord Delaware brought some winegrowers, along with some of their vines, from France to establish the Virginia vineyards. The project failed, however, and the French viticulturists were blamed for poor treatment of the vines. We can assume today that the failure was most likely due to the Phylloxera root louse and not the French *vignerons*.

In a London text describing the Colonies, Beauchamp Plantagenet described four grape varieties:

Thoulouse Muscat, Sweet Scented, Great Fox and *Thick Grape*; the first two, after five months, being boiled and salted and well fined, make a strong red Xeres [Sherry]; the third, a light claret; the fourth, a white grape which creeps on the land, makes a pure, gold colored wine.

These grapes, as depicted, are surely from *V. rotundifolia*, or Muscadines. This may be one of the sources for the term, "foxy," typified by the huge values of aroma and flavor indigenous to the Muscadines. "Foxy" is often used as a rather loose organoleptic term to evaluate the flowery, fruity character that is dominant in the *V. labrusca* cultivar; Concord and Niagara serve as good examples.

A descendant of Pocahontas, Colonel Robert Bolling of Virginia, wrote the first technical literature of grape growing in America during the late 1600s. Entitled *A Sketch of Vine Culture*, in it Bolling commented upon unsuccessful European methods used for *V. vinifera* propagation in the Colonies. The manuscript was copied and distributed but never published.

The seventeenth century continued with unyielding hope and promise that the European vines would find a satisfactory place upon American soil to grow. Laws were passed in Virginia requiring every household to plant ten vines, and in New York for wine to be made and sold without tax. Massachusetts Governor John Winthrop planted wine grapes on Governor's Island, and Lord Baltimore established vineyards in Maryland. William Penn tried to grow the European vine in his vineyard plot near Philadelphia. Portuguese vines were exported to Georgia and King Charles II ordered vines to be planted in Rhode Island. The accounts go on and on, all to no avail.

The lack of significant progress in American viticulture during the early and mid-1700s may reveal a widespread discouragement to continue planting the European *V. vinifera* cultivars, or, for that matter, discouragement for winegrowing at all.

Use of Native Cultivars

A Swiss colony located in Jessamine County, Kentucky, also abandoned their *V. vinifera* efforts, but instead of surrendering altogether, they moved across the Ohio River to the Indiana side. The new settlement was called Vevay, named for the important wine village back in their native Switzerland. At Vevay, the native American vine was commercially cultivated and southeastern Indiana grew to become one of the largest viticultural areas in America prior to the mid-nineteenth century. The effort was led by one John Dufour who attracted the interest of such men as Thomas Jefferson and Henry Clay.

John Alexander, a Pennsylvanian, is given credit for the first cultivation of native American grapes, although on an insignificant scale when compared to the Swiss effort in Indiana. Alexander's efforts in the latter part of the eighteenth century brought forth a selection, unabashedly named Alexander, that became famous in the potential to resist disease and bear good crops of fruit. The wine from the Alexander grape was received with somewhat less enthusiasm.

Thomas Jefferson, who was a grape and wine enthusiast, stated in 1809, as a result of some personal native vine experiments:

. . . it will be well to push the culture of this grape (Alexander) without losing time and effort in the search of foreign vines which it will take centuries to adapt to our soil and climate.

Later John Adlum, from Georgetown, D.C., introduced the variety Catawba, named for the Catawba River in Buncombe County, North Carolina where the cultivar supposedly originated. It is a natural hybrid cross of *V. labrusca* and *V. vinifera* parentage. The Catawba gained much attention as Adlum less than humbly exclaimed to the world that he had performed a greater service to America by introducing this new cultivar than if he had paid the national debt.

No doubt, Adlum, Alexander, and Jefferson did not know that to the west, in what is now New Mexico, the *V. vinifera* had been grown by the Spanish explorers along the Rio Grande River more than a century before. Apparently the hot desert sand provided a natural barricade for the Phylloxera. The cultivar grown was probably Mission which had been grown in Mexico under the direction of Cortez since the early sixteenth century. In 1769, Father Junipero Serra founded the San Diego Franciscan Mission whereupon he promptly planted the first grapes in sunny California. As the Spanish had succeeded in New Mexico, so did Serra, and his vines flourished. By the 1830s there were more than twenty Franciscan missions growing *V. vinifera* grapes along the El Camino Real.

Unlike Adlum, Serra made no pretext that he had done America any great service and the California missions, along with a few commercial ventures, prospered with the culture of *V. vinifera* in rather obscure fashion until the great Gold Rush of 1849.

SITE OF THE FIRST TAYLOR VINEYARDS ON BULLY HILL NEAR HAMMONDSPORT, NY IN THE FINGER LAKES DISTRICT, JUNE, 1977

THE VILLAGE OF MERSAULT, CLASSIC WHITE WINE PRODUCER IN BURGUNDY, FRANCE. JUNE, 1970

JUNIPERO SERRA

Back east, winemaster Nicholas Longworth had built a very imposing winery near Cincinnati, about fifty miles northward along the Ohio River from Vevay, Indiana. The Catawba grape greatly interested Longworth and he planted a large acreage of it in the early 1820s near his winery. The Longworth site stands today historically as the very first wine cellar in America to have produced "champagne", sparkling white wine from Adlum's Catawba. Longworth correctly named his product Sparkling Catawba.

Longworth was proud to the extent that he declared his sparkling wine superior to that of champagne from France. His wine and his winery were intricate and these may have been what tempted Henry Wadsworth Longfellow, a wine lover, to visit the Ohio winery and taste Longworth's Sparkling Catawba. Convinced that, indeed, Longworth was right, Longfellow penned a tribute in 1854 with a poem entitled, "Ode to Catawba Wine," in which he states:

But Catawba wine
Has a taste more divine,
More dulcet, delicious, and dreamy.

A truly fine testimonial, as Longfellow was comparing the Catawba wine to champagne from Verzenay in France—a very highly touted adversary.

The Beginnings of the New York State Wine Industry

A Baptist deacon, Elija Fay, is given credit for planting the first vineyard in western New York State in 1818. Vineyards and wineries were established in the Hudson Valley—the Brotherhood cellars in 1839. However, more notable are the vineyards of the Finger Lakes, which are more centralized in New York. Old local legend has it that, "when God finished making the earth, He laid His hand upon that particular chosen soil to bless His creation"—hence, the imprint of the New York State Finger Lakes topography. Science, however, asserts that the Finger Lakes region was formed by the clawing effects of the great Ice Age.

In either case, St. James Episcopal Church still stands as the landmark in Hammondsport where Reverend William Bostwick, in 1829, first propagated both Catawba and Isabella cuttings onto the roots that founded the New York State wine industry.

By 1860, the success of the Bostwick work was well known in the region and Charles Champlin formed, along with the efforts of several others, the Pleasant Valley Wine Company, a magnificent stone cellar with walls more than six feet thick.

Champlin's idea was to improve upon the wines of Longworth in Ohio. The *Methode Champenoise* (the French method of making champagne) was put to work once more in America. Champlin also hired a French winemaster, Jules Masson, in order to leave no detail overlooked in his quest for the very finest champagne. In 1871, some of the results were taken to the Parker House Hotel in Boston for scrutiny. Among the prestigious guests was Marshall Wilder, president of the American Horticultural Society, who proclaimed that the Pleasant Valley wine was a "great" sparkling wine from the new "western" world and the name "Great Western" was born, and still lives, as the trademark of the Pleasant Valley Wine Company in Hammondsport.

Acceptance and Expansion

The early 1860s marked the beginning of a very promising era for American winegrowing. Significant vineyard and winery investments had been actively pursued and made in no less than 20 states and territories. Mary Todd Lincoln had been the first to serve American wines in the White House and French winemasters admitted that California could grow the Old World *V. vinifera* grapes such as to be capable of entering serious competition with the wines of Europe.

The notoriety and success of Longworth and Champlin, among others, brought forth renewed commercial winegrowing interests which was followed by considerable vineyard expansion and vigorous competition for grapes and wine in the consumer marketplace.

A few miles north of Hammondsport, on the west side of Lake Keuka, the Gold Seal winery was founded by a group of farmers headed by Clark Bell. To the west another few miles, John Widmer from Switzerland founded the Widmer Winery at the southern tip of Canandaigua Lake. Once again, the

WALTER TAYLOR, FOUNDER OF
THE TAYLOR WINE COMPANY

Courtesy: The Greyton H. Taylor Wine
    Museum, Hammondsport, N.Y.

Church became involved, with Bishop Bernard McQuaid of Rochester, New York having founded the O-Neh-Da Vineyard on Hemlock Lake. These were principal examples of the entrepreneurial ventures of the times in the Finger Lakes vicinity—all of which brought forth the need for many supporting goods and services, such as cooperage manufacture.

The market potential for casks and vats attracted cooper Walter Taylor to move into the Finger Lakes area on Bully Hill, just to the northwest of Hammondsport, in the midst of good stands of oak timber and at the heart of the growing New York State wine industry. Taylor also made wine, but his craft was the creation of fine cooperage, much of which was in use upon credit extension to the many new wineries in the region.

But the growth of the wine industry was faster than the market could bear and Taylor was eventually forced to accept bulk wine from the other wineries as some compensation for his labors and expense. Along with his three sons, Fred, Clarence and Greyton, Taylor promptly found himself in the wine business in a rather extensive fashion—certainly far more than he had originally intended. The rest is history, as the Taylor family sacrificed and struggled for the next half-century and more in building the largest winery in the country outside of California.

From New York and Ohio, the Isabella and Catawba cultivars were taken to Herman, Missouri during the mid-1800s—which adapted to yield huge bounties of fruit. This, much the same as in New York, attracted widespread plantings, often in the hands of inexperienced growers who did not recognize or could not provide any treatment for the spread of black rot and mildew. These diseases sieged throughout the vineyards of Missouri and Indiana causing acute discouragement once again, even among the most dedicated of believers. Many of the Vevay winegrowers moved northward to upper Ohio where they joined immigrants, principally of German extraction, in forming the viticultural region that remains today in the Sandusky and Lake Erie Islands area.

The few grape enthusiasts who remained in Missouri became interested in growing a new variety that had been introduced in Cincinnati by a Dr. Kerr who had brought it from Virginia, named simply Norton's Virginia.

Nicholas Longworth pronounced the grape worthless, but a few regarded the red wine from Norton's Virginia as promising. Among these was one George Husmann, a University of Missouri Professor of Horticulture.

Ephraim Bull from Concord, Massachusetts had succeeded in hybridizing a new disease-resistant cultivar from native American *V. labrusca* blood lines. The Concord was also of timely interest among the few remaining Missouri viticulturists. Both the Norton's Virginia and the Concord flourished in Missouri, once again bringing large grape crops to Missouri and rekindling interest in grape culture in mid-America. Vineyards expanded rapidly, to such proportions as to render that state the largest producer of grapes in the Union.

Having once been the center of disease, Missouri, in the span of only three decades, ironically became the major source of disease-free vine stocks being sent to Europe to save the Old World vineyards from the Phylloxera blight.

The 1860s were the finest years of Missouri grapes and wine, although again, the wine industry grew faster than market demand. In 1870 California took over the top spot in vineyard acreage and has never since relinquished that position. Even Professor Husmann left his Missouri homestead, moving to California where he stated that, ". . . this [California] was the true home of the grape . . .".

Arkansas winegrowing commenced during the 1880s with the efforts of the Post and Wiederkehr families who settled in the hamlet of Altus in the northwestern corner of the state. To this day both families remain in the same locale.

As the Alexander, Catawba, Isabella, Norton's Virginia and Concord had become champion grape varieties in the north, the Muscadine cultivar, Scuppernong, had become the pride of the south. The cultivar had been known simply as White Grape for centuries until James Blount of the town of Scuppernong, North Carolina, named the variety in the early 1800s. Sidney Weller established the first commercial winery in North Carolina in 1835, which was later bought by the brothers Garrett who made many millions of dollars in the grape and wine business. Garrett's empire once spanned six states with seventeen facilities producing Virginia Dare grape juice and wine.

The Montevino vineyard, owned by Dr. Joseph Togno of Abbeville, and the Benson and Merrier Winery of Aiken were active examples of wine-growing during the 1860s in South Carolina.

Since the mid-1700s, Georgia had become one of the top ten wine producing states, and by the 1880s accounted for nearly a million gallons per year.

J. M. Taylor grew grapes and made wine in Rienzi, Mississippi, during the 1870s. The Magnolia state boasted of no fewer than 31 wineries by the turn of the century.

A British Consul report made by E. M. Erskine to his government in 1859 stated:

The banks of the River Ohio are studded with vineyards, between 1,500 and 2,000 acres being planted in the immediate vicinity of Cincinnati, with every prospect of a vast increase.

. . . in Kentucky, Indiana, Tennessee, Arkansas, and generally, in at least 22 out of the 32 states now constituting the Union, vineyards of more or less promise and extent have been planted. . .

Another report, resulting from a special U. S. Department of Agriculture study in 1880, revealed that Alabama and Texas combined for nearly 2000 acres of vineyards while Kansas and Missouri accounted for about 11,000 acres. New York and Ohio had swelled to more than 22,500 acres and Pennsylvania had operating wineries in 57 of its 67 counties.

Much of the Texas activity was generated by the enthusiasm of T. V. Munson, from his Denison, Texas, vineyards. Munson was an avid breeder of grape hybrids and his name is still revered by wine historians in America. Few of his cultivars remain in vineyards of any size but the Munson name has been immortalized by the grape genus *V. munsoniana* given to the blood line of the native southern grapes that he worked with much of his life.

Winemaking in California

The most significant growth of American vineyards and winemaking during the nineteenth century was in California, which became the largest grape producer in the country when it overtook Missouri in 1870. Having more than 15,000 acres of grapes prior to 1849, the Gold Rush resulted in even further stimulation of new California vineyards.

Much of the credit for the founding of the now gigantic industry in California is given to Count Agostin Haraszthy, a Hungarian exile, who first immigrated to Wisconsin in the early 1800s. Haraszthy built a beautiful stone winery near what is now the village of Prairie Du Sac and planted vines. But the vines died, presumably either from the rigors of the Wisconsin climate or Phylloxera, or both. In any event, he migrated to the Sonoma Valley north of San Francisco about a year before the Gold Rush. The stone

GEORGE YOUNT

Courtesy: The Bancroft Library,
University of California, Berkeley

cellars in Wisconsin remain to this day and are now operated by the Wollersheim family who grow French-American vines that are much more adaptable to their confines.

By the mid-1860s, the Haraszthy influence had developed the Buena Vista Vinicultural Society and he had brought from Europe some 100,000 cuttings of approximately 300 prized *V. vinifera* cultivars. The impetuous nature of Haraszthy may have led to the legendary mislabeling of his cuttings. While it took years to untangle the puzzle of proper varietal labeling, it was nevertheless, Haraszthy who first brought the precious *V. vinifera* stocks to California.

The Buena Vista estate and winery were extravagant and soon failed, and without renewed financial support, Haraszthy left America. The vines remained, however, and became the foundation for quality winegrowing in California.

Nearby Sonoma is now the famed Napa Valley, probably the single most honored wine region in all of the North American continent. George Yount, a transplanted North Carolinian, first planted Mission grapes in the valley during the 1840s. The village of Yountville in the valley marks his honor.

By the end of the 1850s choice European cultivars were numerous in the little valley, some as the result of Haraszthy from Sonoma, and others from Samuel Brannan who had collected vinestocks on a personal tour through Europe.

Robert Louis Stevenson wrote lovingly of the Napa Valley in the late nineteenth century praising the wines there as "bottled poetry" and loosely comparing the local products with the Burgundian Clos de Vougeot and the

PAUL MASSON

Courtesy: Paul Masson Vineyard

equally classic Chateau Lafite of Bordeaux. But the Phylloxera root louse invaded Sonoma and Napa in much the same manner as it had in Europe. From more than 15,000 acres, the blight brought such destruction as to leave only about 3000 acres of vineyards under cultivation by the end of the century.

Born to Burgundian ancestry, Paul Masson came to California in the late 1870s and took employment with Charles Lefranc. Lefranc, along with his father-in-law, Etienne Thee, had maintained vineyards a few miles from Los Gatos since the early 1850s.

Masson was proficient in the making of sparkling wines and, by the end of the 1880s, the Masson "Champagne" was a popular local item. After Lefranc's death, Masson married his daughter, bought out the interest of his brother-in-law and formed the Paul Masson Champagne Company. By the turn of the century, Paul Masson wines were winning medals for excellence, even in Europe. In operation since the 1852 plantings of Thee, the Paul Masson brand remains as the oldest continuous producer of wines in California.

Prior to the development of affordable irrigation techniques during the late nineteenth century, the huge San Joaquin Valley was, for all intents and purposes, a desert, with rainfall less than 15 inches in an average year and usually no rain at all during the hot summer months. Among the first vintners to brave the conditions there was George Krause, who carved a wine cellar called Red Mountain into a hillside some 20 miles east of Modesto.

Today the San Joaquin Valley of California accounts for about two-thirds of the total winegrowing in all of America. Much of this vast expanse of vineyards and cellars has developed since Repeal in 1933. Most noteworthy are Ernest and Julio Gallo who, since renting an old warehouse in Modesto shortly after Repeal, have developed their business to occupy about one-third of the shelf-space available in American wine retail outlets. Gallo, the largest single winery in the world, is so large that it even makes its own wine bottles.

The year 1880 marks two important milestones in the history of American winegrowing. First, the legislature of the state of California enacted support for viticulture and enology research at the University of California at Davis, just west of Sacramento. The Davis facility reveals the importance that grapes and wine had gained in the agricultural and economic profiles of California. In modern times, the Davis institution ranks among the world's finest schools for students of the winemastering skills.

PROHIBITION

The other 1880 event took place in Kansas, where it was voted to adopt "prohibition" of the sale of beverages that contained alcohol. Behind this was one Carrie Nation, who took it upon herself to be judge, jury and executioner of anything that had to do with the "devil's brew". The Kansas incident was the first outbreak in America of the "dry movement"—a threat more terrible to the wine industry than any vineyard disease or winery bacterial infection.

It had been a long time building, some eighty years or so. America was changing rapidly and life styles were forcibly altered by monumental advances in agricultural and industrial technology. The population explosion compounded urbanization. The Victorian era, Protestant sectarianism, the aftermath of the Civil War, by-products of the Industrial Revolution and other conditions all contributed to the idea that alcohol of any manner or type was responsible for setting the country astray.

Most of Canada adopted such prohibition in 1916 and this, no doubt, set the stage for the eventual passage of the Volstead Act in America in 1919, national Prohibition.

Many of the more farsighted vintners, who had foreseen Prohibition, had geared up production and marketing machinery for sacramental, medicinal and salted cooking wines which had remained legal for production. Other winemasters went into the grape juice business. Many businesses, under Prohibition, were forced to close down altogether. Growers sold what grapes they could in the fresh-fruit market for table consumption and home-wine-making (the head of a household could make up to 200 gal. of wine per year for family consumption). Other growers converted their vineyards from wine grapes to juice and table grapes, or ripped them out completely in favor of other crops. But, for all of the ingenious ideas for makeshift operations, few vineyards and wineries could survive.

The beautiful Stone Hill Wine Cellars at Hermann, Missouri became caves for mushroom propagation. Garrett's California Virginia Dare winery became a motion picture set and The Glen winery in Hammondsport, New York became a storehouse for Glenn Curtiss' World War I airplane parts.

Some wineries desperately tried to make a go of it in secret, as bootleggers but, sooner or later, pictures of "T-men" standing amidst the rubble of complete destruction would dominate the front pages of local newspapers as an example to anyone else who dared ferment grape juice.

Alva Dart of Hammondsport remodeled an automobile so as to be fueled from small containers of gasoline under the seats of the vehicle. This left the fuel tank available for the transportation of brandy distilled by him and his father from area wines. Quite inconspicuously, Dart's car would pull up to one or another gasoline filling stations, but instead of filling the gasoline tank, the station pumps were reversed to empty the payload of this remarkable automobile into the hidden underground "gasoline" storage tanks. Later the brandy would be rectified and distributed by means of other clever schemes.

Many attempts were made for reversal of the Volstead Act, as well as some proposals for modification. It may have been Thomas Jefferson's old, but sound advice ringing over and over again that brought the proponents of the dry movement to its senses: "No nation is drunken where wine is cheap." Bootlegging was making wine expensive, and even worse, Prohibition was placing the grape and wine business into the hands of the underworld.

REPEAL

Surely the forces behind Repeal were not from the wine-growing industry, or for that matter, from the beer and whiskey industries; most of the legal vintners, brewers and distillers in America were long since broke. The illegal purveyors were doing a huge business and surely discouraged ideas of repealing the Volstead Act. But Repeal came, nevertheless, early in December of 1933, bringing to an end fourteen years of disastrous experiment.

Many states, especially in the south, voted to remain prohibitionist. Mississippi continued until 1966. Other states voted for rigid state government control of the purchase and sale of alcoholic beverages. Still others enacted laws regulating alcoholic strength and rigid marketing statutes.

Repeal came in the midst of the great Depression and the vintners who were impoverished by Prohibition were relieved. Many of the same winemasters who had foreseen the emergence of the Volstead Act were equally astute about the impending Repeal. Consequently, they had equipped their facilities once more for fermentation functions.

Many of these vintners were in California, where most of the vineyards which had been converted for table grapes awaited in ready resource. Unfortunately, table grapes are not adaptable to fine winemaking, especially delicate dry table wines, but they provided some semblence of a fresh start. Nearly a whole generation of wine culture, education and research had been lost to Prohibition. Per capita wine consumption at more than six gallons annually had been reduced to less than one gallon per person after Repeal.

The stage was set in the late 1930s for the popularity of dessert wine types. The table grapes of California and the juice grapes of the east lent themselves to sweet, high-alcohol ports, sherries and muscatel quite adequately.

The Wine Institute was formed in 1934 by Leon Adams, a noted wine authority both in America and Europe. This association of winemasters has contributed ever since in the stabilization of the United States wine industry. Mr. Adams also authored *The Wine Study Course* which has probably been the largest single educator of wine consumerism in history.

By 1940 about 110 million gallons of wine was sold in America, but still more than three-quarters of the total was comprised of the sweet dessert wines. The phenomenon of America's demand for sweet wines continued through World War II and on through the 1950s and most of the 1960s—a period of creative innovation in the industry. New products such as "Cold Duck" (a mixture of champagne and sparkling burgundy) became very popular. Then "pop" wines (artificially flavored low-alcohol wines associated with "soda-pop") stormed the market place.

Concepts of public relations brought more and more people to the wineries for touring and tasting which stimulated highly profitable retail sales.

Of particular note was the winery of George Lonz, which stood on the island of Middle Bass in Lake Erie, some seven miles out from Port Clinton, Ohio. Despite its remote setting, it was a favorite excursion. Lonz was a very colorful sort who would publicize his harvest and promote his sales of wine by posing among the local ladies. The winery still stands in grim contrast to the activities that once took place there.

Although there was some wine made in Michigan prior to Prohibition, it did not become a major winegrowing state until after Repeal. And even then, few attempts were made in variance from the sweet and "innovative" wine types that were favored.

With the promise of Repeal instantly opening up huge latent markets, winegrowing in Michigan commenced again with aggressive abandon, quickly creating an oversupply. A protective discriminatory tax was enacted, levying higher taxes for wines sold in Michigan that were produced outside the state. Despite the tax break, few Michigan wine cellars survived, and most of those still operational today produce post-Repeal types of sweet wines.

Varietal Labeling

Perhaps the most significant event during the 1950s and the early part of the 1960s was an idea first suggested by the late Frank Schoonmaker, a devoted and respected American wine authority. Schoonmaker felt that renewed interest and pride would be freshly generated in both the wine industry and the wine consumer if wineries would label their wines as *varietal* wines rather than continuing to exploit the more traditional generic names. This brought the name of the grape variety, or cultivar, used to make a wine onto its package instead of the European title. In other words, instead of Burgundy, a place that makes dry red wines in east-central France, the same wine type produced in America would be labeled Pinot Noir or Gamay. Rhine wine became much more properly Johannisberg Riesling or White Riesling or Sylvaner. Chablis became much more meaningfully correct as Chardonnay or Pinot Blanc, etc.

This idea met with limited success at first, but within a decade, varietal labeling was a major factor in the 1960s wine boom.

There was a relatively easy conversion to varietal labeling in California, especially in the Napa, Sonoma, Livermore, Santa Clara and other northern regions where many of the European classic varieties were growing from the Haraszthy influence a century earlier. But in the east, varietal label conversions were difficult as many of the native American grapes such as Concord, Niagara, Delaware, Ives, etc., were not internationally known. Many eastern wineries continued with the old labels: New York State Burgundy, Ohio Rhine Wine and Michigan Chablis. The most dynamic winemasters chose to search for more advanced eastern wine products.

In 1957, Dr. Konstantin Frank became one of the first to successfully grow the European *V. vinifera* in the east, as a viticulturist employed by Gold Seal Vineyards in Hammondsport, New York. Frank started and continues with his own Vinifera Wine Cellars winery nearby. However, the methods of Dr. Frank with his special rootstocks are not common knowledge and, along with considerable skepticism by his peers, the *V. vinifera* acreage in the east remains very small and no competition for the huge California production.

The most significant salvation for quality modern winegrowing in eastern American has evolved from the importation and development of the French-American hybrid varietal selections. Many of these cultivars thrive in the Phylloxera soils through harsh winters to yield fruits, in varying degrees, similar to their European parentage.

The first to import the hybrids into America was Philip Wagner, a retired newspaper editor from Baltimore, who became interested in quality winegrowing early in life, advancing through amateur and hobby ranks.

One of the most successful winemasters in eastern America producing wines from the French-American hybrids is Walter Taylor in his Bully Hill winery. This facility is located at the site of the original Taylor Wine Company in Hammondsport. During the last decade or so Taylor's altercations with the Taylor Wine Company over the use of his family name has drawn national publicity. In Lonz fashion, Walter Taylor is an innovator in the wine industry. An enthusiast of many arts, he is the founder and curator of the first wine museum in America, an unimposing structure named after his father, Greyton H. Taylor.

By 1968, table wines had surpassed dessert wines in both production and sales in the United States. Much of this was due to the production from small new wineries in many states renewing, at last, the pulse that was building in American winegrowing during the latter part of the nineteenth century.

By 1972, the wine-boom had accounted for a per capita wine consumption rate approaching the two gallon mark. But the boom subsided and growth has been more marginal ever since.

The wine industry has not stagnated, however. Technology advances with both viticultural and enological functions yielding wines of better quality at lower cost. Federal and state regulations are being reconsidered from a more realistic attitude and purpose, yet even higher standards are being developed. Market acceptance is becoming more and more receptive, especially in the southern sun-belt that has been traditionally "dry". The public and private enthusiasm about good wines grows with each new chapter organized in the Brotherhood of the Knights of the Vine, Les Amis Du Vin and The American Wine Society, among other wine-oriented consumer groups. It would seem that America is viticulturally healthy once again and that the devoted winemaster can pursue his art with renewed vigor and tenacity.

WORLD VITICULTURE

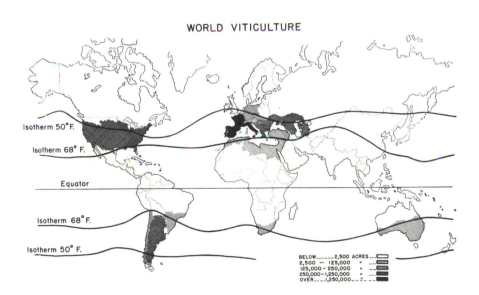

Courtesy of "World Viticulture"

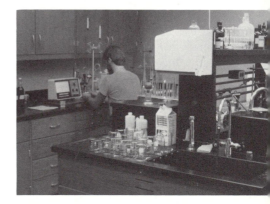

Courtesy: The A. B. McKay Food and Enology Laboratory at Mississippi State University

2

The Winery and the Laboratory

In most of the great European wine districts that were mentioned in the previous chapter, winemaking remains a traditional process, often based on national and district laws. The experience and practice, of the European winemasters span many generations, with several generations possibly having used the same equipment, in the same winery. The questions of winery site or construction materials may never be seriously considered in a lifetime, and the selection of a major piece of winery equipment may occur only several times during the career of a typical European vintner.

The regulation of specific types of wines, grown from particular cultivars of grapes in distinct geographical areas, is strictly enforced in Europe. Despite the fact that American winemaking is also closely regulated, no such factors regarding winemaking styles or grape cultivars limit our vintners. Geographic limitations are, however, beginning to be established and enforced in the United States.

This freedom allows for a multitude of different wine types to be manufactured in almost any area of our country. This, in turn, renders possibilities virtually endless for plant and equipment, but this chapter shall include only some of the more common elements found in American wineries. Small wineries are defined in this text as those with less than a 100,000 gal. production capacity annually.

Wine processing functions are those operations performed in the winery, while wine quality controls are the analyses that take place in the laboratory. However, some of these functions overlap; for example the testing methods employed by a cellarmaster insure that the cellar and equipment remain clean, and the winemaster prepares yeast and fining agents in the laboratory for use in wine processing. The laboratory is usually contained in one of the winery buildings. Despite this, we shall consider the laboratory a separate and distinct facility.

All activities that take place in the winery and laboratory are regulated by the Bureau of Alcohol, Tobacco and Firearms (ATF). It is important that the prospective vintner become fully familiarized with the regulations set forth by the ATF. Wine production is authorized under the provisions of Part 240 of Title 27 Code of Federal Regulations, while labeling and advertising of wine is permitted in compliance with regulations under the

Federal Alcohol Administration Act, Title 27, Code of Federal Regulations. Both of these publications are essential to commercial winemaking and are provided free by regional ATF offices. The vintner should also be informed of the state and local statutes that are applicable to the operation of the wine producing facility.

The establishment of a winery is regulated in Subparts C through N, CFR, including Sections 240.120 through 240.345. The difference between bonded winery and bonded wine cellar is slight. Generally, the bonded winery is considered a wine manufacturing, bottling and warehousing operation, while the bonded wine cellar is primarily just a bottling and warehousing facility. There are exceptions, but for the sake of consistency, this book shall consider a winery to be the entire wine manufacturing concern, complete with buildings and equipment sufficient to receive grapes, process, and ship packaged wines.

The term, *bonded*, refers to wines for which excise taxes have not been paid. A *bond* is a legal instrument purchased by a vintner from a bonding company which will protect the maximum tax liability to the ATF. The rates levied are discussed in Chapter 4. The bonded winery is a facility within which bonded wines may be legally manufactured and stored. The bonded premises is that exact portion of real property limitation upon which the bonded winery may operate.

LOCATION

The site upon which a new winery is to be established is very important, requiring proximity to suppliers as well as to prime wholesale markets. Harvested grapes can decay rapidly, rendering wines that are poor in quality.

The volume and profit of retail wine sales can be the difference between success and failure, so careful considerations must be made of traffic flow and ease of accessibility in selecting a site.

Most readily available winery machinery is powered with high voltage electric motors which can save significantly on energy costs. Therefore, a 220-volt power source must be available to the proposed winery site. The new vintner should also investigate the availability of power and calculate the cost of utility service.

A rule-of-thumb figure for calculating water usage in a winery is that about ten gallons of good quality water will be needed for each gallon of finished wine, so a plentiful supply of clean water must be available. Most of this large requirement is used in cleaning.

The contemporary concern for ecology and the environment is reflected in ATF F 1740.1 and ATF F 1740.2, forms that are designed to investigate the impact of effluents from a proposed winery. These questions must be dealt with in depth before construction or remodeling of a winery building.

Apart from obvious concerns such as local alcohol control statutes, zoning, real estate taxes and insurance, the winery entrepreneur should consider other matters with regard to the proposed winery site. Governmental,

DEPARTMENT OF THE TREASURY – BUREAU OF ALCOHOL, TOBACCO AND FIREARMS	ESTABLISHMENT IS BONDED
APPLICATION BY PROPRIETOR OF BONDED WINE CELLAR OR BONDED WINERY *(Complete in triplicate - See instructions on last page)*	☐ Wine Cellar ☐ Winery

ESTABLISHMENT IS BONDED ☐ Wine Cellar ☐ Winery

REGISTRY NO.

SERIAL NO. 1/

TO: REGIONAL REGULATORY ADMINISTRATOR,
BUREAU OF ALCOHOL, TOBACCO AND FIREARMS, AT

DATE

APPLICATION MADE BY *(Complete appropriate section below)*

INDIVIDUAL OWNER	CORPORATION
FULL NAME	NAME
TRADE NAME *(If any)*	STATE UNDER LAWS OF WHICH INCORPORATED
ADDRESS *(No., street, city, county, State, and Zip Code)*	ADDRESS OF PRINCIPAL OFFICE

PARTNERSHIP OR ASSOCIATION *(List below names and addresses of each person interested in operation)*

TRADE NAME *(If any)*

PURPOSE FOR WHICH FILED *(Such as "Original Establishment," "Extension of Premises," "Change in Firm Name" - Describe briefly)*

APPLICATION IS MADE TO OPERATE

☐ Bonded Winery For Standard Wine   ☐ Bonded Winery For Sub-Standard Wine   ☐ Bonded Wine Cellar For Standard Wine   ☐ Bonded Wine Cellar For Substandard Wine

LOCATION *(State exact location of bonded premises. If located within a city, give street, number, and city. If located elsewhere, give name of county, nearest post office with distance and direction therefrom, and name and number of road or highway on which situated)*

NAME OF OWNER OF PREMISES

DATE OPERATIONS TO COMMENCE

PART A - CAPACITY

ITEM	APPROX. STORAGE CAPACITY *(Gallons)*	EST. ANNUAL PRODUCTION *(Gallons)*	ESTIMATED QUANTITY OF WINE SPIRITS TO BE WITHDRAWN ANNUALLY FOR ADDITION TO WINE *(Proof gallons)*
STILL WINE *(Including distilling material and vinegar stock:* 14% & UNDER			
OVER 14%			ESTIMATED QUANTITY OF WINE SPIRITS THAT WILL BE USED ANNUALLY FOR PRODUCTION OF SPECIAL NATURAL WINE AND EFFERVESCENT WINE *(Proof gallons)*
SPARKLING WINE			
ARTIFICIALLY CARBONATED WINE			

ATF Form 698 (5120.25) (2-77) PREVIOUS EDITIONS ARE OBSOLETE.

THE APPLICATION BY PROPRIETOR OF BONDED WINE CELLAR OR BONDED WINERY—ATF FORM 698 (5120.25)—THE BASIC APPLICATION FORM NEEDED TO COMMENCE A WINERY

	DEPARTMENT OF THE TREASURY – INTERNAL REVENUE SERVICE	CONTINUING BOND		
FORM **700** (REV. MARCH 1969)	BOND FOR BONDED WINE CELLAR OR BONDED WINERY *(File in duplicate. See instructions on back)*		REGISTRY NO.	F.A.A. ACT PERMIT
		BONDED WINE CELLAR		
		BONDED WINERY		

PRINCIPAL *(See instructions 2, 3, and 4)*	ADDRESS OF BUSINESS OFFICE *(Number, Street, City, State, ZIP Code)*
	ADDRESS OF BONDED PREMISES *(Number, Street, City, State, ZIP Code)*

SURETY(IES)	AMOUNT OF BOND	EFFECTIVE DATE

KNOW ALL MEN BY THESE PRESENTS, That we, the above-named principal and surety (or sureties), are held and firmly bound unto the United States of America in the above amount, lawful money of the United States; for the payment of which we bind ourselves, our heirs, executors, administrators, successors, and assigns, jointly and severally, firmly by these presents.

This bond shall not in any case be effective before the above date, but if accepted by the United States it shall be effective according to its terms on and after that date without notice to the obligors: *Provided,* That if no date is inserted in the space provided therefor, the date of execution of this bond shall be the effective date.

Whereas, the principal is operating, or intends to operate, a bonded wine cellar or bonded winery at the bonded premises specified above;

Now, therefore, the conditions of this bond are such that if the principal —

(1) Shall in all respects, without fraud or evasion, faithfully comply with all requirements of law and regulations relating to the operation of the bonded premises specified above; and

(2) Shall pay all penalties incurred and fines imposed on him for violation of any of the said requirements; and

(3) Shall pay, or cause to be paid, to the United States, all taxes, including all occupational and rectification taxes, imposed by law now or hereafter in force (plus penalties, if any, and interest) for which he may become liable with respect to operation of the said bonded premises, and on all distilled spirits and wine now or hereafter in transit thereto or received thereat, and on all distilled spirits and wine removed therefrom, including wine withdrawn without payment of tax, on notice by the principal, for exportation, or use on vessels or aircraft, or transfer to a foreign-trade zone, and not so exported, used, or transferred, or otherwise lawfully disposed of or accounted for: *Provided,* That this obligation shall not apply to taxes on wine in excess of $100 which have been determined for deferred payment upon removal of the wine from bonded premises or transfer to a tax-paid wine room on the bonded premises;

Then this obligation is to be null and void, but otherwise to remain in full force and effect.

We, the obligors, also agree that all stipulations, covenants, and agreements of this bond shall extend to and apply equally to any change in the business address of the bonded premises, the extension or curtailment of such premises, including the buildings thereon, or any part thereof, or in equipment, or any other change which requires the principal to file a new or amended application or notice, except where the change constitutes a change in the proprietorship of the business or in the location of the premises.

And we, the obligors, for ourselves, our heirs, executors, administrators, successors and assigns, do further covenant and agree that upon the breach of any of the covenants of this bond, the United States may pursue its remedies against the principal or surety independently, or against both jointly, and the said surety hereby waives any right or privilege it may have of requiring, by notice or otherwise, that the United States shall first commence action, intervene in any action of any nature whatsoever commenced, or otherwise exhaust its remedies against the principal.

Witness our hands and seals this . . . . day of _____ , 19____

Signed, sealed, and delivered in the presence of—

_____   _____ SEAL

_____   _____ SEAL

_____   _____ SEAL

_____   _____ SEAL

_____   _____ SEAL

| KIND OF BOND *(Check applicable box.)* | ORIGINAL | STRENGTHENING | SUPERSEDING |

FORM **700** (REV. 3-69)

BOND FOR BONDED WINE CELLAR OR BONDED WINERY—ATF FORM 700—THE BOND FORM WHICH IS REQUIRED BY THE ATF TO BE FULLY EXECUTED BEFORE A BASIC PERMIT IS ISSUED TO OPERATE A COMMERCIAL WINERY

Department of the Treasury - Bureau of Alcohol, Tobacco and Firearms

Environmental Information

(See Instructions on back)

1. Name and Principal Business Address of Applicant	2. Description of Activity	3. Number of Employees

4. Location Where Activity is to be Conducted *(Be specific. Number, Street, City, State, ZIP Code, describe locations of buildings and outside equipment and their situation relative to surrounding environment including other structures, land use, lakes, streams, roads, railroad facilities, etc. Maps, photos, or drawings may be provided.)*

5. Heat and Power:

A. Describe types of heat and power to be used and their sources. If they are to be produced in connection with the proposed activity, estimate type and quantity of fuel to be used for each purpose. *(Example: 40 tons/yr. anthracite coal for heat, 20 million cu.-ft./yr. natural gas for power generating.)*

B. Describe any air pollution control equipment proposed for use in connection with fuel burning equipment, boilers, or smokestacks.

6. Solid Waste:

A. Describe amount and composition of all solid waste to be generated.

B. Discuss proposed methods of disposal *(Incineration, open burning, landfill, government or commercial garbage collection, etc.)* Specify whether on-site or off-site.

C. Describe any air pollution control equipment proposed for use in connection with any incinerators.

ATF Form **4871** (1740.1) (6-77)   PREVIOUS EDITIONS MAY BE USED

ATF FORMS (1740.1) and F 1740.2—ENVIRONMENTAL INVESTIGATION FORMS REQUIRED BY THE ATF PRIOR TO ISSUING A BASIC PERMIT TO OPERATE A COMMERCIAL WINERY

	APPLICATION RELATED TO THIS RIDER	
DEPARTMENT OF THE TREASURY Bureau of Alcohol, Tobacco and Firearms **SUPPLEMENTAL INFORMATION WATER QUALITY** **CONSIDERATIONS** — Under 33 U.S.C. 1341 (a)	1. FORM NUMBER	2. APPLICATION DATE
	3. SERIAL NUMBER	

INSTRUCTIONS:

1. COMPLETION. Answer all items in sufficient detail if applicable to your activity. If necessary, continue on reverse side of this form or a separate sheet. Your answers are evaluated to determine if a certification or waiver by the applicable State Water Quality Agency is required under Section 21 (b) of the Federal Water Pollution Control Act (33 U.S.C. 1341 (a)).
2. FILING. Submit an original and copy of this form with the related application or other document, to the Regional Regulatory Administrator, Bureau of Alcohol, Tobacco, and Firearms. This form must be completed and submitted even though three copies of the required certification or waiver has been sent to the Regional Regulatory Administrator or is attached to this form.
3. DISPOSITION. After final action is taken on the related application or other document, the copy of this form will be returned to the applicant.

4. NAME AND PRINCIPAL BUSINESS ADDRESS OF APPLICANT *(Number, street, city, county, State, and ZIP code)*	5. PLANT ADDRESS*(If different from address in item 4)*

6. DESCRIBE ACTIVITY TO BE CONDUCTED IN WHICH THE BUREAU OF ALCOHOL, TOBACCO AND FIREARMS HAS AN INTEREST.

7. DESCRIBE ANY DIRECT OR INDIRECT DISCHARGE INTO NAVIGABLE WATERS WHICH MAY RESULT FROM THE CONDUCT OF THE ACTIVITY DESCRIBED IN ITEM 6, INCLUDING THE BIOLOGICAL, CHEMICAL, THERMAL, OR OTHER CHARACTERISTIC OF THE DISCHARGE AND THE LOCATIONS AT WHICH SUCH DISCHARGE MAY ENTER NAVIGABLE WATERS.

8. GIVE THE DATE OR DATES ON WHICH THE ACTIVITY WILL BEGIN AND END, IF KNOWN, AND ON WHICH THE DISCHARGE WILL TAKE PLACE.

9. DESCRIBE THE METHODS AND MEANS USED OR TO BE USED TO MONITOR THE QUALITY AND CHARACTERISTICS OF THE DISCHARGE AND THE OPERATION OF EQUIPMENT OR FACILITIES EMPLOYED IN THE TREATMENT OR CONTROL OF WASTES OR OTHER EFFLUENTS.

10. I certify that I have examined this rider and, to the best of my knowledge and belief, it is true, correct, and complete, and that copies of this rider may be furnished to the applicable State Water Quality Agency and the Regional Administrator, Environmental Protection Agency.

11. APPLICANT	12. BY *(Signature and title)*

ATF F 1740.2 (5-79) REPLACES ATF FORM 4805 WHICH MAY BE USED

political and social attitudes can prove very costly where establishment of a winery is unpopular. A site selection carefully made, however, may be a welcome addition to the local business community.

DESIGN

A winery operation that expects to enjoy a good tourist business and repeat customers should be designed to be attractive. It should go without saying that the premises must offer maximum safety to visitors, especially as small children may accompany their parents. A floor plan that allows self-guided tours is expense-saving, but must be managed very carefully in order to keep accidents, damage and losses at a minimum.

Winery construction today ranges from the bare simplicity of modular steel buildings to the most romantic laid-up stone and hand-hewn wooden beams. For the maximum return on investments, durable materials are required, as well as efficient use of space. Such planning must involve careful marketing projections, taking into account both short-term and long-term production goals. Future expansion with minimal cost is another important consideration. Figures 2.1 and 2.2 exhibit an elevation and a simple floor plan for a commercial winery capable of producing a maximum of 50,000 gal. (about 21,000 cases) per year.

Remodeling existing facilities can be an effective way to avoid the high cost of new construction. A remodeled winery may not be as efficient as a newly designed facility, but financially may be more feasible. In any case, professionals should be consulted about the design of a proposed winery.

THE RECEIVING AREA

The receiving area in the winery must be designed for the efficient handling of grapes as well as other supplies and equipment. Figure 2.3 illustrates a commonly observed scene in a receiving area. Note that this facility is covered with a roof, yet is actually out of doors. This allows for receiving grapes during inclement weather, and also provides for easier handling of waste product removal. Insect and microbiological control inside the winery is also made easier to handle.

Scales

It is imperative that incoming fruit be accurately described and measured so as to establish the identity and amount of grapes to be processed into wine. A tally sheet for this purpose will be discussed in Chapter 3.

Relatively low cost portable electronic scales, as shown in Fig. 2.4, are readily available for weighing grapes. Some of these will automatically subtract a standard tare weight which has been determined and entered into the device. For instance, common plastic grape lugs generally weigh 2.8 lb. If one lug at a time is being weighed, the tare of 2.8 is entered into the electronic scales, and this will automatically be deducted from each weight.

FIG. 2.1.   ELEVATION OF SMALL WINERY BUILDING

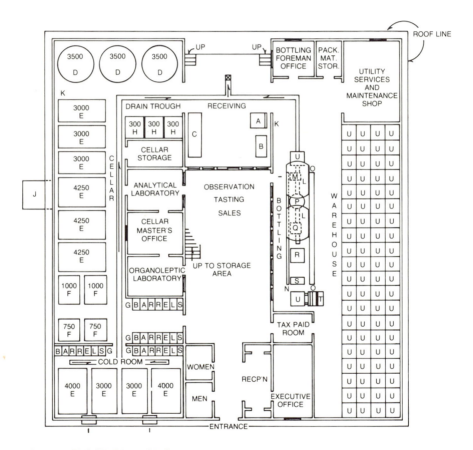

FIG. 2.2.   FLOOR PLAN OF SMALL WINERY BUILDING

A—PORTABLE SCALES B—CRUSHER-STEMMER C—PRESS D—SS MUST-STORAGE TANKS E—SS PROCESS-FERMENTATION AND AGING TANKS F—OAK CASKS G—CRIADERA H—SS MIX TANKS I—COLD ROOM REFRIGERATION J—TANK REFRIGERATION K—PUMP AND FILTER STORAGE AREA L—SS OVERHEAD BOTTLING TANKS M—BOTTLE RINSER N—BOTTLE CONVEYOR O—CASE CONVEYOR P—FILLER Q—CORKER-CAPSULER P—LABELER T—FORK TRUCK U—PALLET STORAGE

FIG. 2.3.   RECEIVING AREA AT A SMALL WINERY

FIG. 2.4.   ELECTRONIC SCALES

The scales can also be used to weigh more than one lug at a time. In a small winery, the weighmaster might weigh ten lugs at a time. A tare of 2.8 lb each, or 28 lb, would be deducted from each gross lot of 10 lugs received.

In larger wineries, where grapes may arrive in one ton bins, each container should be weighed for gross weight before the grapes are emptied into the crusher-hopper. The tare is then taken for subtraction from the gross weight, in order to compute the net weight of the fruit received. A variety of conventional scales of this type are available.

In the largest wineries, containers arrive in one ton bins, or in dump-bodies containing eight tons or more. Gondolas carrying 20 tons are also found. Such volumes require large platform scales where the entire truck or gondola may be weighed, emptied, and then reweighed to find the tare and net weight. However in the small winery, portable scales are more appropriate so that the dock and receiving area can be used for other purposes when grapes are not being received.

Lug or Bin Rinser

The use of clean containers for transporting grapes from the vineyards is essential for minimizing contamination. Immediately after grapes are dumped into the crusher-hopper, the container that held the grapes should be adequately rinsed. Lugs are rinsed in a small portable device that is supplied with a ¾-in. heavy-duty garden hose. Figure 2.5 portrays a lug rinsing device that is suitable for a small winery.

The large one ton bins can be held inverted over a seasonal or permanent spray device that can be easily operated by the lift-truck operator who handles the grape dumping operation. The device usually has a bin rinsing spray head mounted in a convenient and effective manner.

Crusher-stemmer

The first step in wine processing is the breaking or crushing of the grape skins, and the removal of the stems. Shredding or grinding the grapes is not necessary. In fact, this is not desired as the seeds will also be broken, which can release bitter oils and astringent tannins to the *must* (crushed, de-stemmed grapes). The gentle crushing of the skins facilitates the release of the pulp inside the grape berries. This, in turn, allows easier release of the juice in the press. In the case of red wines, crushing the grapes allows the cultured yeasts added by the winemaster easy access to the fermentable sugars stored inside the fruit.

Stem removal also reduces the astringency that can be contributed by natural tannins. In some wine districts a portion of the stems is allowed to remain with the must in order to obtain a "steminess" from the tannins. Tannins are complicated phenolic compounds that inhibit wine aging by reducing oxidation. Red table wines are believed to "live" longer when some, or all, of the stems are processed with the must. Today, however,

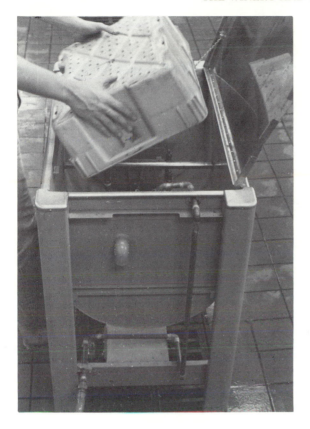

FIG. 2.5.   PLASTIC LUG RINSER

specific amounts of tannic acid are added at some particular time during the processing of the young wines in order to better control these reactions. The stemming procedure can be virtually eliminated in the case of Muscadine cultivars as these grapes do not have stems of any significant consequence.

Figures 2.6 and 2.7 show two views of a popular crusher-stemmer that is found in small wineries. This type of crusher-stemmer ranges in capacity from about one ton to more than ten tons per hour. Similar machines with much larger capacities have compound mechanisms performing the same function. Two rollers rotating in opposite directions, fed by an auger, crush the grapes as they are dumped into the hopper. The perforated stainless steel shroud around the tooth-shaft allows seeds, pulp, juice and skins to pass into the lower stainless steel must hopper. Stems, however, cannot easily pass through these perforations and are picked up by the teeth of the rotating shaft and rifled out the end of the device.

Another type of crusher-stemmer device gaining popularity in larger wineries is a vertical machine that is more efficient that the horizontal type, but is much more expensive.

Courtesy: Budde & Westermann, Montclair, New Jersey

FIG. 2.6.   SMALL CRUSHER, STEMMER AND MUSTPUMP

Must Pump

With appropriate hoses and/or pipelines, the must pump transfers the must from the crusher-stemmer must hopper to further processing vessels. Some crusher-stemmer models have centrifugal must pumps built into the bottom of the hopper, and may be fed by an auger.

Must pumps can be either centrifugal or positive displacement devices of various designs. Normally, inlet and outlet minimums are of two inch inside diameter. Smaller capacities are quickly clogged with must. A common type, is a "quad-chamber" positive displacement must pump that can be reversed with a handle mounted at the front of the machine for ease in unclogging.

Centrifugal must pumps are not as popular with winemasters, because of concerns for oxygen pickup during this segment of processing. These types, properly designed with large outlet diameters and low speeds can, however, function effectively.

Piston-type must pumps are generally regarded to be the most reliable for pumping crushed, destemmed grapes. However Fig. 2.8 shows that piston pumps may be too bulky and heavy to be efficient devices for the small winery.

A new type of must pump (Fig. 2.9) can be used both as a must pump beneath the crusher-stemmer, and also for assistance in emptying red wine fermenters.

Press

The press is a device used to extract juice or wine from musts. It is often employed in the receiving area for pressing juice from white grape musts, and in the cellar area for pressing wine from fermented red musts.

There are a number of press types available, and the capital outlay for a press still remains one of the largest single items that a vintner encounters. Because of this, some vintners initially purchase only one small press and then add another machine when production growth warrants a move. Having two machines also lessens the possibility of breakdown totally crippling pressing operations during the vintage.

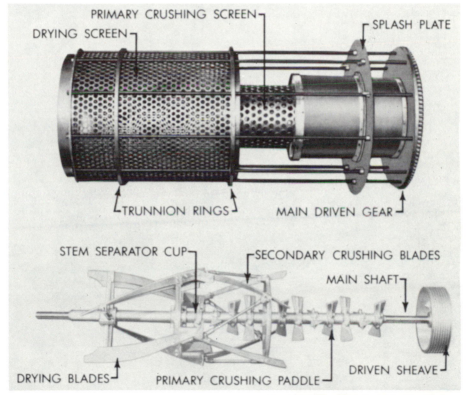

Courtesy: Valley Foundry & Machine Works
Fresno, California

FIG. 2.7.   CRUSHER-STEMMER—INTERNAL OPERATION DIAGRAM

FIG. 2.8. A STATIONARY PIS-
TON-TYPE MUST PUMP

Courtesy: Valley Foundry And Machine
Works
Fresno, California

Courtesy: Menestrina
Scott Laboratories, San Rafael, California

FIG. 2.9.   PORTABLE MUST PUMP WITH HOPPER AND AUGER

Perhaps the most common type of press is the horizontal "basket" as displayed in Fig. 2.10. This ranges from about one ton in capacity to more than 20 tons. The basket rotates so that juice can pass through the slots, of a stainless steel type, or between the staves of a wooden model. After the free-run is exhausted, pistons riding on a screw-shaft inside the press converge, creating a pressure on the must and extracting the "first-press", "second-press", etc.

Figure 2.11 illustrates a vertical basket press which, although the basket remains stationary, operates in a similar fashion to the horizontal basket press just discussed. Vertical basket presses are primarily found in the very smallest wineries. Although they are smaller and less efficient, these are among the most modestly priced types.

An unusual type of horizontal basket press is the "bladder" press (Fig. 2.12). This device employs an inflatable bladder, which exerts pressure on the must inside the basket. While these operate somewhat faster and more effectively than the standard horizontal basket press, the cost is often more than double.

Recently a continuous screw press has been developed and introduced. These are also known as dejuicers or auger presses. The principle of these machines is to force the must against a plug of pomace. This plug is

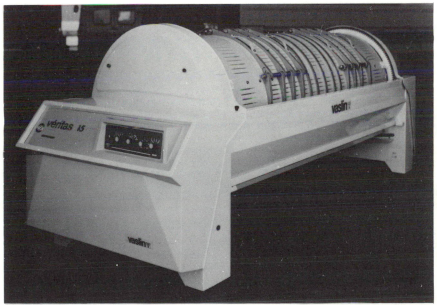

Courtesy: Budde & Westermann
Montclair, New Jersey

FIG. 2.10.   SMALL HORIZONTAL BASKET PRESS

Courtesy: Rossi S.p.A. Presque Isle Wine Cellars
North East, Pennsylvania

FIG. 2.11.   VERTICAL BASKET PRESS

A-IN POSITION FOR PRESSING   B-IN POSITION FOR POMACE REMOVAL

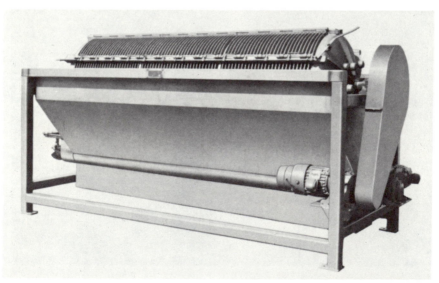

Courtesy: Valley Foundry And Machine Works
Fresno, California

FIG. 2.12.   A "BLADDER"-TYPE HORIZONTAL BASKET PRESS

developed by a weighted door at the pomace discharge outlet. The auger rotates within a heavy-duty stainless steel screen that allows the juice or wine released to pass through to collector hoppers mounted beneath the device. These machines reduce labor requirements and have high product yields. However, continuous presses of this type usually render products with a high proportion of suspended solids. Screw presses are very expensive, even more than bladder presses of the same capacity. Fig. 2.13 displays a popular continuous press found in some American wineries.

THE CELLARS

The juice and must pumped from the receiving area are sent to the cellars to be fermented, racked, clarified, stabilized, primary filtered and aged. Details concerning these aspects of wine processing are discussed in later chapters.

The cellars should be located on the bottom floor of the winery, because of the heavy weight. In addition, for aesthetic appeal, visitors expect the winery "cellars" to actually be in the cellar. If cellars cannot be excavated, the processing and bulk storage facilities can be effectively designed and constructed on grade level. However, in the crusher building, a better flow can be achieved by starting processing at a higher level, and allowing downhill flow.

One of the most common errors found in wine cellar design is that of inadequate drainage of processing water. As was stated earlier in this chapter, about ten gallons of water are required for every gallon of finished

Courtesy: Budde & Westermann
Montclair, N.J.

FIG. 2.13.   CONTINUOUS PRESS—SCREW TYPE

wine, although this can be significantly reduced in larger wineries. In any event, a large amount of water must be discharged from the cellars and receiving areas. This is best accomplished by a good system of pitched floors and drain troughs feeding an approved effluent system of sufficient capacity.

Another design flaw found in some cellars is insufficient insulation for maintaining a reasonably constant temperature in the cellars. Most winemasters prefer the processing portions of the cellars to be kept cool, generally in the 65−70°F range. Of course, special areas are kept cold, such as that used for detartration, while others may be kept as warm as 80°F, for aging some red table wines and dessert wines.

The most practical and economical materials for cellar construction are masonry and/or concrete, although neither of these are good sources of insulation. Other materials such as vermiculite, fiberglass and treated cellulose can be applied effectively to increase the insulative "R-factor". Grade-level facilities can be fabricated from a variety of efficient and durable new materials. Floors should be carefully engineered with steel-reinforced concrete to withstand the heavy loads. It is also recommended that several applications of acid-resistant coating be applied, to preserve floors and masonry walls from the long-range effects of juice, must and wine acids.

Heat Exchanger

In larger wineries it is common to find separate tube-and-shell heat exchangers; one for decreasing temperatures of the product, and another for increasing temperatures. The inside diameter of the tubes carrying the product is usually 1 to 2 in. Smaller diameters can clog easily and larger diameters are less efficient. The shell portion of the exchanger is supplied with a circulated refrigerant such as propylene glycol. The shell can also be used for heating, with hot water or steam. Figure 2.14 illustrates a stainless steel tube-and-shell heat exchanger.

The plate type of heat exchanger is not the most efficient device for use in a winery. The small passages between the plates are easily clogged with seeds, skins and pulp from the must. Also, the buildup of tartrate crystals during low-temperature chilling operations can quickly foul the system.

Used dairy farm storage tanks are found in many small wineries. These vessels usually are equipped with a refrigeration device that can directly chill juice, must or wine in the tank. These vessels are also used to chill water that is circulated through heat exchanger shells or the jackets of fermentation tanks.

Because capital investment is especially critical to small-volume vintners, a mixing tank in the cellars usually also serves as a heat exchanger when necessary. This is accomplished by furnishing the jacket of the mixing tank with a hot water supply. Used creamery process vessels can be modified for these purposes. Figure 2.15 portrays a stainless steel process tank used as a mixing-heating tank.

Courtesy: Valley Foundry & Machine Works
Fresno, California

FIG. 2.14.   STAINLESS STEEL TUBE-AND-SHELL HEAT EXCHANGER

Centrifuge

During the past several decades the use of the centrifuge in American and European wineries has rapidly increased, particularly the continuous types of centrifuge. These machines are, however, extremely expensive, and even a very small continuous centrifuge can cost more than $50,000.

The purpose of this device is to separate the major portion of suspended solids (mostly pulp) from freshly pressed grape juice by the use of centrifugal force. The machine is similar to a milk separator, although the latter cannot be used with safety in a winery. The removal of suspended solids from the juice or new wine saves time, energy and space in the later process of racking (decanting) from the lees (sediment) and may improve quality.

While larger wineries may have the capital necessary for a continuous centrifuge, most smaller wineries continue to use traditional methods of juice and wine clarification. Figure 2.16 illustrates a modern continuous juice centrifuge.

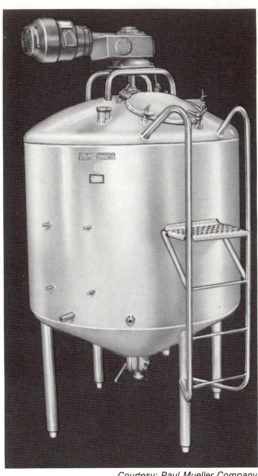

FIG. 2.15. STAINLESS STEEL
CREAMERY PROCESS TANK

Courtesy: Paul Mueller Company
Springfield, Missouri

Must Tanks: Red Fermenting Tanks

In many small or medium-sized wineries, must tanks serve the same purpose as red fermenters. Such vessels allow musts to be treated or fermented before pressing. White grape musts may be treated with press aids such as paper fiber and/or pectic enzymes to assist in the release of juice in the press. Jacketed must tanks may be used as heat-exchange devices when refrigerants or hot water are properly circulated. Must tanks are usually set up in "batteries" providing flexibility: a portable press can be moved back and forth beneath these vessels as is required. The basic idea is to have the vessels supported so that the outlet valve is higher than the inlet to the press. This, of course, allows gravity to empty the treated or fermented must

FIG. 2.16.   AN AUTOMATIC CONTINUOUS CENTRIFUGE FOR USE IN A SMALL WINERY

into the press. Must service valves should be six inches in diameter, or larger. Ideally such tanks should also be sealable, so that they may be used as storage tanks after the vintage season is completed. Agitating devices provide for simple mixing of pressing aids and also allow must tanks to serve as mix vessels during other processing periods. Figure 2.17 offers a sketch of an ideal must, or red fermentation, tank.

Tank Boards

While tanks may be used for more than one purpose in a winery, they must be marked in accordance with Part 240, CFR, Sections 240.163 through 240.165. Such markings are best made on a board such as that in Fig. 2.18. The "S", denoting a storage tank can be covered with an "F", for fermenting tank, during the vintage season.

Storage Tanks

Stainless steel storage tanks have become widely accepted during the past several decades. Durability, ease in cleaning and maintenance head the list of reasons for this response.

The expense of stainless steel tanks can be significantly reduced by purchasing good used vessels, particularly creamery storage types as pic-

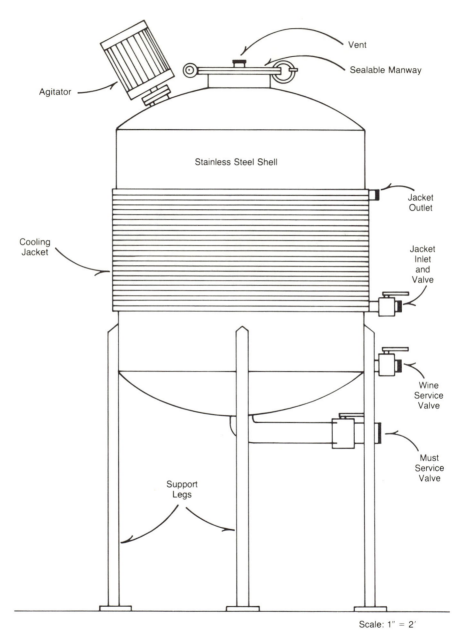

FIG. 2.17.   DESIGN SKETCH FOR A 4-TON RED MUST FERMENTER

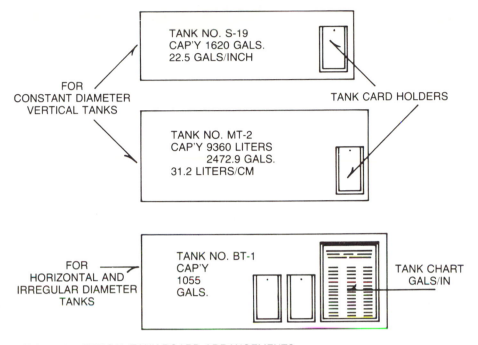

TANK NO. S-19
CAP'Y 1620 GALS.
22.5 GALS/INCH

FOR
CONSTANT DIAMETER
VERTICAL TANKS

TANK CARD HOLDERS

TANK NO. MT-2
CAP'Y 9360 LITERS
        2472.9 GALS.
31.2 LITERS/CM

TANK NO. BT-1
CAP'Y
1055
GALS.

FOR
HORIZONTAL AND
IRREGULAR DIAMETER
TANKS

TANK CHART
GALS/IN

FIG. 2.18.   TYPICAL TANK BOARD ARRANGEMENTS

tured in Fig. 2.19a. These have the added benefits of high-grade stainless steel construction materials, insulation, manways, and sight-glasses, and often are equipped with agitators and heat-exchange jackets. Usually only minor alterations are needed before these containers can be used. With the addition of a racking valve installed just below the manway, these tanks can also serve nicely as white grape juice fermentation tanks. (Fig. 2.19b).

Lined mild steel tanks are found in wineries, but the cost of maintaining the lining materials is reducing their popularity. The lining must not become deteriorated or else exposure to the mild steel can cause iron contamination—a casse that can greatly reduce wine quality.

Large wooden wine tanks have become so expensive to purchase and maintain that new oak and redwood wine storage vessels are more the exception than the rule. Smaller wooden wine containers such as 50-gal. barrels have become popular during the past several decades in America. These smaller barrels have the advantage of mobility: they can be moved from place to place in the winery so that seasonal operations can be more efficiently performed. Being so small, the unit volume of barrels offers more surface area of wood exposure per unit volume than larger tanks. Consequently, aging is faster in the smaller wooden vessels. New barrels, espe-

FIG. 2.19a.   USED CREAMERY STORAGE TANKS

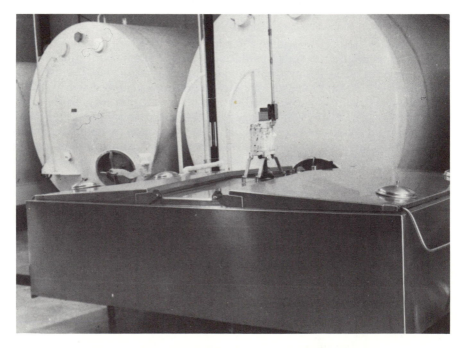

FIG. 2.19b.   A 500-GALLON DAIRY STORAGE TANK IN FOREGROUND

cially those made from imported Limousin and Nevers oak, are expensive, but still relatively much less expensive than larger handmade wooden tanks. The ability of the cellarmaster to handle cooperage repairs in-house is also significantly increased with the use of barrels.

Until recently, fiberglass and plastic tanks were not commonly used. Some winemasters report that tanks made from these materials impart odors and/or flavors that foul and contaminate the wine.

Mixing Tank

The primary purpose for a mixing tank is to mix ingredients or agents used in the bulk winemaking process. Cane sugar added during the vintage season, when juice and must are cold, is very difficult to dissolve without the use of a mixing tank. The preparations of dehydrated yeast cultures for inoculation into juice and must is another tedious task facilitated by a well equipped mixing vessel. This device can also be used as a heat exchanger, a measuring tank, a holding tank for filtration, and a constant agitation device that is needed for making fining agent suspensions.

Many wineries have two mixing tanks. This is helpful during the vintage season when many operations are taking place at the same time, or when several agents are required for a wine clarification. Larger wineries may have a half-dozen or more mixing vessels, each with a specialized purpose. Most small and medium-sized wineries have mixing tanks with capacities that range from 200 to 500 gal. The 500-gal. size is common, but there are tanks that hold 3000 gal.

High-grade stainless steel dairy process tanks can be easily converted to winery mixing tanks. The agitators must be heavy duty, turning at relatively low revolutions per minute. The vessels should be adequately insulated and have ample jacket area for heat-exchange efficiency. Many vintners choose tanks that can handle live steam in the jackets, and adapt a small steam generator to the unit to heat the product. This can save considerable time and energy (the steam generating device can, of course, also be used as a very effective portable cleaning tool). The lids should fit securely to keep dust and insects out of the vessel.

Figure 2.20 illustrates a dairy process tank altered to serve as a winery mixing tank, and exemplifies a custom-designed mixing vessel.

Transfer Pumps

The ability of a winery to handle large volumes of a product is dependent upon adequate pumping capacity. Transfer pumps are placed in operation as soon as the juice or wine is released in the press, and continue to be used in nearly every facet of bulk wine processing. Dependable high-grade stainless steel transfer pumps are of paramount importance for operating efficiency in the cellars.

There are a number of different types of pumps that can be considered, although only centrifugal, rotor and stator models shall be discussed in this section. These are, by far, the most qualified types for use in wine cellars.

Courtesy: Walker Stainless Equipment Co.
New Libson, Wisconsin

FIG. 2.20.   MIXING TANK

Centrifugal transfer pumps are among the most popular types found in wineries today. This is due primarily to their high-speed outputs and low cost. The best of these operate at low revolutions per minute in order to minimize the churning of the product in the impeller housing. The advantages of centrifugal operation also include smooth continuous flows and a high tolerance for suspended solids. Figure 2.21 depicts a common centrifugal wine transfer pump.

Rotor pumps have lobes that are filled with a product during the inlet phase and then positively displaced by the rotor(s) during the outlet phase. This positive displacement feature can be dangerous if restricted by closed valves or plugged hoses and lines. Unlike centrifugal pumps, which can be valved and restricted so as to limit output, rotor models should be equipped with a bypass unit to allow for restricting outflow of product and pinpointing pressures for feeding filters and other sensitive uses. Rotor pumps are not recommended for use with products that have suspended solids, especially abrasive solids such as wine tartrates. These materials will quickly damage the pumping mechanism. Rotor pumps are very gentle but are also expensive, although good quality used machines are often available.

Stator pumps are also called progressing cavity pumps. They operate by virtue of a stator, or helical screw, that gently, but positively, discharges product from inlet to outlet. These models have the best features of both centrifugal and rotor pumps: high output, smooth flow, good tolerance for suspended solids and, when equipped with a bypass unit, they can maintain a constant pressure. Stator transfer pumps are even more expensive than rotor models, and are not as easily available in the used machinery market.

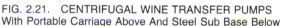

FIG. 2.21.   CENTRIFUGAL WINE TRANSFER PUMPS
With Portable Carriage Above And Steel Sub Base Below

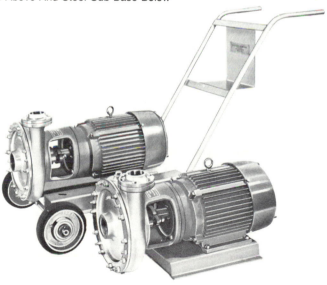

Courtesy: Valley Foundry And Machine Works
Fresno, California

Refrigeration Systems

Refrigeration is a process whereby compressed refrigerant (usually ammonia gas or Freon) is allowed to expand in a specially designed chamber, such as an ice-builder. The expansion of the refrigerant is endothermic (requires heat) and with the loss of heat, the chamber is cooled.

Refrigeration methods in American wineries are varied, not only as to the source of refrigeration, but also in the method.

As mentioned earlier in the chapter, vintners in some small wineries make good use of converted milk coolers designed for bulk farm storage (see Fig. 2.19b). These can be very effective for the chilling of small lots of product for a short period of time, such as cooling juice from the press during the vintage season. They cannot, however, handle the larger refrigeration requirements that follow during the course of wine processing.

The aging of some wines, such as for some of the more delicate white and rosé table wines, should be performed in cellars that are kept cool—generally between 40 and 55°F. Detartration processing requires temperatures of about 25°F. In warmer climates, the entire winery may need air-conditioning.

There are three basic designs for refrigeration systems. In the first, refrigerated air is forced into insulated cellars within which wine storage tanks and other vessels are placed. In the second, a refrigerated liquid (water and propylene glycol mixtures, usually) is circulated through insulated and jacketed storage tanks. In the third system, used in conjunction with the cold room or alone, wine is circulated through heat exchange tubes to insulated storage tanks or to the cold room. There are, of course, advantages and disadvantages for each type of system.

Some winemasters prefer the "cold-room" approach to refrigeration, especially in smaller wineries, as most any wine storage vessel can be placed in that area for cooling. This allows for 5-gal. carboys or 50-gal. barrels to be refrigerated, along with the larger tanks that may be stationary in the cold cellar. This advantage is offset, however, by a lack of flexibility in temperature control: each wine storage vessel in the cold room will be the same temperature. Customarily, cold-room installations are somewhat less expensive than other refrigeration systems. Figure 2.22 illustrates 3 remodeled semi-truck refrigeration unit employed in a small winery cold room.

Other winemasters prefer individual refrigerated tank systems. It is common to find several tanks supplied with a single refrigeration unit, each tank being temperature controlled by the amount of refrigerant that is allowed to circulate through its cold-wall, or jacket. Figure 2.19a depicts used tanks that can be employed for such a purpose. While individual tank systems are flexible as to temperature, they do not serve well when small or odd-sized lots of product are involved. Figure 2.23 depicts a "manifold" system of refrigerant applied to an independently refrigerated storage tank system.

In larger wineries both of these systems may be found, each with specific purposes. In some instances, both cold-room and individual tank systems

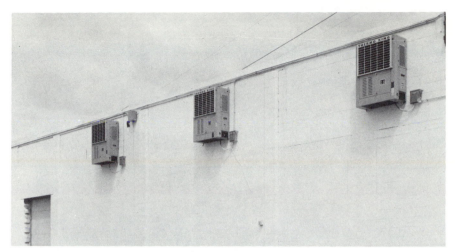

Courtesy: Thermo King Corporation
Minneapolis, Minnesota

FIG. 2.22.   SEMI-TRUCK REFRIGERATION UNITS ADAPTED TO WINERY REFRIGERA-
TION REQUIREMENTS

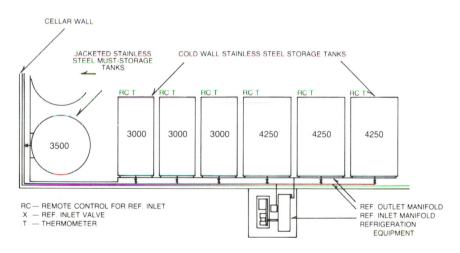

FIG. 2.23.   MANIFOLD SYSTEM OF TANK REFRIGERATION

FIG. 2.24.   DIATOMACEOUS
EARTH DOSING UNIT

Courtesy: Seitz-Werke GmbH,
W. Germany
SWK Machines, Inc.,
Bath, New York

are operated from the same refrigeration generation system. Engineering is so advanced that a system can be designed with elements to fit the requirements of almost any winery purpose.

Filters

Filtration is one of the most critical and technical operations in the winery cellars. In order to determine the proper size and type of filter to be used, flow rate, porosity and condition of product must be considered.

Most winery operations make use of the plate-and-frame type filter, especially when sub-micron porosity filtration is required. These devices operate by the insertion of pads between the plates and frames. Filter pads are available in many different porosities, for removing particles of various sizes (see Appendix B). Wine is pumped into the inlet side of the filter at a constant pressure, usually at less than 26 psi, to avoid rupturing the pads. The inlet frames receive the wine, but in order for the wine to reach an outlet plate, it must pass through a pad. This, of course, removes any suspended solids that are larger than the porosity of the filter pads being used.

Plate-and-frame filters for use in a winery should be specially designed and fabricated for cellar use—conversion or makeover from related-type filters is not recommended. The market for used high-grade wine filters is practically nonexistent.

Some vintners choose to adapt the plate-and-frame filter for very coarse wine filtration with special frames that are designed to hold the diatomaceous earth media. This powder-type filter media is available in a number of different grades and porosities. Filtration with diatomaceous earth may be made with wines that have not responded adequately to fining agents or that are difficult to filter through coarse pad media. In larger wineries, where wine processing time is critically maintained, coarse filtration with diatomaceous earth may be a routine treatment. Figure 2.24 illustrates a dosing unit that is commonly used in diatomaceous earth filtrations. This device regularly measures and "doses", or pulse-feeds, a specific amount of the filter media for maximum efficiency of the surface area. Figure 2.25 portrays a modern plate-and-frame filter. The same filter can be fitted with diatomaceous earth frames for coarse filtration duty.

In larger wineries one can usually find diatomaceous earth filters specially designed and fabricated for coarse filtration. With the advent of centrifuges and more efficient clarification materials, such filter types are, however, not nearly as commonplace as was the case a decade or so ago.

Several plate-and-frame filters may be employed in different areas of the winery. Even in some small wineries, one plate-and-frame filter may be reserved for the bottling area, while another remains full time in the cellars. In larger wineries, several such filters may be employed, each having a specific purpose. However, the heavy capital outlay required for a top quality stainless steel plate-and-frame filter (some heavy-duty plastic plates and frames are acceptable) usually limits most small wineries to one device. Therefore, the choice of manufacturer, filter capacity, potential for increased capacity (addition of more plates and frames), adaptability for diatomaceous earth filtration, and ease of absolutely thorough hot-water (or steam) cleaning must be careful considerations.

Ion Exchange Columns

The concept of ion-exchange is an intricate subject in inorganic chemistry and shall not be considered in depth here. The application of this process to the stabilization of wines, however, has brought dramatic savings of time and energy to wine production.

The reaction involves the exchange of potassium ion from unstable potassium bitartrate for the sodium ion from sodium chloride (common table salt). The reaction forms a stable (soluble) sodium bitartrate compound in the wine, and potassium chloride, which is discarded. In actuality, a large number of reactions take place in the exchange operation, involving such elements as calcium, magnesium, iron, copper and other natural wine constituents. The chemical composition of the wine can be significantly altered. The principle is similar to that of water softening. In general, the ion exchange being done currently is not of this type. This method has been given up by most wineries because of the addition of sodium ions to the wine. Many use combinations of two and three column exchange using anion and cation columns.

Winemasters interested in the highest quality wine production usually do not choose this method of stabilization, and favor instead the traditional process of cold-storage precipitation of potassium bitartrate salts. Ion-exchange is a rather controversial topic among winemasters: good cases can be made to support either pro or con. In any event, ion-exchange columns are expensive and require very sophisticated laboratory controls for proper operation. Figure 2.26 illustrates a modern manual ion-exchanger for use in a larger winery.

BOTTLING, PACKAGING AND WAREHOUSING MACHINERY

The operations that take place in the bottling and packaging facilities change the status of wine products from bulk inventory to cased goods. Clean bottles are filled with mature wine, adjusted, preserved, final filtered, checked and double-checked. Then they are immediately corked or capped, and usually packaged with capsule and label(s). A few winemasters choose to remain with the more traditional bottle-aging of newly bottled wines for several months or years (in cases, bins or stacked in piles) before packaging materials are applied.

Bottling, packaging and warehousing operations may be located away from the receiving and cellar portions of the winery, if necessary. Well

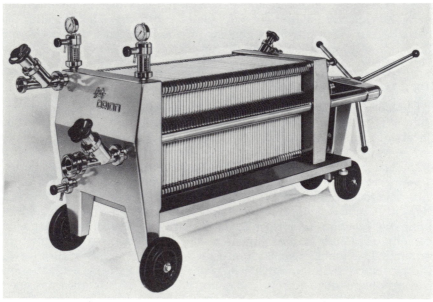

Courtesy, Seitz-Werke GmbH, W. Germany
SWK Machines, Inc., Bath, New York

FIG. 2.25.   STAINLESS STEEL PLATE AND FRAME FILTER

FIG. 2.26.   A WINERY ION-EXCHANGER

Courtesy: Valley Foundry And
Machine Works
Fresno, California

designed and protected pipelines can be used to move the final product from the cellars as needed.

Bottling machinery is not as heavy as the large bulk wine storage vessels in the cellars, although floors must still be properly engineered for vibration and loading. Bottling cellars should also have sufficient drainage. A well engineered discharge system for the cleaning water should include pitched floors and drain troughs, feeding into the same approved effluent system as is utilized by the cellars. Bottling cellar floors should be protected from being corroded by wine acids by several applications of appropriate sealant.

Larger and more automated bottling and packaging operations normally require much more electricity than the smaller semi-automatic facilities. Good planning should include a "busway" or some other accessible source of electricity installed directly over the line (usually mounted upon the ceiling). This allows for conduit and wiring to be flexible and manipulated as needed, while being out of the way of other functions.

A major design parameter of warehousing that is sometimes overlooked is sufficient floor load capacity. Stacked cases of wine have high load requirements (Fig. 2.27). For example, two loaded and stacked pallets, each with eight layers of filled wine cases, will need floor support of at least, 400 lb/sq ft. The operation of a fork-lift truck on warehouse floors can easily double this load factor.

Temperature control is also very important to consider in the warehousing areas. Today finished packaged wines are commonly stored at 72°F— somewhat warmer than traditional temperatures prior to modern-day energy shortages. Even so, this function still requires significant insulation and air-conditioning investments.

FIG. 2.27. PALLETIZED
WINE CASES

*Courtesy: Wiederkehr Wine
Cellars, Inc.
Altus, Arkansas*

Winery tourists are often very interested in the bottling, packaging and warehousing operations. Watching grapes being received is also popular, but the vintage season lasts only a short time. The cellars are customarily quite picturesque, but there are few actual operations taking place for visitors to watch. Bottling is an active operation: wine can be seen entering the bottles and the final product taking shape. A wide inventory of cased goods can also be interesting to visitors. It should be remembered, however, that there are many hazards in the bottling, packaging and warehousing areas of the winery. To avoid accidents, tourists should be protected by screens, windows, and other restraining devices which still allow for observation.

The market for bottling and packaging machinery is keen, with many different types and sizes of equipment available. Cost, speed and output are factors to be considered when choosing this machinery. Semi-automatic bottling and packaging output are usually restricted by labeling, the slow-est operation to be performed without total automation. A speed of 5 to 10 bottles per minute can be expected from most semi-automated equipment, providing sufficient labor is available. This averages about 300 cases per day. The diversification of labor that is usually necessary in small wineries limits bottling and packaging operations to about 10 weeks per year, or 15,000 cases annually. At 2.37781 gal. per 9-liter case, those 15,000 cases amount to just over 35,667 gal. of production yearly. Automated bottling and packaging lines can be designed and fabricated to produce from 30 bottles per minute to more than 10 times that amount. The calculation is, of course, modeled after production requirements—large wineries are often designed with several lines for the finishing operations.

Descriptions of each of the hundreds of modern bottling and packaging machines available would, in itself, comprise a large text. This discussion is meant only to acquaint the reader with some of the more common and popular items operating in American wineries. Figure 2.28 provides a graphic illustration of a compact single-unit bottling and packaging system now available.

Courtesy: Seitz-Werke, GmbH, West Germany
SWK Machines, Bath, New York

FIG. 2.28.   A COMPACT SINGLE-UNIT BOTTLING AND CORKING MACHINE

Final Filtration

The very last operation prior to bottling is final filtration. Generally, the task involved is to remove the yeast and bacteria that remain in the wine. Sometimes there are also light colloids or other minute suspended solids that require removal.

As stated previously, the final filter in a small winery may well be the same filter that is used for more coarse work in the cellars, and is transported back and forth as needed. This can be an effective way to economize, but unfortunately, it can also lead to microbiological contamination. Under such circumstances, checked and double-checked hot-water cleaning programs are most essential.

In plate-and-frame filters, final filtration can be adequately accomplished with pads having a porosity of less than .75 microns. Properly performed, such filtration should remove all yeast cells and most harmful bacteria (see Appendix B for the relative sizes of these microbes). Despite this technology, it is highly recommended that adequate preservative be added, even when very low porosity pads are used. Without the protection of a preservative, the unchecked passage of just one viable cell could spoil the product, especially with residually sweet wines.

Extremely low porosity filtration can be made with the newer membrane filter types. Measurements are often made at the millimicron level. Application of such devices usually requires pre-filtration of the product through some other type of filter to remove all solids other than microbes.

Such equipment is very expensive to purchase, to supply and to monitor in the laboratory. Where bottling facilities are maintained under careful standards of sanitation, perhaps even sterile conditions, membrane filters are justified. In some newly designed medium and large winery operations, such equipment is becoming increasingly accepted. Figure 2.29 illustrates a modern membrane filter applied to final filtration requirements in a winery.

Bottling Tanks

A well designed bottling line will have at least two bottling tanks: one to contain the wine being bottled, and the other to hold the wine being prepared for the following day's production. In some wineries a third tank is utilized, to contain the wine which has been transferred from the cellars prior to adjustments, preservation and final filtration. In any event, each tank should be of sufficient volume so as to hold the bottling capacity of one entire day or shift.

Custom designed high-grade stainless steel bottling tanks should be provided with adequate insulation if ambient temperatures in the bottling cellars vary more than 3°F from 72°F. In some instances, such as where bottling tanks are mounted above the bottling line so as to gravity-feed a filling machine, refrigerant may need to be circulated through the bottling tank jacket in order to control the temperature. In other cases, wines may arrive from the cellars too cold, requiring jacket hot water circulation prior

MEMBRANE BEING PACKED INTO FILTER

FIG. 2.29.   MEMBRANE FILTER

Courtesy: Millipore® Corporation
Bedford, Massachusetts

to bottling. Low-speed agitation is necessary so that adjustments and pre-servatives can be uniformly dissolved prior to final filtration. Each bottling tank should be identical in size and style to others being used in the same line so that volume comparisons before and after final filtration are easily checked. The final qualifications for bottling tanks are that they be in-stalled and located so that steam cleaning may be done safely and easily, and also that the tanks drain dry with an open outlet valve. Both of these features can significantly reduce microbiological contamination.

In smaller wineries, where semi-automated bottling lines can produce less than 400 cases per shift, the application of used farm bulk storage tanks can, again, provide significant cost savings.

Fully automated bottling lines require much larger bottling tanks. Such operations are usually sophisticated enough to warrant the installation of custom designed bottling tanks, although converted vessels of various de-signs can be used effectively. The larger, high-speed lines do not normally rely upon gravity in order to supply filling machines, but rather, use low-speed and low-head centrifugal pumps.

Bottle Cleaner

New glass bottles are shipped from manufacturers in paperboard cases, standardized by ATF regulation to exactly 24 ⅜-liter half-bottles per case, 12 ¾-liter bottles to the case and 6 1½-liter jugs per case. Several other sizes are also available. The bottles are clean, but usually are contaminated with case dust. This must be removed prior to filling the bottles, in order to prevent the wine from also becoming contaminated.

The most simple device for performing this task is a spray tube upon which a bottle is placed upside down. A hot water spray is applied, and the bottle is allowed to drain dry, the case dust having been removed by the rinsing action. While these spray cleaners, as illustrated in Fig. 2.30a, are inexpensive to purchase, they can be rather expensive to run due to the slow manual operation required.

A variation on the inverted spray principle is shown in Fig. 2.30b. This custom-built hot water bottle rinser is capable of several times the output realized by the device described in the previous paragraph. Many other variations on this rinser exist.

Another type of cleaning mechanism is illustrated in Fig. 2.31, whereby rinsing water containing free sulfur dioxide disinfectant is used in the rinse operation.

Of course, the faster bottling lines cannot operate efficiently with small-capacity bottle cleaners. Three major types of high-speed bottle cleaners are commonly found in American wineries: the air-blast type, the multi-cycle bottle washer and the simple bottle washer with one spray.

The air-blast machine (Fig. 2.32) is usually fed a number of bottles simul-taneously. The containers are held securely and "pulsed" with high-velocity jets of air directed inside each bottle. This can be an effective means of removing case dust from bottles, but such systems can be inadequate at

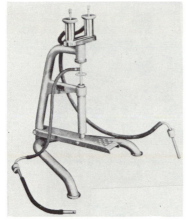

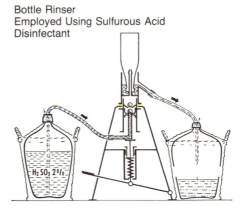

Bottle Rinser
Employed Using Sulfurous Acid
Disinfectant

Single-Valve Bottle Rinser
With Dual Stations

Courtesy: SWK Machines, Inc.
Bath, New York

FIG. 2.30a.   SINGLE-VALVE SPRAY-TUBE BOTTLE RINSER

FIG. 2.30b.   INVERTED BOTTLE SPRAY WASHER RINSER

times. For example, when the bottles are cold, moisture condenses inside, and because of surface-tension, contaminants are not released. Figure 2.32 provides a graphic description of an air-blast bottle cleaner.

FIG. 2.31.   SULFUR DIOXIDE
(H₂SO₃) BOTTLE RINSER

Courtesy: Seitz-Werke GmbH, W.
Germany
SWK Machines, Inc., Bath, New York

The multi-cycle bottle washer can be a very large, expensive machine. However, there is currently no better way to clean bottles. Most bottle washers provide a mechanism whereby the bottles are held securely on a track that transports them through the various detergent washing and final rinsing cycles inside the machine. Such thorough cleaning is obtained that several manufacturers guarantee sterility of the bottles when the last cycle is completed. These machines are designed primarily for European bottling lines where used wine bottles are refilled. In the United States, multi-cycle bottle washers are used primarily for "transfer-process" sparkling wines, where bottles used for fermentation become heavily exposed to yeast. Figure 2.33 depicts a modern multi-cycle bottle washer.

Bottle Conveyors

The bottle conveyor transports clean bottles from one station to another during the bottling and packaging process. In small wineries there may be no actual conveyor system, only an extended table or some other flat surface upon which bottles can be passed from one station to another.

Powered bottle conveyors are standard devices which are almost always custom designed and built with stock components. Some of the new plastic track is less expensive and has less friction than traditional metal tracks. Newer units have widely diversified guide rails so that a multitude of sizes and shapes of bottles may be handled on the conveyor. In nearly every instance, bottle conveyors are powered with a unit that has a variable-speed transmission. This allows the bottling line foreman to adjust the line speed for the type and size of product being produced. Figure 2.34 portrays a modern variable-drive bottle conveyor.

Case Conveyor

After the cases have been emptied and the bottles are being cleaned, the voided cases are normally placed high on a roller-type conveyor so that gravity can move them to the end of the packaging line. There they can be marked, stamped or labeled and refilled with finished product. Adjustable legs under the case conveyor are necessary in order to make a proper incline. On high-speed bottling lines, where machinery can be spread out to lengths to 75 ft or more, a powered conveyor may be needed in order to raise the cases up high enough so that they can be passed on to the gravity roller conveyor and inclined down to the final station.

Case conveyors serve both for unloading new bottles from transport trucks and loading finished product cases into vehicles for delivery.

Courtesy: McBrady Engineering
East Hazel Crest, Illinois

FIG. 2.32.   AIR-BLAST BOTTLE CLEANER

FIG. 2.33.   A MULTI-CYCLE
BOTTLE WASHER

Courtesy: SWK Machines
Bath, New York

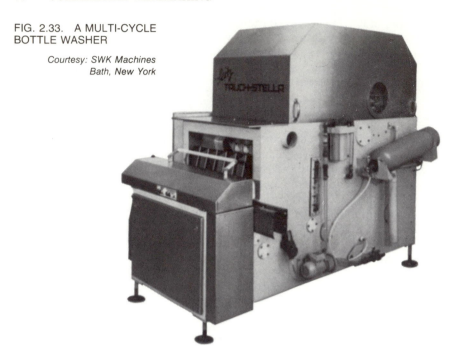

Filler

Unless filling is done quickly, efficiently and with a minimum of exposure, delicate table wines can be adversely affected. Excessive foaming, or "surge", for instance, may reduce free sulfur dioxide preservative levels to the point that browning (oxidation) may result; or worse, inhibited yeast cells may be rendered viable, causing refermentation. Spillage from poorly engineered bottling operations may cause significant product shortages and microbiological contamination.

One of the highest quality devices for bottle filling is illustrated in Fig. 2.35. This machine is usually supplied by overhead bottling tanks so that filling is done by gravity. It can also be used directly with the outlet of the final filter.

Properly operated, such machines can be acceptable for all still (nonsparkling) wine bottling operations in the small winery.

The faster bottling lines are usually fitted with automatic multi-spout fillers of either the vacuum or pressure type. The number of spouts is calculated from the speed by which the line is to be operated. It is characteristic to find 20 to 30 spout fillers in bottling lines that operate at about 60 bottles per minute.

Vacuum fillers draw a vacuum inside the bottle as the filling cycle commences. The spout valve then opens and a precise amount of wine (⅜ liter,

¾ liter, etc.), measured according to a standard fill height, fills the vacuum. These machines are popular with many vintners, as modern pressure fillers are much more expensive. Figure 2.36 depicts a common vacuum wine filler in a winery.

Pressure filling machines provide more protection to the wine in all phases of the filling process than do vacuum fillers. Normally, wines are pressurized in the filler bowl with nitrogen gas, in order to reduce exposure to oxygen and contaminants. Bottles are also flushed with nitrogen gas prior to being filled. The actual filling process is performed with special types of valved spouts that do not permit oxygen exposure. As the filled bottles leave the machine, the remaining head-space is filled with nitrogen to await immediate closure with a cork.

Corker

Immediately after filling, properly soaked corks (see Chapter 6) are compressed and driven into the necks of the bottles. In addition to the

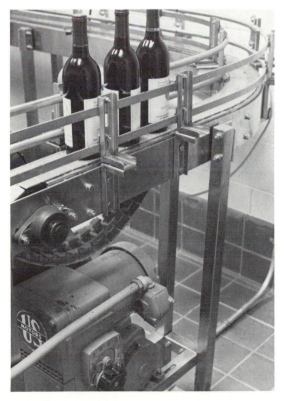

FIG.   2.34.   VARIABLE-SPEED BOTTLE TRACK

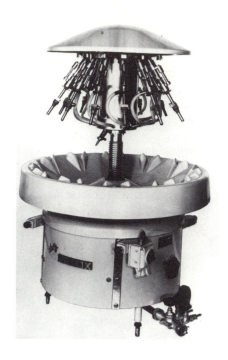

FIG. 2.35.   SEMI-AUTOMATIC BOTTLE
FILLER

Courtesy: Seitz-Werke GmbH, W. Germany
SWK Machines, Inc., Bath, New York

natural cork cut from the bark of the cork tree, there are now agglomerated corks, which are pieces of cork compressed with adhesive to standard shapes and sizes. A further advance in technology has led to the development of synthetic corks which can represent significant cost-savings.

Very simple and low-cost corkers are available for small winery requirements. They are, however, tedious to operate, and semi-automated corkers, such as shown in Fig. 2.37, are preferable for most low-speed bottling lines. The operation of these devices is dependent on a supply of properly prepared corks which are gravity-fed from the hopper to the "jaws" of the machine. The insertion of a cork in each bottle and the activation of the cork compressor subsequently drives the cork to a proper depth in the neck of the bottle.

Fully automatic corking machines are similar. As each bottle enters the corker, it is secured in a position whereby the cork is compressed and driven. Models are available which have several driving heads. Figure 2.38 illustrates a popular model of automatic corker.

Capsuler

Apart from enhancing the appearance of wine bottles, custom-decorated capsuling can provide pilfer-proof sealing.

Some of the earliest capsuling efforts were performed by dipping bottle necks into melted wax. Metal capsules have become most popular, but are very expensive. The cost of lead capsules has been somewhat reduced by the introduction of lead-alloy materials. Nevertheless, some vintners choose to absorb these ever-increasing costs, to maintain the image of top quality. Single-head semi-automatic roll-on machines for metal capsules are available. Multi-head automatic machines, such as pictured in Fig. 2.39 are commonly used in high-quality packaging operations. Figure 2.40 provides an illustration of a properly applied lead-alloy capsule from a roll-on machine.

There are two principal types of plastic shrink capsules marketed to vintners. A traditional type is shipped and stored in formaldehyde. Application

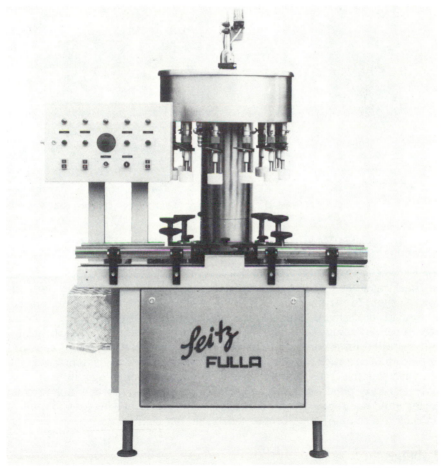

Courtesy: Seitz-Werke GmbH, W. Germany
SWK Machines, Inc., Bath, New York

FIG. 2.36.   AUTOMATIC VACUUM-TYPE BOTTLE FILLER

Courtesy: SWK Machines
Bath, New York

Courtesy: SWK Machines
Bath, New York

FIG. 2.37.   SEMI-AUTOMATIC
CORKER

FIG. 2.38.   AUTOMATIC CORKER

of these capsules requires a few minutes' exposure to the atmosphere, allowing them to shrink to the shape of the bottles on which they are placed. A new kind of shrink capsule is offered which requires heating in order for proper shrinkage to occur. This is normally managed by a heating tunnel installed on a portion of the bottle conveyor. Both types of capsules can be applied either by hand or by machine. Figure 2.41 demonstrates the finished product from the application of a heat-shrunk capsule.

Heavy-duty plastic capsules (Fig. 2.42) are also marketed, which provide a clean and neat appearance. These are, however, nearly as easy to remove as they are to apply and, consequently, do not seal securely.

Labeler

Gluing a label to a wine bottle seems simple enough, but is often the limiting factor for output from hand-operated or semi-automated packaging lines. Despite the fact that a number of semi-automatic labeling machines are offered to vintners, few are consistently effective. The handling of die-cut labels, one at a time, through a glue-roller, then positioning them exactly on each bottle, is an operation that is difficult to engineer in a low-cost semi-automated machine. Figure 2.43 depicts an acceptable low-speed labeler of this type, but even rebuilt used models are very costly.

Rather than making the relatively large investment for a labeler, most small-winery vintners employ one or more gluing machines. Labels inserted into this device are promptly glued so that they can be immediately applied, by hand, individually on wine bottles. Most operators soon develop the skill for uniform label positioning, and a "jig" can be made to insure proper label placement.

Totally automatic labelers are, like all high-speed bottling and packaging equipment, serious investments. Today, there are methods of leasing

Courtesy: Otto Sick KG, W. Germany
SWK Machines, Inc., Bath, New York

FIG. 2.39.   AUTOMATIC ROLL-ON CAPSULING MACHINE

FIG. 2.40. LEAD ALLOY
ROLL-ON CAPSULE

FIG. 2.41. HEAT SHRINK
PLASTIC CAPSULE

FIG. 2.42. HEAVY DUTY
PLASTIC CAPSULE

FIG. 2.43. SEMI-AUTOMATIC LABELING MACHINE

that may be more economical than the usual purchase of such equipment. In either event, most types of automated labeling machines are efficient and dependable, and are often available with supplemental devices which provide for back and neck labeling. Figure 2.44 illustrates a modern, totally automatic labeler in a large winery.

Fork-lift Truck

During most of the year, the fork-lift truck is assigned to duty in the warehouse: receiving cases, taking them to and from the bottling line and loading them on transport vehicles in shipment. Figure 2.45 portrays a fork-lift truck in operation.

The fork-lift is an indispensible tool at the receiving dock, especially during the vintage season. A rotary-head unit can be installed and utilized for dumping grapes into the crusher-hopper. There are, of course, many other uses for the fork-lift truck, such as in the receiving and placement of machinery and equipment.

Miscellaneous

As explained earlier, there exist many different types of devices and machines that may be used in the winery, only a few of which have been introduced in this text. Among the machines which have not been discussed

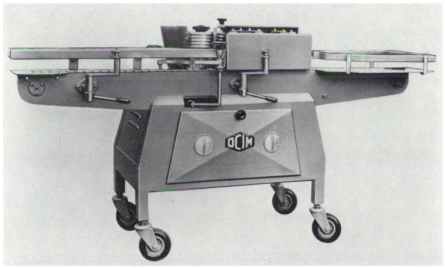

Courtesy: OCIM INTERNATIONAL INC.
Chicago, Illinois

FIG. 2.44.   PORTABLE AUTOMATIC LABELING MACHINE

are automatic and semi-automatic pomace removal equipment, remote-control cellar systems and single-unit bottling and packaging systems. The purpose here, however, has not been to be exhaustive, but simply to provide a basic description of elements most commonly found in small and medium-sized wine processing facilities.

THE LABORATORY

The laboratory is normally the control base for the winemaster (Fig. 2.46). Here information is gathered, decisions are formulated, responsibilities are directed, and actions are recorded.

The winemaster is usually the liaison between the small winery and the Bureau of Alcohol, Tobacco and Firearms (ATF). The establishment of a new winery in the United States is regulated by the ATF in accordance with Part 240 CFR, as described earlier in this chapter.

State and location regulations may also be applicable to the operation of a winery. The winemaster is the communication link with these authorities.

In some large wineries one may find several laboratories, each divided into separate functions. The disciplines of agronomy, botany, plant pathology and horticulture support viticultural operations in a specific area. Another laboratory may specialize in the chemical and physiological analyses necessary for the maintenance of winemaking quality control. In a

FIG. 2.45.   MOTORIZED FORK TRUCK WITH ROTATING HEAD

FIG. 2.46.   TASTING THE COMPONENTS IN THE PREPARATION OF A BLEND

separate facility one may find an elaborate tasting room, free from distractions which may inhibit sensory evaluations. Yet another laboratory may be employed for research and special problems, as well as product development.

Most small wineries, however, combine most, or all, technical functions within one facility, often separating only the tasting room. This concept can serve adequately if the vintner fully understands technical requirements and follows through with a laboratory floor plan that does not create confusion. In other words, a laboratory can be designed to operate effectively and efficiently if specific analytical operations are located in separate areas so as not to cause complication. Such design considerations may well justify the employment of a competent winery consultant to assist with the project.

The breadth of small winery laboratory activity cannot be accurately measured by the number of different wines that are produced in the winery. The analytical parameters for the quality control of only white table wine production, for instance, would not have to be greatly expanded to include rosé table wines, red table wines and even sparkling wines. The volume of laboratory work can, of course, be projected by the numbers of wines produced and the frequency of their production.

With the growth of modern research and technology, the chemical constituency of grapes and wine has become a listing of several hundred elements and compounds, most of which occur in very small concentrations. Normally, the small winery laboratory will be concerned, on a day-to-day quality control basis, with only about a dozen major constituents. For research and special problem areas, many more compounds may be encountered.

There are as many different chemical compositions of grapes and wine as there are different kinds of grapes and wine. The following tabulation serves as a very general idea of natural grape and wine composition:

Component compound	Approximate percentage in grapes	Approximate percentage in wine
Water	75.0	86.0
Sugars	22.0	0.2
Alcohols	.1	11.2
Organic Acids	.9	.6
Minerals	.5	.5
Tannins	.2	.3
Nitrogenous Compounds	.2	.1
Polyphenols	.1	.1
Total	99.0	99.0

It is interesting to note that the very finest of grape cultivars can be distinguished from the most disreputable by differences in only about 3% of total grape composition, considering that water and sugars will not affect aroma and taste.

Sampling

Poor sampling technique is one of the most frequent sources of error and confusion. Whether it is while gathering grape petioles and soil in the vineyard, or sampling juice, must, and wine in the cellars, the main thing to remember is that samples must be representative.

Nowadays most wineries in the U.S. rely upon private laboratories or an extension service to analyze soils and leaf petioles. The high cost of equipment and materials coupled with infrequent testing has resulted in better results for a lower price when independent testing laboratories are employed. The requirements for petiole and soil sampling are usually very definitive.

Sometimes growers will establish rather elaborate systems for gathering grape samples from a vineyard prior to harvest. Thus, a vineyard may be cross-sectioned with samples from the first vine in the first row, the second vine in the second row, the third vine in the third row, etc. This can be further refined by alternating sides of the vines from which the samples are taken, as well as substituting higher and lower levels on the trellis. The wincmaster should insure that the methodology does not exceed sensible levels of expense and practicality. (Building a file on the relation between your sampling results and your actual harvest data is important).

The proper labeling of samples is another way to help insure minimum sampling error. Typical small winery sample labels are illustrated in Fig. 2.47. Of course, the best of laboratory labels will serve no useful purpose unless the personnel gathering samples in the vineyards and cellars are meticulous in providing all pertinent information.

Samples should also be current and timely. Heavy damage can take place very quickly in an infected area when samples are delayed. It may be helpful to frequently review sampling philosophy with all production personnel, as sampling is done for good reason and should not be delayed.

The simple commonsense rule of sampling is that the laboratory must receive representative samples as soon as possible after gathering.

Thief

A thief is a device by which juice and wine samples are taken in the cellars. Crushed grapes (must) are normally sampled in beakers or some other wide mouth container. A thief, or beaker, should be constructed of stainless steel, although polypropylene or glass may also be used. The important item, as regards construction material, is that no chemical alteration of the sample will take place. It follows that the thief should not introduce any contaminants to the wine or juice being sampled.

Simple types of thieves draw from just beneath the wine surface to a depth of only several inches. There are also submerged-action devices that will not allow sample intake until remotely activated by the operator triggering an inlet valve. Figure 2.48 provides sketches of two common wine thieves employed by small wineries.

Thieves are usually no smaller than 300 ml capacity and not larger than 1000 ml in volume. A complete wine analysis will not typically require more than about 700 ml or so. Most small winery laboratories use readily available 750 ml wine bottles as sample containers. (However, recleaning such sample containers may create problems.) This represents much less of an expense than if precision laboratory glassware is used for sample containers.

The thief, the sample container and their storage spaces can easily contaminate samples when not properly cleaned and stored after each use. It is imperative that this be given more than just casual attention. Unless there is an unsightly stain or some other unusual problem that would require brushing or scrubbing, the thief and sample container should be thoroughly cleaned with hot water and stored in an area that is free of dust, dirt and other influences. This is a simple requirement, but one that must always be properly attended to. If more than one sample is to be gathered in the cellars, a container of hot water large enough to allow a complete submerging of the thief should be carried along. The thief can then be thoroughly rinsed between samples, and this will help to avoid any infection.

ABC WINERY Lab Sample—For Analysis Only		
Date	Tank No.	Temp.
Lot No.	Type-Variety	
Remarks:		

XYZ WINERY 702 Vineyard Road Eno, State 12345		
Lot No.	Type-Variety	
Date	Tank No.	Temp.
Alc.	Ball.	V.A.
Remarks:		
LABORATORY SAMPLE FOR ANALYSIS ONLY		

FIG. 2.47. TYPICAL SMALL WIN-
ERY LABELS

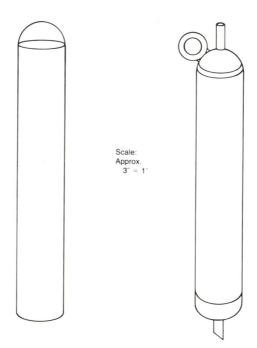

Scale:
Approx.
3" = 1'

FIG. 2.48. COMMON WINE THIEVES

Laboratory Analysis Log

The most basic administrative form in the laboratory is the *laboratory analysis log*. A suggested format is displayed in Figure 2.49. Every sample taken from the vineyards or cellars, whether fully analyzed, partially tested, or not checked at all, should be entered in this log. This practice will result in a complete record of every item offered and considered in the pursuit of quality control.

Complete identification is essential when entering a sample in the log. The little extra time necessary to fill in the spaces for tank number, lot number and type is very small in comparison to the disastrous results that can take place from receiving information late. Consider, for example, the consequences of having added sulfur dioxide preservative to one wine tank twice and none at all to another tank. The double-treated wine may, in time, recover to become a second-class or third-class product. The forgotten wine may well have been lost to oxidation or bacterial spoilage. Obviously, the single error may carry double jeopardy.

Lot Number Systems

A *lot number* system is very helpful in the administration of both laboratory and accounting functions in the winery. ATF inspections will proceed much more smoothly when good identification procedures have been implemented. The general idea is to develop a simple code that will condense critical information into just several digits and/or letters. One common method is the three digit system.

First digit denotes year or vintage:

    0—1980, 1990, 2000, etc.
    1—1981, 1991, 2001, etc.
    2—1982, 1992, 2002, etc.
    3—1983, 1993, 2003, etc.
    4—1984, 1994, 2004, etc.
    5—1985, 1995, 2005, etc.
    6—1986, 1996, 2006, etc.
    7—1987, 1997, 2007, etc.
    8—1988, 1998, 2008, etc.
    9—1989, 1999, 2009, etc.

Second digit denotes origin of grapes or wine:

    0—grapes grown on winery estate
    1—grapes from rented or leased vineyards in the local appellation
    2—grapes purchased that were grown within the local appellation
    3—grapes purchased that were grown within the regional appellation
    4—grapes purchased that were grown within the state appellation
    5—purchased wine grown within the local appellation

LABORATORY ANALYSIS

DATE 9-3-79

TANK NO	LOT NO	TYPE-VARIETY	GALS	pH	BRIX-BALL	ALC	EXT	TA	VA	FSO₂	TSO₂	CU	FE	LT	O₂	CC	N	T	A
		Grape samples: 9-3-79																	
Vyd	T-7	Cabernet Blanc		3.25	15.1			1.03	schedule harvest for 9-5										
Vyd	T-2	Sauvignon Blanc		3.05	10.9				very green yet — new samples 9-6										
Vyd	T-11	Chardonnay		3.10	11.4			"	"	"		"			"		"		
		Stability samples: 9-3-79																	
	819	Sauvignon Blanc	296 c.s.			-1.1	11.3	2.7	.510	.027	84	128	OK	stability seems OK so far			✓	✓	✓
	749	Cabernet Sauvignon	197 c.s.			-1.4	11.4	2.5	.540	.054	48	72	slight sediment — recheck			✓	✓	✓
	849	Aurora Blanc	219 c.s.			0.0	11.8	4.0	.533	.020	88	144	OK	stability seems OK so far			✓	✓	✓
		Tank checks: 9-3-79																	
S-19	818	Cabernet Sauvignon							.060	last check on 8-5 = .054 — recheck 9-4								
S-21	818	Chardonnay							.054	"	"	"	"	" = .054 OK					
		Grape samples: 9-4-79:																	
XYZ Farm		Sauvignon Blanc		3.10	11.6				very green yet — new samples 9-6										
Todd Hanover Farm		Sauvignon Blanc		3.05	11.1			"	"	"		"		"		"		

FIG. 2.49.   LABORATORY ANALYSIS LOG IN USE

6—purchased wine grown within the regional appellation
7—purchased wine grown within the state appellation
8—experimental
9—other

Third digit denotes status of production:

0—fermenting juice or must
1—new wine in first racking storage
2—new wine in second racking storage
3—new wine in third racking storage
4—blending completed
5—clarification agents applied
6—wine in detartration storage
7—wine primary filtered from clarification and detartration lees
8—wine in aging storage
9—wine approved and ready for bottling (or further processing)

Using the above code, a wine with a lot number of 925 Chardonnay, for instance, would instantly provide a vintner with the information that this Chardonnay wine is from the vintage of 1979, made from purchased grapes grown within the local appellation and presently awaiting detartration storage.

A Vidal Blanc from the vintage of 1980, grown from the vintner's own vineyard and aging in the cellars would carry a lot number of 008 Vidal Blanc. When two or more wines qualify for the same lot number, but the winemaster wishes to maintain separate sub-lots, the lot number may be extended as 008-1 Vidal Blanc, 008-2 Vidal Blanc and 008-3 Vidal Blanc. In this way one may also distinguish between a 925A Chardonnay, a 925B Chardonnay and a 925C Chardonnay. There are, of course, many other devices for differentiating between sublots.

Wines that require extensive processing, such as sparkling wines, or extensive aging, such as dessert wines, may require a more elaborate system of lot numbers. The important thing to keep in mind when designing number coding is that the system must be simple and practical. It should not require lengthy, time-consuming bookkeeping, but it should fully serve the purposes needed.

Laboratory Organization

The furniture and fixtures preferred by one winemaster may be far different from those of another. For instance, one winery may be more concerned with one particular analytical area than another.

Most new or remodeled buildings being considered for winery operations are planned with ample space set aside for the quality control functions. An analytical laboratory that is 11 ft wide and 16 ft long (176 sq ft) can serve adequately. Figures 2.50, 2.51, 2.52 and 2.53 are drawn to depict just such a laboratory plan. Note the large expanse of window on either side of the service door in Fig. 2.53. This permits visitor observation. These windows also serve as a barrier in order to protect tourists from the many hazards that exist in such a facility, while providing minimal distraction and inter-ruption for personnel working inside the laboratory.

The quality control area is of interest to many people and should be readily accessible on the winery tour. The prime considerations for location, however, must remain quality control and space utility.

Organoleptic analysis can be performed in an analytical laboratory. However, due to the interference of chemical reagents, cleaning supplies and other materials that can influence the human sensory organs, it is advised that a separate area be reserved for tasting evaluations. It is typical to find the winemaster's office dividing the analytical laboratory and the organoleptic laboratory. A floor plan illustrating such an arrangement is provided in Fig. 2.54.

The tasting laboratory should be designed so as to provide ample space and aesthetic appeal. As mentioned previously, wine can be properly evalu-ated only when the environment has a neutral effect on the human senses. Among the most common contaminants are those of color, light, odor, sound, taste and temperature. Both the analytical and tasting laboratory should have stark white countertop surfaces, so that wine colors and analyt-ical reaction end-points can be observed from an accurate perspective. Walls, ceilings and floors should be covered with neutral colors such as beige or eggshell. Incandescent light of the "daylight" type is preferred to the "blue" of some fluorescent sources. Many winemasters choose to have an unfrosted bulb close by the tasting bench while others still use a candle—both of which are aids in wine clarity analysis.

An exhaust fan may be needed to bring in fresh air from outdoors so as to free the organoleptic laboratory of odor contaminants before each use. Storage of anything other than wine in the tasting facility should be avoid-ed. There should be relative freedom from distracting noise when wines are being judged, although absolute silence is not usually necessary. Soft music may be preferred in order to "drown out" outside interference. Taste con-taminants are often found in glassware that has not been properly cleaned or stored from prior use. (Even containers of distilled water often become contaminated with odors.) Cupboard storage should be carefully evaluated as wine glasses can easily harbor paint and varnish vapors. Hot water-cleaned tasting glasses hung inverted in open racks should relieve this problem. The tasting laboratory should be temperature controlled so that the taste panel is comfortable. The organoleptic laboratory should be a facility that optimizes the winemaster's abilities.

Chapters 5 and 6 will introduce the need for stability analysis and a storage library for retained samples from the bottling line. Figures 2.55, 2.56, 2.57 and 2.58 illustrate typical elevations in an adequate organoleptic laboratory that includes a stability testing area and wine storage library.

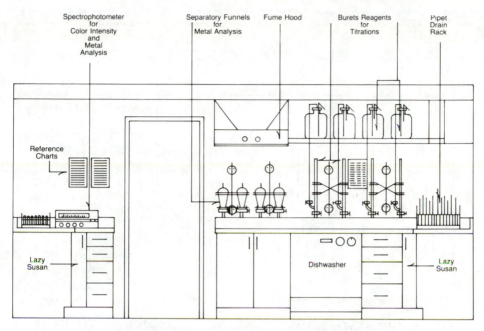

FIG. 2.50.   INTERIOR ELEVATION OF ANALYTICAL LABORATORY

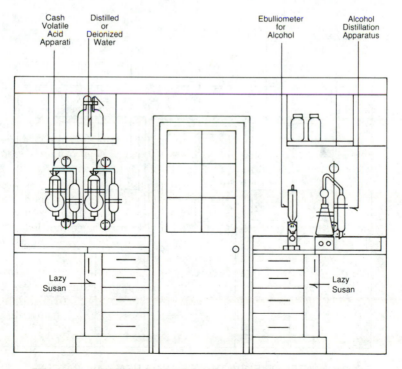

FIG. 2.51.   INTERIOR ELEVATION OF ANALYTICAL LABORATORY

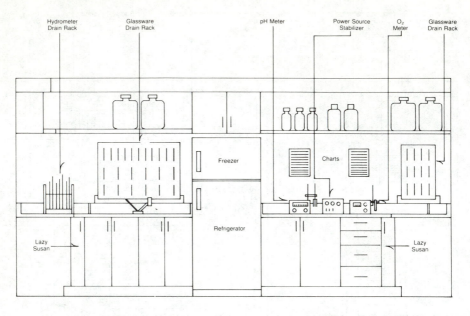

FIG. 2.52.   INTERIOR ELEVATION OF ANALYTICAL LABORATORY

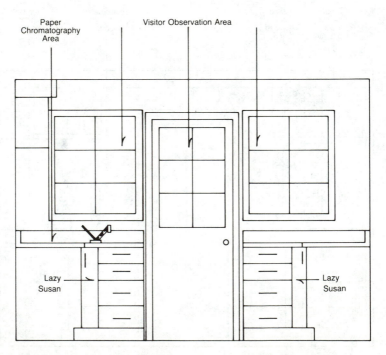

FIG. 2.53.   INTERIOR ELEVATION OF ANALYTICAL LABORATORY

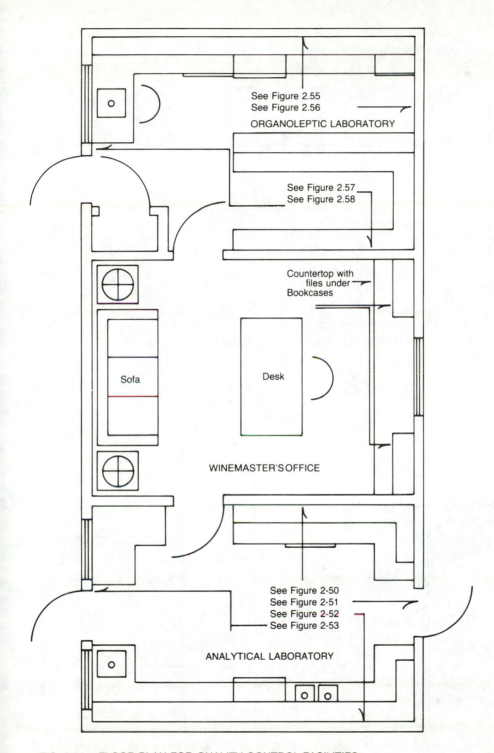

FIG. 2.54. FLOOR PLAN FOR QUALITY CONTROL FACILITIES

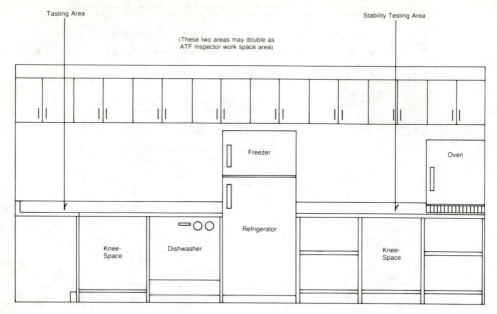

FIG. 2.55. INTERIOR ELEVATION OF ORGANOLEPTIC LABORATORY

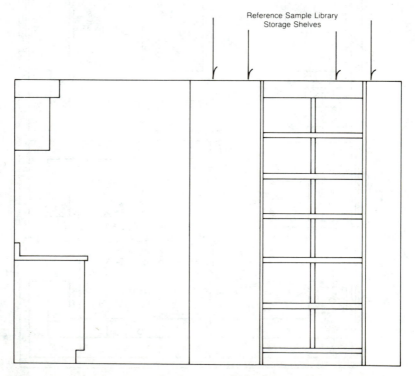

FIG. 2.56. INTERIOR ELEVATION OF ORGANOLEPTIC LABORATORY

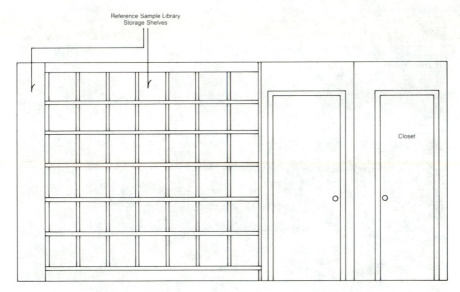

FIG. 2.57.   INTERIOR ELEVATION OF ORGANOLEPTIC LABORATORY

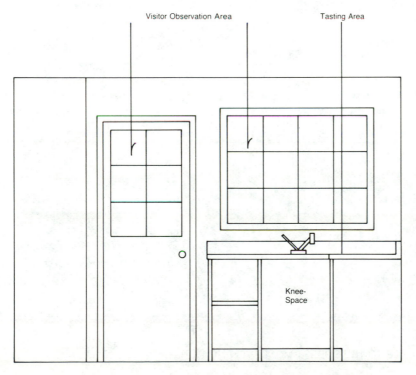

FIG. 2.58.   INTERIOR ELEVATION OF ORGANOLEPTIC LABORATORY

ENTRANCE TO THE SCHLOSS JOHANNISBERG WINERY, RUDESHEIM, W. GERMANY

ENTRANCE TO THE ROBERT MONDAVI WINERY, NAPA VALLEY, CALIFORNIA

The use of quality control facilities should be fully understood through a consistent policy as prescribed by the winemaster. The nature of a laboratory is such that it can become a kitchen and a social gathering place unless persistent measures are taken.

In many small wineries, one person can handle all of the administrative tasks of the office and still have time to assist with some of the tasks in the laboratory. The winemaster should be eager to enlist such help and learn to maximize other sources of labor that may become available. This is particularly important during the harvest season, when there is an abnormally heavy load of analytical quality control work. It should go without saying that the people involved in quality control should have the ability and attitude to produce accurate results.

Equally as often, there may be enough work in the laboratory to require the full-time services of an analyst. This person may be trained to perform many of the more repetitious day-to-day analyses in a "recipe" style, overseen by the winemaster. Record-keeping and other duties can also be assigned in order to relieve some of the "taskmastering" otherwise looked after directly by the winemaster.

Colleges and universities should not be overlooked as a source of labor. Students may often sacrifice higher salaries in order to get on-the-job experience. There are programs where some of the expense is borne by the institution sponsoring the student. Another source of analytical help would be through state extension services and laboratories.

3

Grapes, Juice and Must Quality Control

The capture of natural essences and flavors can be elusive to even the most experienced of winemasters. Most wine experts agree that the largest single variable in winemaking is the fruit, both in terms of variety and condition.

It is imperative that grapes arrive at the winery as soon as possible after harvesting, especially grapes intended for more delicate types of wine such as whites and rosés. Once the fruit leaves the vine, decomposition commences, permitting oxidation, spoilage and other problems to develop. Some further guidance in this area is provided in Chapter 9.

HARVESTING

The harvesting of grapes manually has become a difficult task to justify for the vineyard manager. Ever-increasing labor costs and unavailability of seasonal labor have resulted in machine harvesting becoming a widely accepted alternative. In whatever manner one may choose to evaluate the question of economics, it remains that machine harvesting strips the berries from the vine and bunch, leaving wet scars (lesions on the berry where it has been detached from the pedicel). Such injury can provide the opportunity for negative elements to significantly detract from quality. Furthermore, the question of MOG (material other than grapes such as sticks, stems, leaves, trellis remnants, etc.) remains controversial, although new inspection methods and trellising techniques have greatly reduced the infection from MOG.

None of this is meant to imply that acceptable wines cannot be made from machine harvested grapes—there are plenty of examples to the contrary in the marketplace. Many growers take extraordinary care in the "good manners" of machine harvesting. Small wineries, however, usually find their best niche in the marketplace by insisting upon high fruit quality.

Hand harvesting, properly done, renders only the bunches of grapes, relatively intact, with little of the juice prematurely extracted. The hand harvesting of Muscadine cultivars, some more than others, will normally

*Courtesy of Chisholm-Ryder
Company, Niagara Falls, N.Y.*

FIG. 3.1.   A MODERN MACHINE HARVESTER FOR GRAPES

result in some improvement of fruit quality, although wet scars will continue to be a problem unless maximum attention is given to rapid transit from vineyard to winery.

VINEYARD SPRAY MATERIALS

It is highly recommended that a vintner pay close attention to vineyard spray schedules—both his own and those of independent suppliers. It should be common practice for growers to submit copies of their complete vineyard spray programs for inspection, approval and file prior to any fruit being accepted at the winery. Certainly, each grower ought to know about spray schedule requirements long before the growing season commences. Particularly important is whether or not any sprays were applied within specified periods prior to harvest. Also, it is advisable to compare each grower's spray schedule along with previous programs he has submitted in order to calculate any radical changes that may have taken place which could affect fruit and/or vine quality. The winemaster may wish to inquire if the grower's permit to use spray materials is up to date, so that complications with federal and state environmental protection authorities can be avoided.

SAMPLING

Representative samples of fruit should be taken from vineyards daily after the grapes are past véraison and showing signs of maturity. The winemaster should insure that the best combination of pH, Brix and total acidity will be attained. These analyses are discussed in more depth later in this chapter. Testing methods are described in Appendix A. Currently, these three determinations, in conjunction with the winemaster's best sensory judgement, are the most affordable and meaningful tools available for measuring grape maturity.

Samples of grapes should be closely examined visually for evidence of damage, disease, insects, mechanical injury and other defects. A typical method of standard control is to take a percentage of berries that exhibit particular types of defects and post that figure in the record of that vineyard. Abnormally high results would cause the winegrower to examine for a trend that might need immediate attention. Standards can be set for defective fruit such as "more than 2% diseased and/or damaged fruit will be subjected to reduced price or rejection . . ."—a common condition in grape contracts. In some instances larger wineries will employ an inspection team from the U.S. Department of Agriculture to inspect fruit as it arrives at the winery during the course of the vintage season. While this is expensive, it does relieve winery quality control personnel of the added burden of grape inspection during an already very busy time. The USDA serving as an impartial quality control inspector can also remove internal bias and subjectivity that may occur because of personal relationships between winery and grower personnel.

CELLAR PREPARATION

One of the most common factors contributing to poor results in winemaking is that of inadequate cellar hygiene. Apart from the aesthetic benefits, a clean winery improves efficiency and working conditions and protects the equipment. Sanitary conditions minimize the potential for development of off-flavors which could easily contaminate and destroy wine products. Title 21, CFR, Food and Drugs, Part 110, *Current Good Manufacturing Practice in Manufacturing, Processing, Packing, or Holding Human Food,* is absolutely essential reading for the new vintner. A periodic review of this FDA regulation by established winery personnel is also recommended.

County health laws differ from place to place, but most are very explicit in their requirements for cleanliness in food and beverage production areas. State and federal regulations are also definitive in the enforcement of environmental conditions for the manufacture of consumables. Several quotations from federal codes are common references permitting action to be taken against food and beverage producers who are in violation.

Well maintained sanitary conditions will not generally be criticized often. Innocent violations are normally met with simple directives for correcting the discrepancies.

An entire separate text could be prepared to cover the major sources of filth and infestation possible in a winery. It becomes a matter of common sense for the winemaster to be alert and investigative in his cellars. Failure to properly dispose of waste materials such as stems, pomace, lees, spent filter pads and other such production items can quickly cause contamina-

FIG. 3.2.a. DISEASED GRAPES—
BLACK ROT

FIG. 3.2.b. BLACK ROT ON AGA-
WAM GRAPES

tion. Insecure storage of materials and supplies may attract insects and rodents. Toilets, washrooms, sanitary drainage and effluent processing are often sources for difficulties. Residues from insecticides, improperly used rodent poisons, and even cleaning agents, when inadequately used, are common problems. However, the largest single contributor to unsanitary conditions is human neglect.

CLEANING AND CLEANING MATERIALS

Clean, insofar as this text is concerned, shall be defined as free from harmful contamination. Again, common sense must be applied. The precise nature of steam-cleaning a sub-micron filter pack in order to process wine for bottling is a far cry from greasing the screw-shaft of a grape press, yet both are "cleaning" operations. The former operation protects the product from microbial activity once the wine is bottled. The latter protects the product from rust by the use of a special vegetable food-grade lubricant.

The most effective cleaning agent in the winery is hot water. It is a good solvent, sanitizes and leaves no meaningful residue unless the water is excessively high in calcium, iron and magnesium, a condition normally referred to as hard water. In such a case, the winemaster may be advised to install a deionization apparatus to soften the water supply.

A detergent in solution is designed to act as a magnet which will lift dirt and other contaminating particles from unclean surfaces, dissolve or suspend them, and carry them off as effluent. Most detergents are complex mixtures of ingredients in different concentrations formulated for specific cleaning purposes. One common component of detergents is sodium dodecylbenzene sulfonate, a surfactant that cuts grease. Another typical compound found in such cleaning agents is sodium tripolyphosphate, a water softener and one of a number of phosphates that are pollutants. Carboxymethyl cellulose helps to prevent soil particles from being redeposited back onto the cleaned surface and benzotriazole is an antitarnishing agent.

Strong alkali detergents such as lye, ammonia and other caustics can be good cleaning agents, but are hazardous to use (eye splash may cause blindness) and are corrosive to many surfaces, even glass and metals. Mild alkalis such as TSP (trisodium orthophosphate) and TSPP (tetrasodium pyrophosphate) are very good detergents but are also very serious pollutants. Wetting agents such as quarternary ammonium and sulfonate compounds have the added benefit of being germicidal cleaning compounds. These detergents, in particular, are heavily used in winery cleaning programs.

Inorganic acids are not normally used as cellar and laboratory cleaning agents due to hazards in handling and their extreme corrosive properties. Some organic acids can be used effectively for cleaning and are not normally considered dangerous to manage.

The relatively new organic chelating agents, such as EDTA (ethylene diamine tetra-acetic acid), are detergents that can offer good cleaning results. Some are complex phosphates compounded with other constituents, and are, therefore, pollutants.

FIG. 3.2.c. DOWNY MILDEW ON CONCORD GRAPES

FIG. 3.2.d. DOWNY MILDEW ON FREDONIA GRAPES

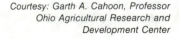

Courtesy: Garth A. Cahoon, Professor
Ohio Agricultural Research and
Development Center

SANITIZING AND SANITIZING MATERIALS

Sanitizing, as differentiated from *cleaning*, describes efforts towards minimizing populations of microbes that may be contaminants, rather than simply removing filth.

Again, hot water is one of the most effective sanitizers in the winery, especially when maintained at a temperature in excess of 165°F, and when combined with steam. These may not be sufficient in open areas, however, and the judicious use of chemical sanitizers may be required.

Apart from the quarternary ammonium compounds mentioned in the previous section, the most common chemical sanitizers commercially available are the halogen-based compounds. Hypochlorous acid is readily available, but is hazardous to use and in the presence of steam can be exceedingly corrosive. Sodium and calcium hypochlorites are common and chlorinated isocyanurates are also found. Iodine sanitizers can have a very strong germicidal action, but can break down very quickly when used with hot water or steam. Bromine compounds can be excellent sanitizing chemicals, but may have only limited use and are difficult to handle.

Commercial competition for the cleaning agent market continues to be keen, and provides the winemaster with an overwhelming choice of materials and methods for the sanitary maintenance program. The following is a suggested list that may be helpful in determining some basic parameters:

Detergents for surfaces in contact with grapes, must or wine

Wood or concrete surfaces:	Soda ash or sal soda
Metal surfaces:	Dilute caustics or citric acid
Sanitizing:	Sulfur dioxide or halogen compounds

Detergents for winery interior surfaces—non-contact with grapes, must or wine

Cleaning:	Soda ash or sal soda
Sanitizing:	Sulfur dioxide, halogen compounds or quarternary ammonium compounds

Detergents for winery exterior surfaces

Cleaning:	Unslaked lime, chloride of lime or sodium bisulfite
Sanitizing:	Sulfur dioxide, halogen compounds or quarternary ammonium compounds

Sulfur dioxide, although listed as a cleaning agent, is even more important to the winemaster as an antioxidant and preservative, and will be further discussed as such in Chapter 5. It is assumed that, where application will permit, hot water and/or steam are the primary materials for any cleaning or sanitizing requirement.

It may be advisable, especially in a new winery, to consider the expenditure for a qualified winery consultant prior to the vintage season. Quite often an expert can offer significant time and money-saving suggestions, even to the most accomplished winemasters. There can be dividends in spotting problem areas before they develop and finding ways to upgrade product quality.

Whether a winery is new or old, several "dry runs" of water should be processed through the entire crushing, stemming, pressing and cooling system to insure the proper functioning of equipment and to flush the system free of any contaminants that could remain from cleaning and maintenance. This should be done each year in plenty of time prior to the vintage season so as to allow for shipment and installation of new parts that may be found necessary.

PROJECTING THE VINTAGE

While a vintner may anticipate 150 to 180 gal. per ton for pressed white grape juice, or 125 to 150 gal. per ton of grapes for white Muscadine juice, almost all red wine grapes will normally yield 175 to 190 gal. per ton if fermented on the skins (fermentation of grape must rather than grape juice), depending upon the volume taken up with stems and MOG. Normally a good cellar plan will insure that space for 275 gal. is available for each ton of grapes scheduled to arrive at the winery. No less than 10 days should be reserved for each fermentation of must, while a minimum of 45 days should be alloted for the cooler and, therefore, slower fermentations of white juice.

The total tonnage expected to be received will determine the volume and time needed so that all operations are properly handled. Scheduling should allow for careful handling of the fruit and proper use of machinery. Supplies, in adequate amounts, must be close at hand so that processes can be completed in timely fashion. Most important is a fully trained, and well-manned labor force. The smaller winery often has an advantage over the larger facility in this regard, in that processing the vintage develops a very close fellowship of hard work among family and friends. The winemaster should be fully prepared with all the resources necessary to see the vintage through in a controlled and efficient manner.

As will be mentioned many times in this text, good records are an absolute necessity in any winery. Recording procedures may be reviewed before the vintage season each year in order to insure full utility.

GRAPE PROCESSING CONTROL

Incoming fruit should be weighed, crushed and stemmed immediately upon arrival at the winery after each container has been carefully inspected.

Stems can contribute to tannic acid astringency in wines and, therefore, *stemming* (stem removal) is a common and recommended practice. Some winemasters prefer to leave portions of stems in red wine must while fermenting , in order to extract some of the *tannins*. This is done in the belief that tannins offer the wine longevity. Tannins are phenolic compounds that can slow oxidation processes in wines and, therefore, slow the aging dynamics. The tannins are much more precisely controlled, however, when added as tannic acid later in the processing of young wines although still not well controlled as they affect taste. Muscadine grapes, of course, do not provide stems of any significance.

Crushing, stemming and pumping of the juice or must are straightforward operations that can be relatively problem-free when performed under clean and efficient conditions.

SULFITING

Sulfiting is the common process of adding potassium (or sodium) metabisulfite to the grapes just before crushing in order to inhibit natural yeasts and other microbes by the release of sulfur dioxide gas. A more in-depth discussion of this important procedure is provided in Chapter 9.

PECTIC ENZYMES

The most common application of pectic enzymes in grape processing is in the processing of grape juice and concentrates. These enzymes break down the natural pectin in grapes which can reduce juice yields, inhibit clarification and foul concentrating equipment. Many larger wineries utilize pectic enzymes, to increase yield and clarity. The normal application is usually about one ounce of enzyme per ton of grapes, held at 120–140°F for 20–30 min before pressing. The quality-conscious winemaster can quickly see the negative aspects of this treatment in the small winery.

PRESS AIDS

Pressing aids are designed to increase yields of juice from pressing operations while reducing energy requirements, pressure and strain upon the press. Most of the readily available press aids are fibrous materials that help to keep grape must solids and semi-solids separated, so that outflow of liquid is accelerated. Wood pulp, flocculents and rice hulls are common press aids. Again, these materials are used primarily in large winery operations where time and yield may be of more significance than absolute quality. For this reason, the small winery may prefer "free-run" juice and "first-pressing" under more natural conditions without the expense of press-aids.

Day's end at a winery during the vintage season, when everyone is tired, is the time when the winemaster must double-check for careful cleanup and lubrication operations to minimize the risk of infection.

GRAPE, JUICE AND MUST ANALYSIS

The analytical testing of grapes, juice and must are normally performed on samples in the liquid state. Representative samples of juice will already exist as a homogeneous liquid (although there may be some separation of solids in juice samples). Grapes and must will require juice extraction prior to testing. This can be a source of significant error in some cases. It is difficult to recreate in the small winery laboratory the exact crushing, stemming and pressing results that can be expected in the cellars. However, the best technicians will endeavor to pursue this task. Grape samples pressed too lightly in the laboratory are likely to yield Brix and pH values slightly higher, with total acidity somewhat lower, than samples from the same fruit processed in the cellars. Blender pulverization often disintegrates grape seeds into suspension and solution compounds, which will also affect lab results.

Most samples of must should be devoid of stems. Grape samples other than Muscadines will require removal of stems by hand in the laboratory. Likewise, must samples will already be crushed while grape samples from the vineyard or picking container will require crushing by hand. Small presses are available from most supply houses that handle laboratory equipment. Some vintners devise their own laboratory crusher-stemmer and press.

A laboratory centrifuge is helpful in separating suspended pulp, skin and seed solids from freshly pressed juice. The supernatant juice will provide greater accuracy in the testing procedures to follow.

pH

The analysis of pH in the small winery laboratory is most significant during the vintage season when it can be used as an indicator of fruit ripeness. Normally a pH of 3.10 would establish the beginning of ripening for most grape cultivars in most locales. A pH of 3.50 may signal the end of maturity and the onset of overripening. A pH of 3.3 can be considered ideal for most vineyards to harvest, at least for scheduling purposes. This level may not prove to be the optimum for every location and variety, but can serve as a starting point for later adjustments.

The best estimates of ripeness are based on experience in each vineyard, considering past performances of fruit quality and yield. Much could also be learned by experimenting in the lab. For instance, a winemaster may harvest a 6 acre vineyard by taking 2 acres at pH 3.2, another 2 acres at pH 3.3 and the last segment at pH 3.4. By carefully scrutinizing the results from the wines made, the vintner can create his own individual standards.

pH is a measure of the effective acidity or basicity of a solution. Precisely, the pH is equal to the negative logarithm of the molar concentration of hydrogen ions. More practically, pH is a measurement of the strength of acidity or alkalinity. The scale of pH is from 0 to 14 with neutrality at pH 7. The higher the degree of acidity, the more the pH will proceed from 7 towards 0. In other words, a pH of 3.0 has a higher degree of acidity than a pH of 4.0. Basic, or alkaline, solutions are read from pH 7 to pH 14. A solution of pH 10.0 has a higher degree of alkalinity than one of pH 9.0.

In measuring the ripeness of grapes, pH is used to monitor the reduction of acid strength as the vine translocates basic ions such as potassium and sodium into the berries. The more basic ions that are pumped into the fruit, the less the *degree* of acidity (this must not be confused with total acidity which is a measure of the *amount* of acidity). A pH of 3.1 may be generally considered too strongly acid with insufficient basic ions having been translocated for peak ripeness. Conversely, a pH of 3.5 can be considered to be too low a degree of acidity, with excess basic ions having been carried to the grapes.

By far, the best results for measuring pH are achieved by the use of an electronic pH meter. Litmus paper and other similar methods are more qualitative than quantitative. A pH meter and the necessary electrodes are relatively expensive, but a quality pH meter is needed for quality pH measurements. Most popular instruments, such as that pictured in Fig. 3.3, are sufficient. See Appendix A for the methodology for determining pH in the laboratory.

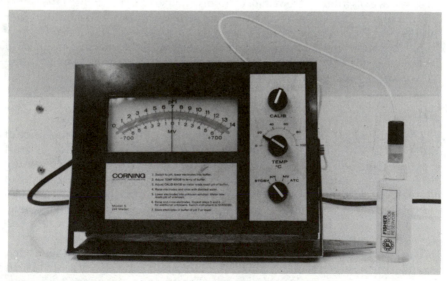

FIG. 3.3.   A pH METER

BRIX

Brix is a very simple and inexpensive measurement of dissolved solids in grapes, must and juice. While Brix testing has become, in the American wine industry, a gauge of sugar content, this is not precisely correct as there are usually other solids besides sugar which affect the testing instrument. In that usually less than 5% of such non-sugar solids are present, few small wineries spend the time or money for precision sugar analysis. For all practical purposes, the Brix measurement is sufficiently accurate for the small winery.

Red grapes and must will generally have slightly higher values of non-sugar Brix solids due to higher concentrations of color pigments and glycerols. Unless the red must is exceptionally pigmented, though, the difference is usually not such as to require alteration of normal amelioration calculation procedures or potential alcohol determinations.

The sugar content of grapes is almost entirely comprised of glucose and fructose. Other fermentable and unfermentable sugars can exist but generally in very small amounts. The level of sugar developed by a given cultivar of grape can vary rather widely from year to year depending upon the growth dynamics of the vine in a particular environment. A hot and dry growing season may yield a light crop of high-Brix fruit. Conversely, a cool and rainy season may result in a heavy crop of low-Brix grapes. Brix cannot be used as a good measurement for ripeness. One year a vineyard may be ready for harvest at a Brix of 17° and the next year not until 21° Brix. Nevertheless, many winegrowers remain steadfast to the Brix measurement as the barometer for ripeness and profit.

Normal Brix results from grapes are usually read within the 12° to 25° range. Wild grapes and some cultivars of V. labrusca, V. riparia and V. rotundifolia may yield fruit of lower dissolved solids. Cultivars of V. vinifera and various French-American hybrids can achieve the upper norms of the Brix scale more consistently. It is common in several winegrowing subdistricts of France and particularly Germany to allow grapes to become overripe and/or to allow molding with Botrytis cinerea—the "noble mold". These practices, when successful, can increase the Brix due to dehydration of the grape berries. This subject is discussed more fully in Chapter 9.

The fermentation of sugar results in the production of ethyl alcohol (ethanol) and carbon dioxide (CO_2) as depicted by the Gay-Lussac equation of 1810:

$$C_6H_{12}O_6 \longrightarrow 2C_2H_5OH + 2CO_2$$

This translates as: one molecule of glucose reacts to give two molecules of ethanol and two molecules of carbon dioxide.

Modern technology, however, has revealed a very long series of biochemical reactions and pathways between the sugar substrate and the yeast waste-products of ethanol and carbon dioxide. This can be only briefly outlined by the following equation:

$$C_6H_{12}O_6 \xrightarrow{\text{(yeast enzymes)}} 2C_2H_5OH + 2CO_2 + 56 \text{ kilocalories of energy}$$

In turn, this can be written as: one molecule of glucose catabolyzed by yeast enzymes gives two molecules of ethanol and two molecules of carbon dioxide plus 56 kilocalories of energy.

Yeast cells are behind the "magic" of fermentation and that fermentation generates energy. In a normal small winery fermentation, about 40% of this energy will be utilized by the yeast cells—the balance of some 60% being converted to heat. Consequently, fermentations of wine are *exothermic* (give off heat) and must be carefully monitored. Higher fermentation temperatures can evaporate precious aromatic flavor compounds, as well as contribute to oxidation reactions.

As would be expected, the sugar, ethanol and carbon dioxide have a direct volumetric relationship to each other in winemaking. One can rather accurately determine how much sugar or concentrate will be required to adjust deficient juice or must to yield a desired percentage of alcohol after fermentation.

One degree of fermentable sugar on the Brix scale (expressed as percent by weight) will ferment to about .535% alcohol by volume. This cannot be precisely calculated in that Brix is not itself a precise measurement of sugar. Nevertheless, the .535 factor is adequate for practical fermentation calculations and is widely used in the wine industry for just such purposes. For one reason or another, some winemasters choose to use slightly different factors such as .525% or .540%, etc.

Figure 3.4 portrays a Brix determination being made with a hydrometer while Fig. 3.5 illustrates a winemaster maneuvering a refractometer towards a light source. Methods for both types of Brix testing are defined in Appendix A.

Bench-type Brix refractometers are available, but are very expensive and have no advantage in the small winery quality control laboratory over a reliable hand-held instrument.

TOTAL ACIDITY

The expression of *total acidity* includes both fixed and volatile acid concentrations in wine quality control. Volatile acidity results primarily from the development of acetic acid (vinegar) by bacterial spoilage of ethanol and other substrates. A more detailed discussion of volatile acidity can be found in Chapter 9.

Confusion can arise as to the difference between pH and total acidity, as both are measurements of acidity. Recall from a previous section in this chapter that pH measures the *strength* of acidity, while total acidity accounts for the *amount* of acidity? Total acidity is, therefore, an expression of tartness.

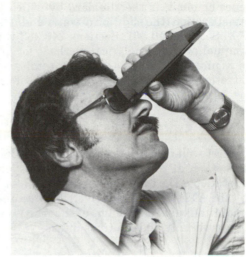

FIG. 3.4. BRIX DETERMINATION
BEING MADE WITH A HYDROME-
TER

FIG. 3.5. BRIX MEASUREMENT CAN ALSO
BE MADE BY MANEUVERING A REFRACTOM-
ETER TOWARDS A LIGHT SOURCE

A vintner may expect total acidity to range from less than .400 g per
100 ml (g/100 ml) to more than 1.400 g/100 ml. There are many factors that
contribute to such a broad variance, including grape cultivar, soil type,
climate, viticultural techniques and many other factors.

Fixed acids comprise the overwhelming majority of acids in wine and can
represent the total acidity in grapes, juice and must. Fixed acidity is composed primarily of tartaric and malic acids, although minor concentrations
of citric and oxalic acids, among others, may be present.

Despite the considerable stability that natural acids can impart in the
preservation of organoleptic qualities, natural grape acids are themselves
quite unstable. Tartaric acid will precipitate, in part, as the acid salt,
potassium bitartrate, sometimes called cream of tartar. A portion of this
precipitate will result from the reduction of specific gravity of juice and
must as fermentation progresses. Another portion will result from the
reduction of temperature in storage. Acids can be utilized by many microorganisms, most of which are spoilage bacteria. The malo-lactic fermentation
is often considered a beneficial reaction and can be induced by inoculation
of new wines with cultures of *Leuconostoc* bacteria. (See Chapter 9).

A total acidity in grape juice of .900 g/100 ml may be reduced to about .700
or even .600 g/100 ml during normal winemaking procedures. Wines made
from the cultivars *V. rotundifolia* (Muscadines and others) often exhibit the
phenomenon of total acidity increasing during fermentation. This synthesis
is not yet fully understood.

Fixed acidity (total acidity less volatile acidity), before fermentation of juice or must, is the standard by which the ATF regulates the limits of amelioration (the addition of water and/or sugar) in Part 240, CFR, Subpart XX, Section 240.961. Sections 240.963 through 240.971 of the same ATF manual provide analytical methods for determining total acidity.

Figure 3.6 depicts a common total acidity test being made by simple titration analysis. Appendix A describes in detail the methodology for performing a total acidity test.

TALLY SHEET

Most well organized records systems in the small winery will include a *tally sheet* at the unloading dock to properly account for grape receiving operations. The tally sheet fully qualifies as a primary supplemental record by the ATF, both qualitatively (source and appellation of the grapes) and quantitatively (net weight of the fruit). Figure 3.7 illustrates an example of a good tally sheet format in use.

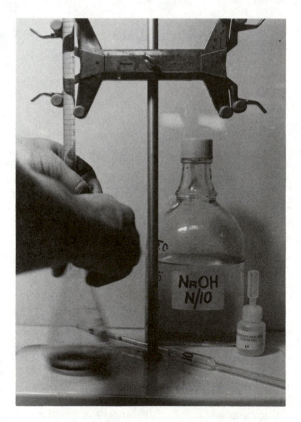

FIG. 3.6. A TOTAL ACIDITY TEST CAN BE MADE BY SIMPLE TITRATION ANALYSIS

TALLY SHEET

CULTIVAR _SEYVAL BLANC_   DATE 9-24-79

NO. OF BOXES	GROSS WT LBS	TARE WT LBS	NET WT LBS	SOURCE DATA	SO₂ ADDITION	MUST TO
227	7,777	636	7,141	XYZ Farms	KMS 14½ oz	PRESS
142	4,541	401	4,140	Todd Yarrow Farm	8½ oz	"
68	2,514	193	2,321	Hilltop Vineyards	4¾ oz	"
120	4,037	339	3,698	Lakeview Manor	7½ oz	"
	18,869	1,569	17,300	= 8.65 tons	35¼ oz KMS	

REMARKS   XYZ and Hilltop grapes had some bird damage — Yarrow grapes had some splitting of skins — all accepted

WEIGHMASTER

FIG. 3.7.   TALLY SHEET IN USE

The beginning winemaker may feel that "grapes are grapes" and that particulars are a waste of time during an already very busy vintage season. It must be made clear that source and appellation of grapes will need to be proven if later statements on wine labels will make mention of specific appellation. In other words, Napa Valley Cabernet Sauvignon may only qualify as California Red Table Wine unless there is proof in the supplemental records that the Cabernet Sauvignon grapes were delivered from a specific Napa Valley vineyard. The tally sheet, along with cancelled checks proving payment to the grower for grapes delivered will provide such a record.

In addition, ATF inspections will include yield data and analytical profiles in order to monitor amelioration volumes and grape concentrate additions. Unless a definite volume of fruit is accounted for, the ATF inspector will have no basis upon which to judge whether or not each lot of wine has been made within the limits of ATF regulation. The tally sheet is vital for establishing a permanent record of grape source and weight data.

Much time and trouble can be saved by the winemaster who insists on a good supplemental records system, of which the tally sheet becomes an important initial part. One might make the analogy of having good records to substantiate returns to the Internal Revenue Service, which is certainly a comfortable feeling during the rigors of an IRS audit.

WHITE GRAPE MUST PROCESSING

The pressing of crushed and stemmed white grapes may, or may not, immediately follow the crushing operation. In either case, the white must should be taken to a temperature equal to or lower than the temperature planned for the eventual fermentation of the white juice.

Unlike reds, white grapes are not normally fermented on the skins since leucoanthocyanin pigments can be extracted. These pigments often contribute undesirable colors and flavors, rendering bitter, foul and heavy white wines, rather than delicate, fresh and fruity.

Hence, much of the flavor constituency remains in the pomace after the white juice is pressed out. Some vintners choose to crush and stem white grapes leaving the must intact from several hours up to several days. While this helps to leach out additional compounds from the skins and enhance flavor intensity, it also has the potential for natural apiculate yeast fermentation, as well as the growth of bacteria and mold. This practice should be attempted with the greatest of caution.

Pressing white grapes, if not completed properly, can result in poor juice quality and, in turn, poor wine quality. A large amount of juice will run through the press without any effects from pressure. This is aptly called "free-run" juice and is sometime kept in a separate lot from juice that has been pressed. Some vintners feel that the free-run juice is superior to that which is extracted under pressure. However, there are probably at least as many winemasters who feel that the juice obtained from lower pressures is every bit the equal to free-run. In any case, most winemasters agree that white grapes cannot be highly pressurized to make high quality juice. For most white wine cultivars, a yield of more than 175 gal. per ton may be considered excessive. For Muscadines, a maximum of 140 gal. of juice per ton would be a limit. After these yield volumes are attained, the pressure necessary for additional gallons would extract oils, solids and other detrimental components.

POMACE MOISTURE TESTING

Pressing efficiency can be monitored in the laboratory by measuring pomace moisture. A sample of the pomace from the press is taken to the laboratory and immediately weighed to at least two decimal places (in grams) in an evaporating pan. The sample is then completely dried and weighed again. After the sample has been discarded, the tare of the pan is taken and subtracted from the "before" and "after" gross weights. The difference in weight is attributed to moisture loss by evaporation from drying. Consider the following example:

Gross weight $-$ 106.657 g (before drying)
Gross weight $-$ 60.252 g (after drying)
Tare weight $-$ 39.873 g (weight of clean evaporating pan)

then:

Net weight $-$ 66.784 g (before drying)
Net weight $-$ 20.379 g (after drying)
Moisture $-$ 46.405 g

$$\text{Pomace moisture} = \frac{46.405}{66.784} = 69.5\%$$

This procedure will not work well for measuring pressing efficiency of white Muscadine grapes. The heavy pulp of these cultivars makes juice extraction very difficult. Consequently, the finest white wines made from Muscadines are a result of free-run and low-pressure juice. The remaining pomace in the press will be very high in moisture.

JUICE CLARIFICATION

Clarification of the freshly pressed juice is highly recommended. Juice with high concentrations of suspended solids can release excessive carbon dioxide during fermentation, perhaps resulting in violent "boiling" which will make temperature control more difficult. Suspended grape solids are also good media for enzyme decomposition, particularly when fermentation nears completion and the solids have precipitated in the form of *lees* (the formation of sediment in the bottom of fermentation vessels).

Centrifugation of fresh white grape juice can be an excellent method of clarifying some species and hybrids, providing the vintner uses a machine designed for the purpose. Poor quality machinery and converted equipment, such as dairy separators, will not perform properly and can be exceedingly dangerous to use. Some cultivars of *V. labrusca* and *V. rotundifolia* have been shown to react negatively to centrifugation, with a marked reduction in available nitrogen as well as pigment distortion. In any event, the centrifuge is a very expensive item—small units are often priced in excess of $50,000. Adequate clarifying can be achieved by allowing the juice to naturally precipitate solids in a refrigerated atmosphere.

Prior to any clarification method used, the new juice should be analyzed, as described earlier in this chapter. This is especially true for the determination of total acidity levels in juice and must that is intended for amelioration. As will be explained in Chapter 4, amelioration limits are regulated by the ATF according to total acidity concentration in juice or must prior to fermentation. Juice in cold storage can precipitate significant amounts of acid salts which can greatly reduce, or even eliminate, the need for amelioration.

Most small wineries will add 50 to 75 ppm of sulfur dioxide (SO_2) at the crusher-stemmer in order to protect the new juice from natural yeast fermentation and bacterial infection. This practice also helps to reduce oxidation. It should be pointed out that pasteurization is emphatically not recommended for microbiological sanitation in musts as this technique can aggravate oxidation potentials and contribute to deleterious changes in natural chemical profiles.

RED GRAPE MUST PROCESSING

Crushing, stemming and pumping of the red must remain rather straightforward operations for red grapes, and can be virtually problem-free when performed under clean and efficient conditions. The winemaster may follow closely the methods for handling white grapes discussed earlier in this chapter.

Extraction of skin color is usually necessary in making the best red table wines. A few of the very darkest cultivars may yield acceptable color from only fermenting pressed juice. However, for the most part, fermentation on the skins is a standard procedure in making premium quality red table wines. Many red grape cultivars, handled as whites with the juice having been pressed from the grapes prior to fermentation, will yield light red or rosé wines, and sometimes even white wines when red pigments are unstable and/or are in low concentrations.

Must samples from the crusher may provide individual results from which a "composite" analysis can be calculated. In other words, if a 25-ton lot of red grapes is to be crushed and fermented, it may be wise to gather a sample from the crusher-stemmer after every five tons have been processed. This would yield five samples for pH, Brix and total acidity analyses. The five results would then be averaged into a single *composite* analysis. This procedure is normally more accurate than taking a single sample from a large fermentation vat. Figure 3.8 displays the formulation of a composite analysis from the laboratory analysis log.

Fermentation vessels for red grape must should be open-top containers, or have a large opening that will allow easy access to the "cap" of skins that will form and float on the juice once fermentation begins.

Stainless steel vessels are best for all wine fermentations—red, rosé, or white. The advantages of stainless steel will be realized in the simplification of cleaning and maintenance operations. Furthermore, stainless steel is a very durable material and is a poor insulator, which helps to dissipate the large amounts of heat that can be generated—especially in red wine fermentations.

Mild steel that has been lined with food-grade surfaces, fiberglass and wooden vessels can be used successfully if they are in sound condition. Metal and fiberglass tanks have the added advantage of permitting sealable manways in the very top of the container so they can be used for wine storage later. Of course, the custom engineering and stainless steel material, especially the high-nickel alloy such as type 316, can be very expensive.

Fermenters should have two outlets with a full-ported valve for each. One outlet should allow juice to be easily withdrawn from beneath the cap during a red wine fermentation. In larger volumes of more than two tons of must, fermenting juice will need to be pumped out and "over-the-top" onto the cap in order to keep the cap wet. This practice will expose more alcohol to the skins which will aid in extracting more color pigments during fermentation. In addition, the fermenting wine will expel the air in the spaces between the skins, discouraging the growth of acetic acid bacteria which are aerobic. In smaller fermentation volumes, the cap is "punched" with a clean device that pushes the cap down beneath the surface of the fermenting juice. The other outlet valve should be positioned at the very bottom of the tank to allow withdrawal of the entire contents when fermentation has run its course. In smaller fermenters a 2-in. outlet may suffice—with considerable difficulty. However, when the volume exceeds 2 tons in a single vessel, it is advisable to consider outlets and valves of 4-in. inside diameter (ID), and preferably 6-in. ID, especially when the fermenters must be emptied to the press by pumping rather than gravity. By far, cone-bottom tanks mounted high enough to allow gravity-feed to the press are the easiest to operate and maintain.

Whatever the combination of devices and materials chosen for red wine fermenters, it is very important that the tanks be cleaned with hot water several times to remove any residues prior to depositing juice or must for fermentation. Cleaning material such as alkali and detergents can withstand several rinsings with water, and can react to contribute noticeable "soapy" values to bouquet and flavor in resulting wines. In addition, large counts of acetic bacteria can be harbored in the pores of fermentation vats, especially those of wood construction, as well as in the irregular surfaces of manways, valves, fittings, unpolished surfaces and other areas of both red and white wine fermentation vessels.

LABORATORY ANALYSIS

DATE _9-29-79_

TANK NO.	LOT NO.	TYPE-VARIETY	GALS.	pH	BRIX-BALL.	ALC.	EXT.	T.A.	V.A.	FSO₂	TSO₂	CU	FE	L.T.	O₂	C	N	T	A
F-11	940	SEYVAL BLANC	1,441	3.35	12.7			.908								✓	✓	✓	✓
c/s	920	DeChaunac	APPROX 3 TONS	3.30	20.7			.855								✓	✓	✓	✓
c/s	"	"	"	3.35	20.2			.863								✓	✓	✓	✓
c/s	"	"	"	3.30	19.9			.818									✓	✓	✓
c/s	"	"	"	3.30	20.3			.818									✓	✓	✓
c/s	"	"	APPROX, 2 TONS	3.65	19.1			.705									✓	✓	✓
c/s	"	"	"	3.35	19.8			.810									✓	✓	✓
DeChaunac COMPOSITE AVG = 15 TONS			APPROX 3.32		20.2			.833											✓
			5)16.60/100.9					4.164											

discarded this sample—water which is from clean-up operations.

FIG. 3.8. COMPOSITE ANALYSIS CALCULATED UPON THE LABORATORY ANALYSIS LOG

SUMMARY

The condition of grapes arriving at the winery is of prime importance to the quality of the resulting wine. Hand-harvested grapes are preferred to machine-harvested fruit and the winemaster must insure that vineyard spray programs have been properly conducted by qualified personnel. Pre-harvest samples should be analyzed and re-analyzed as is necessary to provide sufficient data for determination of maximum potential.

The winemaster should make every effort to insure that his wine cellars are properly prepared to receive the vintage. Sources of filth and infestation should be determined and dealt with in an appropriate manner. The development, or redevelopment, of a complete and sound cleaning and sanitizing program should be made in accordance with public health laws.

Projecting the needs of the vintage season, and following up with assurances that these needs will be met is a responsibility that the winemaster must look after very closely during the course of grape processing.

The methodology used for grape processing is determined by the preferences of the winemaster. Sulfiting is highly recommended, but the use of pectic enzymes and press-aid materials is discouraged.

The three-point ripeness indicator scale is composed of pH, Brix and total acidity, with pH being the most important contributor. Brix is most useful in determining the amount of sugar that may be needed in order to ferment to a predetermined level of alcohol. Total acidity is necessary in order to find the maximum amelioration level that the ATF will allow. All three tests, in the hands of a learned winemaster, are used to determine grape ripeness and harvest scheduling.

Good records are a necessity in the operation of any winery. The tally sheet is a very important initial record that accounts for both source and quantity of the fruit taken for wine production. This is particularly important in the establishment of supplemental records necessary for ATF inspection.

White and red grape crushing, stemming and pressing operations require careful handling, treatment and cleanliness. The use of stainless steel fermenters can aid greatly in the ability to maintain clean, efficient operations.

4

Quality Control During Primary Fermentation

This portion of the text will deal with the most common testing procedures and recording practices that are carried out during primary fermentations in the small winery. The expression *primary fermentation*, as used in this book, refers to the initial fermentation of grape juice or must by cultured yeasts, as opposed to *secondary fermentation* in the vinification of sparkling wines or the bacterial *malo-lactic fermentation*.

Several new analytical procedures and supplemental records will be introduced for support of production quality control and ATF requirements. There are more tests and data that could be considered beyond what will be found here, but for the sake of practicality, the following may be assumed as a minimal program of quality control necessary for consistent quality in table wine fermentations.

It is not the way of things for grapes to have a perfect sugar-acid ratio every year. Such fruit is developed only occasionally even from the best vineyards. Some viticultural areas, such as in California and other western states, will consistently yield fruit that is high in sugar content, but deficient in acidity. Most eastern states have vineyards that ripen grapes of high acidity, but low in sugar. Soil types, climates, cultivar selection, cultivation techniques and other factors contribute to this variance. In almost any winegrowing region the winemaster can predict that uncommonly dry and hot growing seasons will yield grapes that are higher in sugar and lower in acid than normal years. Cool and wet growing seasons can be expected to have the opposite effect. The point is that the winemaster can fully anticipate different grape characteristics each vintage season.

Acidity deficiencies can rather easily be made up with simple calculations and additions of citric acid, fumaric acid, tartaric acid, or some special combination of the three. Each of these acids is readily available from commercial winery suppliers. Juice and musts considered only moderately high in acidity may be adjusted with an application of calcium carbonate (chalk). The addition or reduction of acidity should be attended to during later processing of the wines when more precise requirements can be measured. A more detailed discussion of acidity manipulation can be found in Chapter 5.

The addition of sugar to juice, must or wine is illegal in the state of California. Lower values of Brix in juices and musts are necessarily corrected with additions of concentrate. To calculate concentrate additions precisely, one should refer to the Brix tables in Appendix B under the column entitled, "Lb/Solids per U.S. Gallon". The following formula can then be employed:

$$\text{Gal. juice} \times \frac{(\text{LSR} - \text{LSE})}{(\text{LSC} - \text{LSR})} = \text{gal. of concentrate required}$$

LSR = lb of solids per gal. required
LSE = lb of solids per gal. existing in the juice
LSC = lb of solids per gal. existing in the concentrate

Consider the following example:

A winemaster has 1944 gal. of white grape juice that has 18.1° Brix. Grape concentrate at 68.0° Brix is to be added in a sufficient quantity to raise the entire blend to exactly 22.5° Brix:

$$1944 \times \frac{(2.054 - 1.623)}{(7.586 - 2.054)} = 151.5 \text{ gal. of concentrate required}$$

This same general formula may be utilized to add cane sugar (sucrose) without a conversion of Brix values to solids per gallon. Juice gallons, however, must be converted to weight in pounds by using the same Brix tables mentioned above, but under the column entitled, "Lb Total Weight per U.S. Gallon". (Remember that cane sugar may *not* be added to commerical juice, must and wine in California). At 18.1° Brix, 1944 gal. then becomes 8.967 × 1944, or 17,432 lb of juice:

$$17,432 \times \frac{(22.5 - 18.1)}{(100.0 - 22.5)} = 989.7 \text{ lb of dry sugar required}$$

LOT ORIGINATION FOR WHITE WINES

Every lot of wine should have a "birth certificate" in the form of the *lot origination form*. The lot origination form can also serve in the collection of data and in the calculation of sugar additions.

The lot origination form portrayed in Fig. 4.1 includes the data of juice analysis for Lot No. 940 Seyval Blanc exhibited in the laboratory analysis log provided in Fig. 3.8. With the measurement of net grape tonnage as furnished by the tally sheet given in Fig. 3.7, there remains only the exact quantity of net juice to be determined in order to proceed with the task of lot origination.

LOT ORIGINATION

DATE ___9-27-79___

LOT NO. ___940___

SERIAL NO. ___9___

FERMENTER NO. ___F-11___

NET GALLONS ___1,441 (1,544)___

NET TONS ___8.65___

GALLONS PER TON ___166.59___

pH ___3.35___ T.A. ___.908___

BRIX ___17.7___ ALC. _____

EXT. _____

CULTIVAR _SEYVAL BLANC_

REMARKS: some slightly distressed fruit—all crushed stemmed and pressed on 9-24 — juice held in cold storage cellar no. 2 - Tₑₘₚ #5.4 for 72 hours ± 40°F — 105 gals juice lees destroyed

AMELIORATION = ___15___ % @ ___22.4°___ BRIX = ___12.0%___ ALCOHOL

___1,441___ GALLONS START
___.85___ INVERSE OF
    AMEL. PERCENTAGE
= ___1,695___ RESULTING TOTAL
    GALLONS OF PRODUCT

___1,695___ TOTAL
    GALLONS
@ ___22.4___° BRIX(___2.044___ LBS/GAL) = ___3,465___ TOTAL
    LBS.

___-1,441___ START
    GALLONS
@ ___17.7___° BRIX(___1.584___ LBS/GAL) = ___2,283___ START
    LBS.

= ___254___ AMELIORATION
    GALLONS
___1,182___ ADDITION
    LBS.

___- 87.5___ SUGAR AS
    GALLONS
× .074 (GAL/LB) = ___87.5___ SUGAR AS
    GALLONS

= ___167.5___ WATER
    GALLONS

___1,4.1202___ YEAST ADDITION = RATE OF ___1___ LBS PER ___1,000 gals.___

REMARKS: juice has typical Seyval aroma and flavor plus the usual "milky" consistency — looks good

FIG. 4.1.   LOT ORIGINATION FORM IN USE FOR WHITE JUICE

As mentioned in Chapter 3, most winemasters prefer to ferment white grape juice as "clean" as possible—to the extent that some vintners will invest in a centrifuge. Far more operations will employ cold storage of juice for several days in order to precipitate suspended solids which form the juice lees. The juice is then *racked* (decanted) from the lees very carefully and pumped to a clean fermenter. The remaining juice lees are then immediately destroyed and the cold storage vessel cleaned thoroughly. The resulting juice can then be accurately measured and recorded as net juice.

The calculation of gallons per ton is another indicator of pressing efficiency (along with pomace moisture determinations) and can easily be determined with simple division of net juice gallons by net grape tons.

The balance of the lot origination form provides the format for concentrate, cane sugar, liquid sugar or *amelioration* additions (the addition of sugar and/or water), depending upon the design required. Figure 4.1 is a lot origination form prepared for the calculation of amelioration.

Amelioration, apart from being illegal in California, is a subject of great controversy among many wine consumers and producers. This results primarily from the fact that some wine manufacturers will use amelioration to stretch their production, rather than to reduce acidity and increase Brix.

In calculation of exact percentages for amelioration, the percent of the *resulting product* (must or juice plus amelioration materials combined) is determined rather than the percent of the beginning gallonage. In other words, 1000 gal. of juice ameliorated to 10% yields 111 gal. of amelioration material to be added—*not* 100 gal. The 111 gal. is 10% of the resulting product of 1111 gal. This is defined in Part 240, CFR, Subpart XX, Section 240.960 of the ATF regulations.

Consequently, in calculating the total gallons after amelioration, the inverse of the desired amelioration percentile becomes the denominator and the beginning gallonage becomes the numerator. For example, an amelioration of 10% would provide a denominator of .90 (in the illustration given in the previous paragraph note that $\frac{1000 \text{ gal.}}{.90} = 1111$ gal. of resulting product); an amelioration of 20% would provide a denominator of .80; an amelioration of 25% would require a denominator of .75, etc.

AMELIORATION LIMITS

ATF maximums for amelioration additions are gauged by natural levels of total acidity measured in the juice or must. Figure 4.2 furnishes a table for finding amelioration limits based upon the ATF precepts.

The Federal amelioration allowances are very liberal and usually provide for considerably more dilution than is necessary for quality winemaking. Apparently, this rule was intended to allow eastern vintners to compete with the more bland wine products from the warmer climates of California. The prudent limit of amelioration is that level whereby juice or must is properly adjusted, even though the ATF may legally allow much more

LIMITS OF AMELIORATION*

Minimum Total Acidity g/100 ml			Maximum Percentage (Resulting Product) of Amelioration	Maximum U.S. Gallons Amelioration Material Allowed for Each 1000 U.S. Gallons of Juice to be Fermented
.500			none	none
.510			1.96	20.0
.520	.526—— 5.0		3.846	40.0
.530			5.66	60.0
.540			7.407	80.0
.550	.555——10.0		9.09	100.0
.560			10.714	120.0
.570			12.28	140.0
.580	.588——15.0		13.793	160.0
.590			15.25	180.0
.600			16.666	200.0
.610			18.03	220.0
.620	.625——20.0		19.354	240.0
.630			20.63	260.0
.640			21.875	280.0
.650			23.07	300.0
.660	.666——25.0		24.242	320.0
.670			25.37	340.0
.680			26.470	360.0
.690			27.53	380.0
.700			28.571	400.0
.710	.714——30.0		29.57	420.0
.720			30.555	440.0
.730			31.50	460.0
.740			32.432	480.0
.750			33.33	500.0
.760			34.210	520.0
.769 and over			34.997	538.4

* For grapes, fruits and berries other than currants, gooseberries or loganberries as permitted by the U.S. Bureau of Alcohol, Tobacco and Firearms, Part 240 of Title 26, CFR, Subpart 240.960 and 240.961.

FIG. 4.2. TABLE OF ATF AMELIORATION LIMITS BASED UPON TOTAL ACIDITY LEVELS IN JUICE AND MUST

sugar/water addition. The proper adjustment is normally a level of about .770 g/100 ml total acidity, and Brix sufficient to generate the desired amount of alcohol.

Under most conditions the winemaster can expect to lose between 25 and 40% of the natural total acidity in juice and must by the time the wine product is finished. Most of this loss arises from the precipitation of acid salts during fermentation and detartration. This subject is discussed in more detail in the next chapter. The "ideal" total acidity of .770 g/100 ml, therefore, can be expected to decrease to about .510 g/100 ml in the completed wine. This is considered perhaps a little low for European-style table wines, but is more in line with American tastes. Pre-determining ideal levels of juice and must total acidity can be, at best, only an educated guess by the winemaster, as the same vineyard can yield fruit that varies greatly in total acidity from one year to the next. This is further complicated by the fact that the individual acid constituency will also vary, causing different amounts of acid salts to be formed and precipitated. The level of .770 g/100 ml is simply a rule of thumb.

The example given in Fig. 4.1 portrays Seyval Blanc juice with a Brix of 17.7° and a total acidity of .908 g/100 ml. If the winemaster desires a finished wine with approximately 12.0% alcohol, then the 17.7° Brix will be insufficient as 17.7 × .535 (alcohol generated by 1.0° Brix) = only 9.47% alcohol. In this example the .908 g/100 ml of total acidity is higher than the .770 g/100 ml considered optimal. The method of calculating the proper level of amelioration is then determined as follows:

$$.908 - .770 = .138 \quad \frac{.138}{.908} = .1520 \text{ or } 15\% \text{ amelioration}$$

From this result we can easily see why the 15% amelioration rate was chosen for this lot of juice. Note, however, that according to ATF limits as tabled in Fig. 4.2, this juice could be legally ameliorated to the maximum level, although at the great expense of quality.

In some cases there are other dynamics that need to be considered. Muscadine cultivars, for instance, may have deficiencies in both total acidity and Brix. Normally this situation is corrected with an appropriate calculation for a grape concentrate addition, similar to what is commonly practiced in California. A lot origination form exhibiting the use of concentrate is illustrated in Fig. 4.3. The mathematical formula for calculating this addition was discussed earlier in this chapter.

Muscadines also have the unique ability for acid synthesis while fermenting, as mentioned in Chapter 3. In some vintages this is an intense phenomenon, while in others the reaction is milder, a topic of great concern and research in the south. In any case, a deficient total acidity in the juice or must of a Muscadine lot at, say, .488 g/100 ml may reach a level of .550 or even higher during and after fermentation. At this writing there are no specific ATF regulations to deal with this contingency. The winemaster may be advised to wait until his fermentation is well in progress before calculating and adding his amelioration, particularly in terms of total acidity adjustments.

The wise winemaster will consult with another qualified person to formulate the amelioration plans. Given agreement, then the entire calculation should be rechecked in order to be sure that ATF, state and any other existing regulations are complied with and that mathematical determinations are correct.

LOT ORIGINATION

DATE _9-29-79_   CULTIVAR _MAGNOLIA_

LOT NO. _920_   REMARKS: _<15% green and_

SERIAL NO. _11_   _<20% overripe. hand picked_

FERMENTER NO. _F-4_   _9-26 early am - then_

NET GALLONS _1,682 (1.994)_   _crushed and pressed late am_

NET TONS _13.810_   _and early pm - juice to_

GALLONS PER TON _121.80_   _TK #5-13 for 70 hrs. ± 40°F_

pH _3.35_   T.A. _.488_   _312 gals juice lees_

BRIX _17.1_   ALC. _____   _destroyed_

EXT. _____

AMELIORATION = _____ % @ _22.5°_ BRIX = _12.2 %_ ALCOHOL

none — see below (above AMELIORATION)

_1 lb 3 oz_ YEAST ADDITION = RATE OF _1 lb_ LBS PER _1,000 gals_

REMARKS:

$$1,682 \overline{)2.054 - 1.584} \quad = \frac{790.54}{5.532} = 142.9 \text{ gals}$$

with $.47$ above the division, and $7.584 - 2.054$ over 5.532

= 142.9 gals concentrate @ 68.0° BRIX

1,682 gals juice + 142.9 gals concentrate = 1,824.9 total gals.

FIG. 4.3. LOT ORIGINATION FORM USING GRAPE CONCENTRATE TO INCREASE BRIX LEVEL

WORK ORDER

All operations directed by the winemaster to his cellars should be issued in a written *work order*. This becomes a permanent supplemental record which is vital for maintaining quality control in the wine cellars. An original should be prepared and given to the cellarmaster and a carbon copy retained in the laboratory, preferably in a bound or padded record. This format should be followed even if the person who made the calculations and wrote the work order is the same one who is going to follow through with the actual work in the cellars. Completed work orders are returned to the laboratory where gallonage figures and other pertinent data can be posted to the appropriate records. The duplicate can then be checked off in the work order record and the original filed in that particular month's *file of operations*. The file of operations will be discussed in further detail in Chapter 10. Figure 4.4 portrays a work order issued for the amelioration additions calculated from the lot origination form worked out in Fig. 4.1.

FERMENTATION CONTROL RECORD

The *fermentation control record,* or *fermentation log,* is a rather simple form kept to monitor and record fermentation events and progress. It is usually attached to, or printed on the reverse side of the lot origination form in order to keep the data from each fermenting lot in an organized manner. Figure 4.5 illustrates a good fermentation control record that might be expected to follow the Seyval Blanc juice lot calculated for amelioration in Fig. 4.1 and work order issued in Fig. 4.4.

The tally sheets, lot origination forms and fermentation control records comprise the three major annual supplemental records that should be kept on file in the small winery. When the vintage season has been completed so that all new wines have been entered in the bulk wine book inventory (defined later in this chapter), the annual supplemental records may be kept in a file entitled Vintage Season 1979, or some other appropriate title. More on this topic is provided in Chapter 10.

Fermentation Tube

It is often difficult to determine whether or not a given lot of juice is fermenting in the initial stages. A rather quick and simple answer can be obtained qualitatively with the use of a fermentation tube. Being modestly priced, several glass fermentation tubes are a sensible investment in the small winery's analytical laboratory.

A representative sample of the juice to be tested can be introduced into a clean fermentation tube as indicated in Fig. 4.6. If fermentation is in progress, some of the carbon dioxide gas generated will collect in the enclosed end of the tube, forcing the fermenting juice downward as pictured in Fig. 4.7. If no movement of the juice occurs during 48 h or so after yeast inoculation, the juice can be considered to be fermenting either very slowly or not at all.

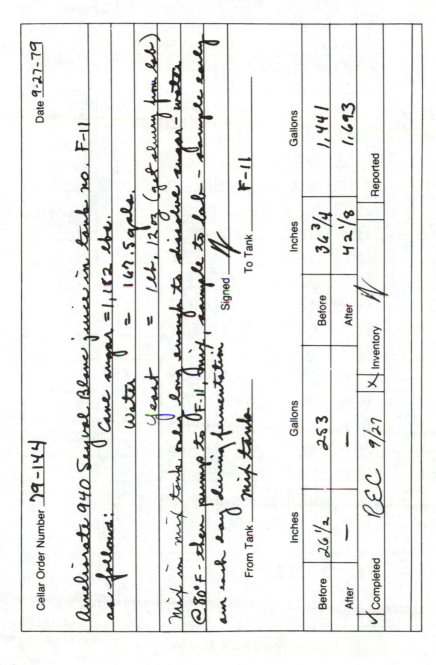

FIG. 4.4.   WORK ORDER FORM IN USE FOR WHITE JUICE AMELIORATION

FEMENTATION CONTROL RECORD

SERIAL NO. 9

TANK NO. _F-11_   LOT NO. _940_   VARIETY _SEYVAL BLANC_

DATE	TIME	TEMP.	BALL.	ALC.	T.A.	V.A.	C	C	N	T	COMMENTS
9-27	9:00 am	52°F	22.6		.780		✓	✓	✓	✓	W.O. #79-144 ameliorated
9-28	8:30 am	52°F	22.1			.006	✓	✓	✓	✓	ferm. tube positive
9-29	8:45 am	53°F	22.4		.780	.006	✓	✓	✓	✓	
9-30	9:00 am	53°F	22.0				✓	✓	✓	✓	first ferm. signs in F-11
10-1	8:45 am	54°F	21.2		.773	.012	✓	✓	✓	✓	
10-2	8:45 am	56°F	20.0				✓	✓	✓	✓	F-11 is "gassy"
10-3	8:15 am	59°F	18.4		.788	.018	✓	✓	✓	✓	cooled to 55°F
10-4	8:30 am	57°F	16.4			.018	✓	✓	✓	✓	
10-5	9:00 am	60°F	13.5	5.1	.780	.021 / .024	✓	✓	✓	✓	cooled to 54°F
10-6	8:15 am	56°F	11.1			.024	✓	✓	✓	✓	
10-7	8:15 am	59°F	8.8		.720	.021	✓	✓	✓	✓	
10-8	9:15 am	55°F	6.7			.021	✓	✓	✓	✓	
10-9	8:30 am	56°F	4.9		.728	.024	✓	✓	✓	✓	
10-11	9:00 am	56°F	1.7		.690	.027	✓	✓	✓	✓	W.O. #79-174 racked to F-6
10-12	8:30 am	56°F	0.8	10.6		.024	✓	✓	✓	✓	
10-14	8:30 am	54°F	-0.4			.024	✓	✓	✓	✓	clearing up slightly
10-16	9:00 am	55°F	-0.8		.660	.030	✓	✓	✓	✓	
10-18	9:15 am	55°F	-1.1			.027	✓	✓	✓	✓	
10-20	9:00 am	56°F	-1.4	11.7	.638	.030	✓	✓	✓	✓	
10-22	9:30 am	55°F	-1.7			.027	✓	✓	✓	✓	
10-24	9:30 am	56°F	-1.6	11.9	.638	.030	✓	✓	✓	✓	
10-26	9:45 am	55°F	-1.7			.030	✓	✓	✓	✓	
10-28	9:30 am	54°F	-1.7	12.2	.645	.030	✓	✓	✓	✓	W.O. #79-198 racked to S-15

CELLAR — *LABORATORY*

FIG. 4.5.   FERMENTATION CONTROL RECORD IN OPERATION FOR WHITE JUICE

FIG. 4.6.   FERMENTATION TUBE

FIG. 4.7. POSITIVE RESULT IN A FER-
MENTATION TUBE

Several items of caution should be discussed when considering the use of
the fermentation tube in the laboratory. First, cleanliness is extremely
important to insure that any yeasts from previous samples have been
completely removed. To this end, the winemaster may wish to boil the tubes
gently in water (they are very delicate), cool and dry before use. Second,
temperature control is necessary so that the fermentation tube environment is reasonably similar, within a couple of degrees Fahrenheit, to that of
the fermenter in the cellar. It will not mean much to observe a fermentation
tube under dynamic progress at 72°F when the fermenter in the cellar
remains at 55°F. Third, samples should be changed at least every 48 h, if
practicable. After extended lengths of time, the samples become unrep-
resentative under the best of controlled conditions. Lastly, fermentation
tubes are usually made from very lightweight, thin-walled glass. As indicated previously, they must be handled with great care in order to avoid
injury and expense from breakage.

Balling

One of the best indicators of fermentation progress is the Balling test.
Balling is a function of specific gravity, or the scaled discrimination between heavy and light-bodied wines. As applied to fermentation control, the

Balling test indicates a somewhat regular regression as the heavy sugars are fermented to ligher alcohol and gaseous carbon dioxide. (Figure 4.5 fermentation control record reveals a beginning Balling of 22.6°—calculated on the Fig. 4.1 lot origination to be 22.4° Brix—the 22.6° is an actual result that is well within the practical error limits that a winemaster may expect). Balling is often confused with Brix in that both analyses may be performed with hydrometers in the same manner. The Brix-Balling hydrometer, properly used, will yield accurate determinations of dissolved solids by weight, in grape juice or must. However, once the juice or must has begun fermenting, alcohol will be produced, having an effect on the hydrometer diametrically opposed to the effects of dissolved solids.

The Balling of water is 0.0°. A dry white wine with an alcohol content of 12% by volume will actually have negative Balling, i.e, will be lighter than water—probably in the area of -1.5 to -2.5° Balling. The "mouth-feel" of such a wine may also be "thinner" than water. This wine may also be referred to as a "light-bodied" wine.

Following the "Ball." column of the fermentation control record in Fig. 4.5 downward, one can see that a very nice curve could be drawn to depict fermentation progress as the Balling regresses.

Professors, winemasters and analysts in America have several different opinions as to the importance of Brix and Balling distinction in winemaking terminology. A good case can be made for clarity in communication if we utilize the term Brix for soluble-solids determinations in grapes, must and juice, reserving Balling for specific gravity measurements once there is alcohol present in the solution. For example, a juice ameliorated and ready for fermentation may be at a level of 22.6° Brix, but the next day fermentation will have reduced this to perhaps 22.1° Balling, as indicated in Fig. 4.5.

Figure 4.8 portrays the simplicity of the Balling test. The entire procedure is described in Appendix A.

Alcohol

Next to the Balling column on the fermentation control record is a column titled "Alc." for alcohol. Of course, the more fermentation progresses, the more ethyl alcohol is going to be produced. Alcohol also will increase proportionately to the reduction of sugar solids.

During the initial stages of fermentation, it may not be necessary to monitor alcohol as closely as Balling unless some special consideration is being made. The Balling test is much quicker to make than an alcohol determination, yet provides an equally accurate measurement of fermentation progress. However, during the latter portions of fermentation, the winemaster may make several alcohol tests on each lot in order to be sure that proper calculations and treatments have been made.

It is the alcohol content on which the rates of Federal excise taxes are based for still (non-effervescent) wines. Wines containing not more than 14% alcohol by volume are subject to a federal excise tax rate of $.17 per gal. The dessert wine category (containing from 14−21% alcohol by volume)

FIG. 4.8. BALLING TEST
BEING MADE WITH A BRIX-
BALLING HYDROMETER

carries a $.67 per gal. tax assessment by the ATF. Wines above 21% alcohol
and necessarily below 24% carry excise taxes of $2.25 per gal. (Alcohol
contents in excess of 24% by volume constitute a spiritous beverage, not a
wine.) Artificially carbonated wines ("crackling" and the like) procure a
figure of $2.40 per gal. while naturally carbonated wines ("sparkling"
wines, champagne, spumante, etc.) incur Federal excise taxes of $3.40 per
gal. All of these rates can easily be converted to "per liter" amount by
dividing the tax rate by 3.785. For example, the $.67 per gal. rate for dessert
wines works out to be $.1770 per liter.

As mentioned in the previous chapter, ethyl alcohol is produced during
fermentation by the action of yeast cells on sugars. Ethanol, in itself,
contributes little taste value, especially in lower concentrations such as in
table, carbonated and sparkling wines that are less than 14% alcohol by
volume. Ethanol is, however, a very good solvent for the many aromatic and
flavor compounds found in some wines.

Methanol, or "wood" alcohol, may be found only at very low concentrations in wines. Generally, methyl alcohol is believed to be formed by pectic

hydrolysis. Consequently, one may expect to find somewhat higher levels of methanol in wines that have been treated with pectic enzymes and in red wines that have been fermented on the skins.

There are some higher alcohols in wine such as propanol, butanol, pentanol and hexanol. In table wines, natural higher alcohols usually do not contribute significantly to the organoleptic profile. In dessert wines, however, higher alcohols may be concentrated due to the added brandy, often detracting from bouquet and flavor values.

Wines that contain no appreciable amount of sugar and have known alcohol contents in concentrations of less than 14% by volume may be analyzed using several types of *ebulliometers*. These instruments measure the difference in the boiling point of water (distilled or deionized) with that of a wine sample. This difference is then calculated to a relatively accurate alcohol content by volume. These instruments are fully defined in Part 240, CFR, Subpart XX, Sections 240.981 through 240.1014. One of the more popular types of ebulliometer marketed in America is the Salleron-DuJardin device as pictured in Fig. 4.9.

Wines that have positive Balling tests (more than a level of 0.0° Balling) and/or alcohols known to be in excess of 14% by volume should be analyzed for alcohol by the distillation method as described in Appendix A. Significant dissolved solids and higher alcohol contents will distort the results obtained by boiling point determinations in ebulliometers.

Total Acidity

Referring once again to Fig. 4.5, the fermentation control record, we will find that, after Balling and alcohol, the next column for consideration is entitled "T.A."—*total acidity*.

The total acidity for fermenting wine may be analyzed in much the same manner as for grape juice or must. The fermentation will render a rather large amount of carbon dioxide gas which must be removed before total acidity can be measured accurately. As with grape juice and must, there may be too many suspended solids for the pipet to handle in the inlet tip capillary. It is advisable to carefully filter the fermenting wine sample in a neutral media so as not to influence the acid level prior to running the total acidity test. Failure to extract the carbon dioxide gas properly may result in the titration of this gas as carbonic acid, making the total acidity measurement inaccurate. Figure 4.10 illustrates a glass funnel lined with neutralized filter paper so that a sufficient sample can be filtered to make a proper total acidity test.

Note that, as the progression of fermentation is followed in the fermentation control record of Fig. 4.5, the total acidity is gradually reduced as acids react and are utilized in the biochemical processes of fermentation. The formation and precipitation of acid salts at cooler fermentation temperatures also contributes to the reduction of total acidity.

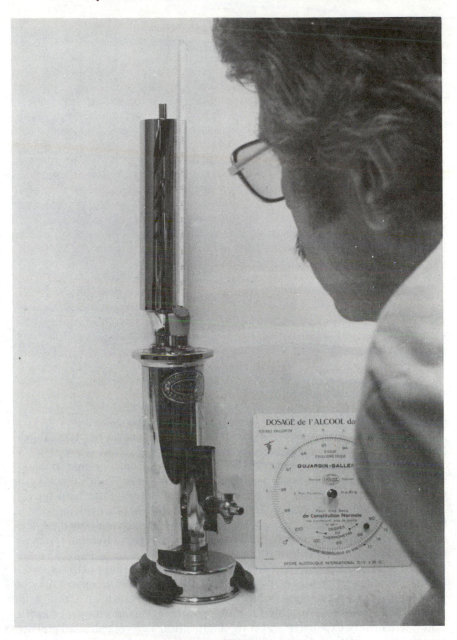

FIG. 4.9.   SALLERON-DuJARDIN EBULLIOMETER IN OPERATION

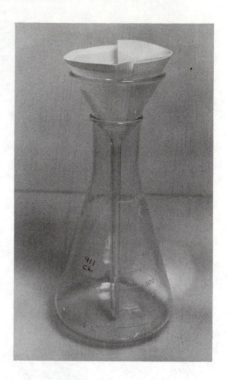

FIG. 4.10.   FILTRATION APPARATUS
FOR A TOTAL ACIDITY TEST

Volatile Acidity

The final chemical analysis on the fermentation control record is abbreviated "V.A.", *volatile acidity*.

For most small winery vintners the term volatile acidity refers to "vinegar spoilage," the acids developed by vinegar bacteria. Volatile acids are distillable acids which are organic in nature and of the fatty acid type.

The most common volatile acid in winemaking is acetic acid, generated by the oxidation of ethanol by acetic acid bacteria, earlier defined as *Acetobacter*. There are a number of different strains of this microbe, the most common being *Acetobacter aceti*.

The traditional test for volatile acidity in the small winery utilizes distillation methodology. While acids such as butyric, formic and others may be present in small quantities, volatile acidity is expressed as acetic acid. Several methods are available for more accurate measurements of acetic acid, but they are expensive and time consuming. The volatile acid method using the cash still as described in Appendix A is more than sufficient for the small winery quality control program, provided the procedures are properly followed and maintained.

The presence of carbon dioxide and/or sulfur dioxide in wine samples can greatly alter results of volatile acidity from distillation in the cash still. In the distillation of volatile acidity from a fermenting wine sample, the analyst is striving to isolate acetic acid (primarily the result of *Acetobacter* activity and spoilage). The analyst should be careful to remove the carbon dioxide which is dissolved in the sample, so that this will not be distilled over as carbonic acid, which would be titrated and determined as acetic acid in error. Normally, there will not be a significant amount of sulfur dioxide in fermenting wines. However, later in the life of a wine, during processing and aging storage, free sulfur dioxide used as a preservative may easily reach concentrations in excess of 40 parts per million (ppm), enough to effect errors in volatile acidity measurements.

Part 240, CFR, Subpart U, Section 240.489 limits volatile acidity content in wines to a maximum of .140 g/100 ml. This is much more than a vintner of high-quality wines will ever tolerate. Normally, white wines should not reach more than .036 g/100 ml by the end of primary fermentation, and not more than .060 g/100 ml by bottling time. Reds may be allowed a little more leeway, say, .048 and .078 g/100 ml.

More regarding the development and nature of bacterial effects in wines can be found in Chapter 9.

ORGANOLEPTIC EVALUATION

As has already been established, the analyses of Balling, alcohol, total acidity and volatile acidity during fermentation are essential for maintaining adequate quality control. The fermentation control record is a log that indicates the completeness and the frequency of each individual vintner's control program. A very important part of a superior control profile is that of *organoleptic evaluation*—the sensory examination of juice and must during fermentation. These analyses are posted under the headings of *C, C, N, T* on the fermentation control record, representing color, clarity, nose and taste, respectively. These are common terms used for sensory evaluation of wines in most wineries.

To the highly sensitive and experienced palate of a master winemaker, the organoleptic analysis can often be more practical, in terms of decision-making, than chemical and physical analyses. This is not to say that a good winemaster can replace an analytical laboratory; on the contrary, the good winemaster's palate serves to expand the capacity of the laboratory and to make it more effective in quality control.

Few positive organoleptic qualities of distinction or superiority are identifiable during fermentation. Most often the fresh fruit aromas and flavors are modified, reduced, or lost, by the time a young wine has begun to develop. However, in the case of young wines made from the heavily flavored cultivars of *V. labrusca, V. rotundifolia* and, in particular, the Muscats of *V. vinifera*, this intensity may not be noticeably weakened.

Sensory evaluations during transformation of juice and must into wine are usually based upon negative quality control parameters. In other words, one should learn to recognize the early signs of hydrogen sulfide production, mold, mustiness, oxidation and other deleterious factors.

Laboratory Wine Glasses

The wine glass is the sole instrument used in organoleptic evaluation. The laboratory may be equipped with a stark white countertop and an unfrosted electric light source in order to minimize and help standardize the influence of wine color and clarity. Use of the all-purpose glasses pictured in Fig. 4.11 is acceptable. The glass should be "flint" (colorless and clear) and without engraving, embossing or any other marking that can cause confusion. Generally, the more simple the glass, the better. The well equipped small winery laboratory may have two or three dozen, or more, wine glasses on hand. They need not be expensive crystal; ordinary types are fine. Whatever choice is made, the entire lot should be of the same size, type and style in order that different wines may be accurately compared with one another, side by side.

In the fermentation control record (Fig. 4.5), there are check marks in each of the columns of color, clarity, nose and taste. These signify that the analyst has performed each of these basic organoleptic tests on the sample and has found nothing out of the ordinary. A question may be raised as to how an "OK" could be given to clarity, as the wine would necessarily be very cloudy during the fermentation. The check mark is always awarded to organoleptic evaluation found to be in accordance with standards. In this case, a cloudy sample during fermentation is expected; a perfectly clear fermenting juice sample would raise immediate questions as to a sampling

FIG. 4.11.   SUITABLE GLASSES FOR ORGANOLEPTIC EVALUATION IN THE LABORATORY

mix-up or some other difficulty. As with every other analysis performed in the laboratory, the organoleptic results are first posted upon the laboratory analysis log and then in any other appropriate record as is necessary—in this case, the fermentation control record.

There is much more detail and discussion of organoleptic analysis in Appendix A. For the purpose of restricting organoleptic considerations to those relevant during fermentation, the commentary on organoleptic evaluation at this point is separated from organoleptic evaluations of the finished wine.

Color

Perhaps the French term *blanc*, "without color", is a more accurate word for the color value of what is referred to in English, as "white" wines. Obviously, white wines are not actually white as is milk, yet neither are they absolutely colorless. Rather, white wines can be so pale as to be nearly without color, ranging to dark golden straw in hue.

The colors found in white juice and must, apart from any oxidation that may have browned or yellowed color values, are contributed by leucoanthocyanin pigments. These are found primarily in the outer layers of the grape skin, and are readily extracted during fermentation (or from heating the must). They are usually rather unstable compounds and are readily oxidized, especially at higher temperatures, into compounds that detract significantly from most quality standards. Consequently, most white wines, certainly the delicate types, are made from the cool fermentation of freshly pressed grape juice.

An experienced wine judge will not normally become concerned with a decrease in the amount of color, or a lighter hue, while the white juice is undergoing fermentation. This indicates that some of the more unstable leucoanthocyanins may be precipitating and/or that some previously oxidized compounds may be growing lighter as oxygen is utilized by the yeast or some other factor. Increased color of white juice during fermentation will cause greater concern, indicating possibly too warm a fermentation temperature or some contaminating exposure to oxygen. As young wine develops from white grape juice it will clarify itself somewhat, precipitating the heavier suspended solids, yeast cells and other organic matter. This may provide the illusion that the wine is becoming darker, when actually it is only becoming easier for the naked eye to see through.

Red wine fermentations (which will be discussed in more detail later in this chapter) progress much more quickly than whites due to the higher temperatures that are normally afforded. The development of color is very important, of course, and the primary reason that red wine fermentations usually take place with the must, "on the skins", is so that optimum color can be extracted from the skins. Experience will provide the best training to determine what cultivars yield acceptable volumes of color in individual climate/soil conditions. Also, certain cultivars will provide different hues of red, such as brownish-reds or tawny reds, while others may be normally crimson, ruby, or even purplish. These differences arise from five main types of red and blue grape anthocyanin pigments—cyanidin, delphinidin, malvidin, peonidin and petunidin. Single-sugar based compounds of these pigments (monoglucoside anthocyanins) are usually much more stable to oxidation and precipitation than double-sugar compounds (diglucoside anthocyanins). Pigment type is totally inherent in the grape cultivar and has nothing whatsoever to do with climate/soil conditions in a vineyard. *Vitis vinifera* red cultivars normally contain monoglucoside anthocyanin pigments while many *V. labrusca* and most all of the *V. rotundifolia* red varieties yield the more unstable diglucoside anthocyanin compounds. As may be expected, the French-American hybrid cultivars may have either, or both, types of these pigment compounds.

The small winery vintner may be concerned with a precise amount of color extraction during fermentation of the must, especially when making rosé wines. Because some cultivars yield more or less color than others, and this color may be of varied stability, a great deal of planning and quality control must be exercised by the winemaster to consistently produce wines within standards. As red color intensity increases during fermentation, the heavy exposure of the wine to the skins and seeds may increase bitterness due to tannin (see Glossary) extraction. When color is at least adequate in intensity to meet quality control standards, and bitterness is not objectionable, pressing the must should be considered so that the new wine is not exposed to the skins any further. In the case of cultivars known to have unstable pigments (those which solidify and precipitate later in the life of the wine), the winemaster may carefully allow the must to ferment beyond the adequate point in order to generate sufficient color to persist throughout the aging process and beyond. For *V. vinifera* and some of the French-American hybrid cultivars, bitterness may not reach an objectionable threshold until fermentation has progressed to about 3.0° Balling, or even less. Cultivars of *V. labrusca* may provide must ready for pressing at approximately 7.0° Balling, or more. *Vitis rotundifolia* reds may well require being pressed even before the fermentation has progressed half-way.

Browning of red colors during fermentation of must may indicate bacterial infection, which would yield a volatile acidity analysis with a higher than normal value. This could be in excess of .060 g/100 ml in the beginning stages of serious infection. Of course, the brown hue may also be the normal color extracted from a particular cultivar. In the case of fermenting hot-pressed red juice (red grape must heated to 130–150°F, then pressed to juice and cooled), heating the must prior to fermentation may have generated a precursor of browning due to oxidation. However, most well known commercial red grapes yield attractive red to purple hues during fermentation on the skins.

Color of juice and must, and its development in the new wine, requires study by hands-on experience in the winery before mastery can be truly achieved.

Clarity

Clarity cannot be a critical factor in the organoleptic evaluation of fermenting juice or must. During normal fermentation, an overwhelming number of yeast cells are undergoing division and propagation. Therefore, the juice and must should have a reasonable degree of turbidity due to the presence of millions of yeast cells per milliliter in suspension of the fermenting juice or must. In addition, there should be some pulp solids from the grape in suspension which serves as a nitrogen source for the yeasts. The nitrogen is available primarily as inorganic ammonium compounds and proteinaceous materials. Yeasts require a sugar-nitrogen ratio of about 10 to 1 in order to properly carry out the dynamics of the fermentation process. Many large wineries nowadays centrifuge fresh grape juice in order to remove much of the suspended pulp that later would form the *lees* sediment. This has been discussed earlier several times. Excessive centrifugation can greatly reduce the amount of nitrogenous compounds necessary for yeast propagation and growth. This may cause fermentations to lag, or perhaps, stop altogether, creating a "stuck" fermentation. The winemaster is then forced to add an expensive nitrogen source material such as ammonium sulfate, diammonium phosphate, urea, or some other acceptable food-grade compound of the type.

Clarity can actually be a disadvantage during fermentation. A clear, or relatively clear, juice may not be optimum for fermentation.

Apart from nitrogen deficiencies, slow or stuck, fermentations can also result from juice being too cold (perhaps less than 55°F), or having too high a free sulfur dioxide level (in excess of 40 ppm) remaining from the crusher-stemmer, or a combination of these difficulties, or some other problem. As fermentations of white juice near completion, however, it is common to observe a slight clearing of the new wine in the upper strata of the fermenter. This may be seen in wine sample containers having been left for a few hours in the laboratory.

In summation, functions of clarity during fermentation are fairly simple: the opposite parameters of those judgments for wine clarity to be found in Appendix A.

Nose

The term *nose* has become familiar in the jargon of the American wine industry as embracing all functions of the nose in the judgment of juice, must and wine quality. Winemasters will commonly refer to the nose of juice and must as the aroma, while the nose of a wine will be its bouquet. The aroma is felt to be a natural constituent of the grape cultivar, and the influence from climate/soil conditions. The bouquet is comprised of the aroma and other aromatic values that have been contributed by the winemaster's particular cellaring techniques, including aging and blending. As mentioned earlier, there is very little that can be determined about how good a wine product may be while that wine is in the stages of fermentation and very early development. Some grape species, on the other hand, have such intense values of aroma that rather early quality judgements can be made. Among these are the Muscats of *V. vinifera* which have heavy concentrations of linaloöl plus geraniol aromatic compounds. The cultivars Concord and Niagara, among others generally categorized with the species *V. labrusca*, exhibit varying high levels of the distinctive "foxy" methyl anthranilate. Most all cultivars within the species *V. rotundifolia* are easily discernable with the overwhelming aroma of beta-phenyl ethyl alcohol.

There are, however, some characteristics that indicate the lack of wine quality after fermentation has been completed. Everyone judges factors and constituents of nose somewhat differently. This makes definitive values difficult, since objectivity is replaced by subjectivity. Again, one may best learn the association of different aromatic compounds by experience. Some winemasters learn by taking different substances such as acetic acid, fusel oils, lactic acid, and others, dissolving them in water and ethanol-water solutions and then evaluating and judging each on an individual basis. These are tested at different concentrations and then taken blind (unknown) at different combinations and intensities of the compounds being considered. Following such programs, the novice wine-taster can form rather keen detection abilities with frequent and regular practice. More on this method of training is detailed in Appendix A.

Taste

Experts agree that taste is less important to quality in wines than smell. However, both are very important and very subjective values. The winemaster may make a wine that he feels is superior, but does not sell very well because others in the marketplace do not share that opinion. The successful vintner does not make wines that only please the in-house palates. The consumer must be satisfied or else the wine will not be successful in the market. It follows logically then that the standards set up in a small winery laboratory may have to be reformulated and constantly updated as market profiles change.

The parameters of taste are somewhat more defined than those of smell, being primarily functions of acidity, astringency, bitterness and sweetness. These are more easily described and introduced as they are sensations that everyone has already experienced in one manner or another. More detail concerning the functions of taste can be found in Appendix A.

Often the term balance is used, associating the degree of acidity with sweetness. Sugars can mask acid intensity and vice-versa. This function is relatively useless in evaluating grape juice or must. During fermentation there will be a large imbalance with very high sugar concentrations masking acidity, astringency and bitterness with sweetness. It is not until the fermentation nears completion that one can make an intelligent judgment for the value of taste.

Using similar methodology to that previously described for developing nose, the newly qualified wine-taster is advised to set up different amounts of common taste compounds found in grape juice and must and dissolve them in water and water-ethanol solutions. Among these may be the constituents mentioned previously for nose, plus some solutions of tartaric acid, hydrogen sulfide, sulfur dioxide, acetaldehyde and others. The concentration, dilution, blind-analysis, and other factors used in developing the palate can be as extensive as dedication permits.

The most important negative tastes to be identified in fermenting juice and must are those of acetic acid, hydrogen sulfide and acetaldehyde. Confirmation of acetic acid will be reflected in a higher than normal result of volatile acidity, as mentioned previously. Hydrogen sulfide has the characteristic nose of "rotten eggs" and a very foul taste sensation, as opposed to the pungent vinegar nose and taste of acetic acid in higher concentrations. Acetaldehyde may form in fermenting juice and must from more intense oxidation reactions—usually accompanied by a browning of color. When the presence of these compounds is confirmed, corrective measures must be taken as soon as possible. In most cases, however, it will already be too late for the wine to survive as a quality product. Stringent quality control of all aspects must be maintained prior to the fermentation.

For best results in grape juice and must organoleptic evaluations, the analyst must educate his or her eye, nose and palate to the skills required. As these skills are then further developed, they will be more meaningfully applied as analytical tools in the small winery laboratory.

Whenever the analyst feels that an item should be logged that is meaningful to the history and quality of that wine, the "comments" column of the fermentation control record may be used for that purpose. Note in Fig. 4.5 that the very last item entered in the comments column is a cross-reference to a work order written ordering a racking (decanting from the lees sediment) of that wine to tank S-15. This signifies that this wine has left the "F" (fermenting) tank and has been transferred to an "S" (storage) vessel. This is particularly important in that this operation removes the new wine from the "juice" account to the "wine" account in the bulk wine book inventory. This change in status makes for full accountability of this lot with the ATF and requires continued close attention in exacting inventory procedures. Figure 4.12 portrays a work order form issued for the racking operation previously mentioned. Figure 4.13 illustrates a common bulk wine book inventory card (usually 5 × 8 in.) kept in a sturdy card file cabinet for desk-top convenience and protection.

LOT ORIGINATION FOR RED WINES

It is apparent that one cannot accurately measure the volume of juice contained within must so that a precise lot origination for red wines can be drawn prior to the must being pressed. Consequently, it is necessary to make a lot origination in two parts for each lot of red wine to be fermented on the skins. Hot-pressed red juice may be treated in the same manner as for white juice.

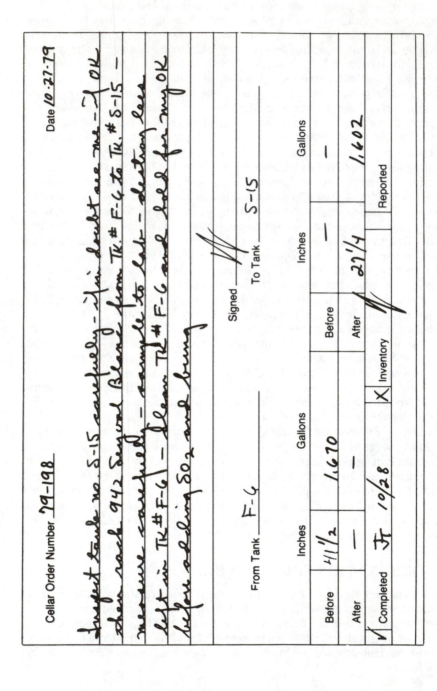

FIG. 4.12.   WORK ORDER FORM IN USE FOR RACKING

Lot No. 942 Type-Variety SEYVAL BLANC Class -14% Color WHITE

Remarks:

Date	Racked From Tank W.O. Gallons	Treatment-Disposition	Racked To Tank W.O. Gallons	PHYS INV	ALC	BALL	T.A.	V.A.	FSO$_2$
10-28 1979	F-6 79-198 1,670	2nd rack	S-15 79-198 1,602		12.2	-1.7	.645	.030	

FIG. 4.13. BULK WINE BOOK INVENTORY CARD INITIAL USE

The first part of the red must lot origination is an estimation of the volume of grape juice contained in the must. The second portion is based upon actual measurements after the eventual press has been made.

Figure 4.14 illustrates the "estimated" lot origination form completed for a sample lot of DeChaunac grapes, as analyzed in Fig. 3.8. This is followed by the normal work order drawn for amelioration and the resulting fermentation control record that would be expected for this lot of DeChaunac grapes, in Figs. 4.15 and 4.16, respectively.

The lot number, serial number, weight, pH, total acidity and Brix are all items that can be immediately assigned or determined, and posted in the estimated portion of the red wine lot origination. The winemaster should have, and will require, a reasonably good estimate as to how much red wine will result from his methods of winemaking to determine the estimated gallons per ton figure. This, as exemplified by the number 180 in Fig. 4.14, can be multiplied by the actual tons of grapes received from the appropriate tally sheet—12.160 tons as posted in this sample lot. This calculates out to be an estimated 2189 (180 × 12.160) gal. of red juice in this lot of must. We can then proceed with the calculations for amelioration percentage and additions in the same manner as described for white juice in Fig. 4.1.

Reviewing the determination for white juice amelioration additions, the *resulting product* is the basis for percentage calculation—not the beginning gallonage. Again, in the language of the ATF, the product that is resulting from amelioration, i.e., juice plus additions of sugar and/or water, is the whole resulting product. For example, 2189 gal. of estimated juice gives a resulting product of 2516 estimated gal. at the rate of 13% amelioration— *not* 2474 gal. One should also be mindful that amelioration is regulated by the ATF in Part 240, CFR, Subpart XX, Sections 240.960 and 240.961. Figure 4.2 provides a table for finding amelioration limits based upon total acidity.

Only that amount of amelioration which is necessary to reach the desired "ideal" total acidity in the must should be considered. The .770 g/100 ml level that was recommended for white juice earlier in this chapter may also be standardized for red must—the calculation made in the same manner. The novice winemaker may be advised to use only a portion of his allowed or calculated amelioration in the red must—especially when this amount is close to that which is legally allowed. Should the yield from pressing be less than estimated, there will remain some credit to adjust the final amelioration to pinpoint accuracy.

RED MUST FERMENTATION CONTROL

The small winery quality control program for fermenting red must is similar to that for white juice except that red must fermentations are usually designed to proceed much faster. This requires more frequent analysis—perhaps three or four samplings of each red must fermenter per 24-h period.

LOT ORIGINATION

DATE _9-27-79_   CULTIVAR _DeChaunac_

LOT NO. _920_   REMARKS: _bunches more_

SERIAL NO. _10_   _compact than most other_

FERMENTER NO. _F-20_   _years - but grapes have_

NET GALLONS _EST, 2,189_   _only traces of any damage_

NET TONS _12.160_   _overall in very good_

GALLONS PER TON _EST. 180_   _condition - crushed, stemmed_

Composite!
pH _3.32_ T.A. _.883_   _and must pumped to_

BRIX _20.2_ ALC. _____   _Tk.# F-20 late pm of 9-27_

EXT. _____

EST.
AMELIORATION = _13_ % @ _22.4_ ° BRIX = _12.0%_ ALCOHOL

EST. 2,189 GALLONS START   = _2,516_ RESULTING TOTAL
   _.87_ INVERSE OF   GALLONS OF PRODUCT
     AMEL. PERCENTAGE

EST.
_2,516_ TOTAL   @ _22.4_ ° BRIX(_2.044_ LBS/GAL) = _5,143_ TOTAL
     GALLONS   LBS.

_-2,189_ START   @ _20.2_ ° BRIX(_1.827_ LBS/GAL) = _3,999_ START
     GALLONS   LBS.

_= 327_ AMELIORATION   _1,144_ ADDITION
     GALLONS   LBS.

_- 84.7_ SUGAR AS   × .074 (GAL/LB) = _84.7_ SUGAR AS
     GALLONS   GALLONS

_= 242.3_ WATER
     GALLONS

_2 lb. 8oz_ YEAST ADDITION = RATE OF _1_ LBS PER _1,000 gals._

REMARKS: _____

FIG. 4.14.   LOT ORIGINATION FORM IN USE FOR ESTIMATING JUICE IN RED MUST

Cellar Order Number 79-144

Date 9-27-80

Ameliorate 920 De Chaunac must in Tk. # F-20 as follows:

Cane sugar = 1,144 lbs.
Water = 242.3 gals.
Yeast = 2 lbs. 8g (get slurry from lab)

Mix with 80°F water until all sugar is dissolved - then pump to
Tk. No. F-20 - mix in F-20 1/2 hr - sample to lab - sample F-20
every 12 hrs. during 06 fermentation   Signed _____

From Tank _Mich Tank_   To Tank _F-20_

	Inches	Gallons		Inches	Gallons
Before	38 3/8	326.5	Before	42 1/2	EST.
After	—	—	After	48	EST.
✓ Completed JT 9/27			Inventory ✗	Reported	

FIG. 4.15.   WORK ORDER FORM IN USE TO AMELIORATE RED MUST

SERIAL NO. 10

FEMENTATION CONTROL RECORD

TANK NO. **F-20**   LOT NO. **920**   VARIETY **DeChaunac**

| DATE | TIME | CELLAR | | | LABORATORY | | | | | | |
		TEMP.	BALL.	ALC.	T.A.	V.A.	C	C	N	T	COMMENTS
9-27	11:30 pm	71°F	22.2		.765	.004	✓	✓	✓	✓	W.O.#79-144 ameliorated
9-28	11:15 am	71°F	22.5			.012	✓	✓	✓	✓	
9-28	10:30 pm	73°F	19.7			.018	✓	✓	✓	✓	" gassy "
9-29	10:45 am	75°F	18.0			.018	✓	✓	✓	✓	
9-29	10:45 pm	77°F	16.4			.021	✓	✓	✓	✓	cooled to 70°F
9-30	11:15 pm	71°F	13.7			.021	✓	✓	✓	✓	
9-30	11:45 pm	76°F	11.7			.027 .027	✓	✓	✓	✓	cooled to 70°F
10-1	9:45 am	72°F	8.5			.024	✓	✓	✓		
10-1	9:45 pm	76°F	5.3			.027	✓	✓	✓	✓	
10-2	10:45 am	77°F	1.0	9.7	.720	.030	PRESSED TO TK #F-8				
10-2	8:30 pm	70°F	0.4			.027	✓	✓	✓	✓	W.O.#79-160 amel. adjust
10-3	8:30 am	72°F	-0.1			.027	✓	✓	✓		
10-4	8:30 am	74°F	-0.2			.030 .033	✓	✓	✓	✓	
10-5	9:00 am	75°F	-0.2			.036	✓	✓	✓	✓	
10-6	8:15 am	76°F	-0.3	11.4		.033	✓	✓	✓	✓	cooled to 70°F
10-7	8:15 am	71°F	-0.5			.036	✓	✓	✓	✓	
10-8	9:15 am	72°F	-0.7			.033	✓	✓	✓	✓	
10-10	8:15 am	72°F	-0.7	11.6		.039 .039	✓	✓	✓	✓	
10-12	8:30 am	72°F	-0.7			.042	✓	✓	✓	✓	
10-14	8:00 am	71°F	-0.9			.042	✓	✓	✓	✓	
10-16	9:00 am	71°F	-0.8			.039	✓	✓	✓	✓	
10-18	9:15 am	70°F	-0.9	11.8		.042	✓	✓	✓	✓	
10-20	9:00 am	70°F	-1.1	11.9		.039	✓	✓	✓	✓	W.O.#79-171 racked to S-21

FIG. 4.16.   FERMENTATION CONTROL RECORD IN USE FOR RED MUST

Red must is normally fermented in the range of 70° to 80°F in order to produce the ethanol as quickly as possible without damaging the organoleptic profile natural to the cultivar. As the skins float to the top of a red fermenter during fermentations, a cap is formed that is an excellent medium for the growth of *Acetobacter.*

White juices are usually fermented in a vessel with a closed top, using an air-lock or some other device to allow carbon dioxide gas to escape, but to keep air out. This all but eliminates oxygen entry into the white juice fermenter. *Acetobacter*, being aerobic, require oxygen in order to develop. The exposure of the red fermenter becomes a profound source of worry for the winemaster.

The cap is commonly "punched" three or four times in each 24-h period of red must fermentation. This operation consists of pushing the cap down so as to be submerged in the fermenting red juice. In larger red must fermenters, the juice may be pumped from below the cap up and over the top so as to drench the cap thoroughly. In either method, the necessity of an open-top fermenter is apparent. Furthermore, the open top allows for somewhat better atmospheric conditions for the cellarmen to work in, although this can be of great danger during violent fermentations.

The potential for asphyxiation is a danger that cannot be emphasized too strongly. Many cellarmasters require that their workers work in teams so as to minimize such hazards. This danger is reduced, but must be dealt with nevertheless, during the racking processes when considerable carbon dioxide gas can be released. However, the open-top fermenter does allow some air to be exposed to the cap and this can be sufficient for the *Aceto-bacter* to cause an infection—often in just a few hours. This can be detected in the early stages by accurate analysis of volatile acidity—considered positive when in excess of .048 g/100 ml.

Figure 4.16 shows what might be considered a normal course of events in the fermentation of a typical lot of red must. Note that the record is developed in two stages, one for the estimated portion of the fermentation as red must prior to pressing and the other for the post-pressing final stage of the process.

Figure 4.17 displays a completed "actual" lot origination form representing the post-pressing phase of the same lot of DeChaunac must as estimated in Fig. 4.14. The "attach to" notation points out that this form should be permanently affixed to the "estimated" portion so that both forms are kept as a single permanent record of the individual lot.

The "actual gallons" realized from the pressing in this example was a net 2217 gal. (2544 total gal. less the 327 gal. of amelioration added in the estimated portion). It can now be calculated that "actual gallons per ton" is 182.319 (2217 gal. divided by 12.160).

The measurement of actual net gallons must be carefully determined in order to accurately calculate the second, or "actual" portion of amelioration adjustment. Some winemasters have become so adept in their estimations of red must potential that they prefer to accept the results generated, when plus or minus to no appreciable degree. This, of course, leaves the records

LOT ORIGINATION

DATE _10-2-79_

LOT NO. _920_

SERIAL NO. _10_

FERMENTER NO. _F-8_

NET GALLONS _ACT. 2,217 (2,544)_

NET TONS _12.160_

GALLONS PER TON _ACT. 182.319_

pH _____ T.A. _720_

BALL
BRIX _1.0_   ALC. _9.7 (18.1)_

EXT. _4.4 + 18.1 = 22.5 pot._

CULTIVAR _De Chaunac_

REMARKS: _pressed late am
of 10-2 – wine dark
and full-bodied – no
browning – heavy CO_2
emphasizes fruitiness –
looks very good_

TOTAL
AMELIORATION = _13_ % @ _22.4°_ BRIX = _12.0%_ ALCOHOL

_2,217_ GALLONS START = _2,548_ RESULTING TOTAL
_,87_   INVERSE OF   GALLONS OF PRODUCT
      AMEL. PERCENTAGE

ACT.
_2,548_ TOTAL   @ _22.4_ ° BRIX(_2.044_ LBS/GAL) = _5,208_ TOTAL
      GALLONS   LBS.

_-2,544_ START   @ _22.5_ ° BRIX(_2.084_ LBS/GAL) = _5,225_ START
      GALLONS   LBS.

= _4_ AMELIORATION   ___ ADDITION
      GALLONS   LBS.

___ SUGAR AS   × .074 (GAL/LB) = ___ SUGAR AS
      GALLONS   GALLONS

= _4_ WATER
      GALLONS

___ YEAST ADDITION = RATE OF ___ LBS PER ___

REMARKS: _____

_____

_____

FIG. 4.17. LOT ORIGINATION FORM IN USE FOR ACTUAL RED WINE AMELIORATION
ADJUSTMENT

such that many different actual amelioration levels with odd decimal fragments remain. This book will make the case for second ameliorations to be calculated and added in order to maintain exacting levels of alcohol, amelioration percentages, yield maximums and standard adherences.

The newly pressed 920 DeChaunac red wine has been determined to have an alcohol content of 9.9% by volume, a Balling of 1.1° and an *extract* of 4.5°.

EXTRACT

Extract is nearly synonymous with Brix in that it is a measurement of total dissolved solids expressed in degrees, approximately equal to percent by weight. The difference is that extract analyses are *extracted* or calculated from wine samples, rather than from juice or must. Consequently, mention of extract implies that any alcohol has been removed (usually by distillation) prior to the measurement of dissolved solids in the wine sample.

Dry white table wines may be found to have approximately 1.5 to 2.5° extract, while dry reds, with obviously heavier pigmentation, and other dissolved solids, may have extract analyses ranging from 2.0 to 3.0°, and even higher in the *teinturier* (see Glossary) varietals.

Extract can be determined rather accurately in table wines by the proper use of the nomograph outlined in Appendix A. There are two nomographs to be found there, one for dessert wine extract determinations and the other for table wine usage. An analyst, knowing the accurate level of alcohol and Balling can easily identify the extract by simply placing a ruler at the proper gradations. For example, a wine with 11.0% alcohol and a Balling of 0.0° will have an extract of about 3.7°. A wine with 12.2% alcohol and an extract of 3.5° has a Balling of −0.6°. A wine with a Balling of 2.0° and an extract of 6.0° has alcohol of approximately 11.8% by volume.

More precise measurements of extract can be made by following the parameters outlined by the ATF in Section 240.975 of Part 240, CFR. Limits for extract can be found in the same manual under Sections 240.363, and 240.368, as well as 240.430.

Methodology for the determination of extract, both by distillation/hydrometer and by nomograph, is detailed in Appendix A.

Referring to Fig. 4.17, the actual portion of the DeChaunac lot origination, note that a distillation-method alcohol analysis is required because of the higher level of remaining dissolved solids (extract), which will render the more simple ebulliometer test for alcohol inaccurate. If the winemaster so desires, the remaining extract from the alcohol distillation test can be measured with a hydrometer to double check. This is recommended if time permits.

At this point we need to determine just how much extract potential exists. This will provide the basis for calculating the correct amount of ameliorating materials necessary to adjust sugar and/or water deficiencies to precise proportions.

The extract potential is found by adding the extract that remains as dissolved solids and the extract which was calculated to generate the alcohol during fermentation. In the example provided in Fig. 4.17, there has been 4.4° extract analyzed. This added to the 18.1° extract calculated from the 9.7% alcohol present (9.7 divided by .535% alcohol per degree Brix) results in a total extract potential of 22.5°. As this example works out, the extract potential is a little more than sufficient to result in the desired 12.0% alcohol. To the beginning winemaker these calculations may seem difficult to perform accurately. However, after several vintage seasons of experience and proper instruction, the apprentice should realize estimates of red must that require very little adjustment. It goes without saying that accurate estimations can eliminate the need for increased amelioration levels of added sugar or water.

A work order should be executed as soon as possible after pressing and sampling so that any amelioration ingredients can be added before the new red wine has completely fermented to dryness. This will help to avoid stuck fermentations.

The quality control program for red wine fermentations is carried out in similar fashion to the testing of Balling, alcohol, total acidity, volatile acidity, color, clarity, nose and taste described earlier in this chapter.

After fermentation of the new red wine has been completed, a work order similar to that illustrated in Fig. 4.12 should be drafted and issued. This work order number should be cross-referenced on both the fermentation control record and the bulk wine book inventory card to facilitate future references. This operation will transfer the new red wine from the fermenting juice inventory to the bulk wine book inventory in a similar manner to the new white wine portrayed in Fig. 4.13.

GENERAL FERMENTATION NOTES

At this point, it may be helpful to consider a few fermentation principles.

Table wines of superior quality require simple, but careful and clean fermentation procedures. Ameliorations, if any, should be added in the bottom valves of fermenting vessels and mixed so that oxygen exposure is minimized. A small amount of the juice or must may be moved into a clean mixing tank where sugar, water, yeast and any other required components can be dissolved and/or suspended gently before redepositing the juice back into the fermenter.

Amelioration materials may be added at any time during fermentation when yeast cells are reproducing. By waiting until fermentation is well under way before sugar and water are added, a winemaker can conserve refrigeration energy. Also, adding cold amelioration water can aid in reducing temperatures of fermenters that have become too warm. Conversely, warm water may be used in amelioration additions to heat refrigerated juice.

A healthy yeast culture, free of bacteria and other contaminants, with a good population of viable cells (5 million cells per milliliter or more), may be inoculated in fermenters using the same careful methods as used for mixing amelioration ingredients with the juice or must. Dehydrated cultured yeast strains are readily available from winery suppliers and are highly recommended. While dried yeasts are easy to use and store, they should be used when fresh, and certainly less than a year old. Viability counts of dehydrated yeast supplies can drop rather significantly over longer periods of storage, even when an optimum cold and dry atmosphere is provided.

It is imperative that fermentations are not allowed to proceed with wild yeasts. There is no substitute for inocula of proven cultured strains of *Saccharomyces cerevisiae*. Wild, apiculate, yeasts are not predictable and can yield very disappointing, even disastrous, results. Unless checked at the crusher-stemmer, wild yeasts can multiply and become dominant in just several hours.

New wines may be racked several times before transfer to the bulk wine book inventory. This is especially recommended in the case of white juice being fermented at lower temperatures. Such fermentations may take several weeks or months during which two or three rackings may be advisable so that the lees sediment can be separated from the fermenting juice. Lees can readily react with enzymes from autolyzed yeast cells causing "mousiness" and other off qualities of nose and taste.

A word to the wise and a lesson to the apprentice: there is no instrument or machine that can replace the keen mind and ambitious spirit of the winemaster, especially during the vintage season at the smaller winery. The very finest of quality control programs cannot possibly be effective unless the work is planned and the plan is worked. Dedication and tenacity are absolute necessities if any hope of achieving top-quality production is entertained.

SUMMARY

The most significant formative period in the life of a table wine is during its fermentation from juice or must. It is at this juncture that the winemaster must be very careful to insure that nature takes a straight and narrow course.

Grapes may, and usually do, vary in constituency each vintage season, even from the same vineyard. The ideal balance of pH, Brix and total acidity may be seen only rarely in the entire career of the winemaster.

In California, where higher Brix and lower total acidity are commonplace each year, acidity deficiencies are made up with the addition of citric, fumaric, or tartaric acid, or a combination of these. Brix shortcomings are made up with the addition of grape juice concentrates. These practices may be used in the south as well, especially when making wine from the Muscadine cultivars.

In the east, where it is common to find lower Brix measurements and higher total acidity levels, small winery vintners also use grape juice concentrate to raise sugar solids to improve their wines. This, of course, may raise high acidity levels even higher. The eastern winemaster may choose to deacidify his juice or must, using calcium carbonate or other approved products. This is usually performed just before clarification and detartration, however, in order that precise calculations can be made. More often, vintners choose to ameliorate juice and must by the addition of sugar and/or water so as to raise Brix and reduce total acidity. This practice is not legal in California, but in most other states and countries, amelioration is looked upon as a legitimate adjustment. Unfortunately, some wineries use amelioration as a stretching agent to increase production volume at the expense of quality.

Good wines result from good planning. The lot origination form allows for combining the natural characteristics of the juice or must with the winemaster's scientific skill. Furthermore, the lot origination becomes a very important permanent supplemental record, often referred to as the "birth certificate" by winemasters.

A serialized record of work orders details the duties and operations in the small winery—exhibiting ingredients, materials, lot numbers, tank numbers, treatments, measurements and the personnel involved so that a complete record is available for both vintner and inspector. It must be remembered that the work order form should be made in duplicate, the original issued to the person performing the assigned task and the duplicate kept in the work order "prescription" book. After the work is completed and all measurements reconciled, the original is returned to the laboratory to be checked off the duplicate pending copy in the work order book. Work order originals are compiled monthly and filed in the file of operations (discussed in detail in Chapter 10) to substantiate production operations for that particular month. Work order duplicates are kept intact in book form so that pending unfinished work can be determined at a glance. The duplicate work order books also represent a permanent chronological record of laboratory-cellar communications and operations.

A fermentation control record should be maintained for every lot along with the appropriate lot origination form to properly monitor fermentation progress. Organoleptic evaluations made frequently on samples taken from fermenters are supported by analyses of Balling, alcohol, total acidity and volatile acidity. These findings are first posted in the laboratory analysis log and then in the fermentation control record so that the perpetual analytical record is maintained and each lot is individually separated for ready reference.

Wines are transferred from the fermentation control record to a bulk wine book inventory card when the winemaster so determines. This officially takes production lots from the fermenting juice category of storage to bulk wine inventory—a status that becomes directly accountable to the ATF. The work order numbers involved in such transfers should be posted so as to cross-reference all movements. This forms bridges that can be traced when necessary to recall the particulars of origin.

Tally sheets, lot originations and fermentation control records should be attached together by lot and in an appropriately tabbed annual or vintage file folder.

Lot origination forms for red wines being fermented on the skins should be initiated in two parts. The estimated lot origination provides the basis for

Courtesy of Wine Institute

TAKING TEMPERATURE OF WINE FERMENTATION WITH LONG STEM THERMOM-
ETER

making amelioration additions when exact gallonage of juice cannot be determined due to the presence of skins, pulp and seeds in the must. The "actual" lot origination is drawn after the newly pressed red wine has been carefully measured and analyzed so that adjustments of sugar and water can be made for amelioration and alcohol content. Extract analysis or determination is required in order to calculate extract potential in partially fermented red wines. This provides the basis for ensuring the precise amount of sugar to be added, if any, in the amelioration adjustment.

Bulk wine book inventory cards should be designed to carry all specific data for each lot of wine in storage as it is received from the fermentation control record.

5

Quality Control During Aging, Clarification and Stabilization

The term *aging* in this text refers to the entire time between fermenter and bottle: blending, clarification, detartration, primary filtration and organoleptic fulfillment. Aging can mean different things to different winemasters: a good case can be made for aging to be defined as the time of "adult maturity" in the life of a wine. Another definition may broaden the aging period to include the entire life of a wine from fermentation to the wine glass. For the purpose of this work, however, aging shall be limited by post-fermentation and pre-bottling as recorded in the bulk wine book inventory, unless specifically excepted.

BULK WINE BOOK INVENTORY

The *bulk wine book inventory* is a perpetual inventory designed to maintain a constant record of wines in aging storage, both as to production status and finite volume. Along with proper identification and descriptive headings, a good bulk wine book inventory card should provide ample space for incoming and outgoing gallonages (or liters), with appropriate tank numbers and work orders. The treatment and disposition of wines while in the bulk wine book inventory should be posted and a column should be provided for reconciling the monthly bulk wine physical inventory, often signified by a simple check mark denoting agreement between the two inventories. The taking of the bulk wine physical inventory each month presents a good opportunity to sample and analyze wines in aging storage to insure that they remain in sound condition.

The ATF requires only a semiannual bulk wine physical inventory (end of June and end of December), but the wise vintner will not allow his inventory to go unchecked that long. Bookkeeping and recording reconciliations can become unwieldy over a six-month period. Monthly sampling and analysis are imperative, and the best winemasters will recheck troublesome or suspicious lots even more frequently.

The bulk wine book inventory card exemplified in Fig. 5.1 illustrates the monthly state of affairs for a white wine that has just been inventoried, found to be in agreement with the bulk wine physical inventory, and

A "chai" (Bordeaux style aging cellar) in which oak barrels are stored for the aging of table wines
Courtesy: Sterling Vineyards

| Lot No. 943 | Type-Variety SEYVAL BLANC | | | | Class −14% | | | Color WHITE | |
| Remarks: | | | | | | | | | |
Date	Racked From Tank W.O. Gallons	Treatment-Disposition	Racked To Tank W.O. Gallons	PHYS INV	ALC	BALL	T.A.	V.A.	FSO$_2$
10-28 1979	F-6 79-198 1,670	2nd rack	8-15 79-198 1,602		12.2	−1.7	.645	.030	
10-31 1979			1,602	✓	12.1	−1.7	.645	.033	16
11-12 1979	5-15 207 1,602	3rd rack	5-31 79-207 750, 5-34 (bbl) 750, 50						
11-13 1979	(63 gals been destroyed – ATF approved – 11-13-79)								
11-30 1979				✓	12.2	−1.8	.630	.036	56
12-31 1979				✓	12.1	−1.75	.608	.038	40
1-31 1980				✓	12.2	−1.8	.615	.036	32
2-29 1980				✓	12.1	−1.8	.608	.039	60

FIG. 5.1.   BULK WINE BOOK INVENTORY CARD IN USE

analyzed for several important quality control functions. Analysis of alcohol, Balling, total acidity, volatile acidity and free sulfur dioxide, plus an intense organoleptic evaluation, comprise a good monthly examination for every lot of bulk wine in the small winery. Of course, as with all testing performed in the laboratory, the analytical results are first posted on the laboratory analysis log and then on the bulk wine book inventory card.

Some of the new lighter and fruitier red wines that have become so popular in America recently may be aged according to programs similar both in style and duration to those of white and rosé table wines: short terms of storage at cool temperatures in large, non-wood vessels. This, obviously, encourages products that are fresh, fruity and tangy. However, many winemasters still make the more traditional dry red table wines that are darker, heavier, more complex and mellow, aged in wooden barrels and casks. It is common to see bulk wine book inventory cards stapled together in the file, recording several years in the long aging program of a red wine.

RECOGNITION OF QUALITY

It is very important that the winemaster be able to recognize superior and inferior traits inherent in young wine, especially as these qualities may apply to his individual product standards. He must first identify what he has before he can design a course for reducing negative and developing positive traits. For example, a winemaster may observe that more of his consumers prefer total acidity in the .488 to .518 g/100 ml range, to which he would, of course, set the standard. If a test for a given wine being made within this particular standard proves to be .578 g/100 ml total acidity, then, obviously, the total acidity level must be reduced. The point is to understand the market demand, set standards of quality, and then strive to meet those standards.

With modern technology there are many analyses that can be made to identify important constituents. Unfortunately, most of this technology is out of reach for the small winery laboratory, as the instrumentation necessary is very expensive. There are, however, some simple and less costly tests that can be made. In addition to high-cost laboratory testing is the intensive training of ones senses towards judging wines by organoleptic analysis. As mentioned before, organoleptic analytical procedures for wines are detailed in Appendix A.

Table wines can easily spoil before any promise and distinction are realized. Biochemical and microbiological damage can occur in a matter of days or hours. Consequently, before wine development and stability are discussed, it is necessary to consider preservation technology in small winery quality control.

SULFUR DIOXIDE

The most commonly used preservative in the small winery is sulfur dioxide. Some winemasters utilize cold storage techniques, which can be very expensive, and they may still require the addition of sulfur dioxide in order to fully protect some wines. Other vintners may utilize ultra-filtration through media of less porosity than the size of spoilage microorganisms. This, too, is quite expensive and time consuming. Furthermore, the "sterile" filtered wine may still require the addition of sulfur dioxide prior to storage in a clean, but not sterile, storage vessel.

Sulfur dioxide has been used both as an antioxidant and as a preservative in wines for centuries. While the ATF limit is 350 ppm (Title 27, CFR, Part 4, Section 4.22), some wineries, with superior cleanliness and filtration techniques, can reduce SO_2 requirements to less than 50 ppm. In the case of red Muscadine wines, the virtual elimination of SO_2 is necessary after fermentation as the highly unstable diglucoside anthocyanin pigments natural to most of these cultivars will oxidize immediately upon contact with sulfur dioxide. White wines made from Muscadine grapes are not affected in this manner.

It is only free sulfur dioxide that is of value as an antiseptic and inhibitor of enzymatic antioxidation in wines. In larger wineries, additions of SO_2 are made from direct applications of sulfur dioxide gas dissolved in water to form *sulfurous acid* (H_2SO_3). (This is not to be confused with the *sulfuric acid* (H_2SO_4), a very dangerous and corrosive chemical.) Sulfurous acid can also be hazardous and cellar workers should be fully instructed in the methods of careful handling of such materials.

Most small wineries use potassium meta-bisulfite (known as "KMS") as the sulfur dioxide source. This compound is less dangerous to handle than other forms of SO_2, but carries with it the disadvantage of being highly unstable. It can lose more than half of its sulfur dioxide content in just several weeks of poor storage conditions. Many wines have spoiled because of the use of old or weak potassium meta-bisulfite.

Fixed sulfur dioxide results from the bisulfite ion (HSO_3^-) having reacted with aldehydes, proteins, pectic compounds, sugars and other substances. The combination of free SO_2 and fixed SO_2 is known as total SO_2, which is normally analyzed in the small winery only to insure compliance with the ATF maximum of 350 parts per million.

Sulfur dioxide is not a cure-all for the many microbial and oxidative problems that can occur in winemaking. Normally about 50 ppm of SO_2 is added to grapes or must as soon as each container is dumped into the crusher hopper in order to inhibit natural yeasts, and to create an atmosphere of SO_2 gas. It may, however, be necessary to reduce this SO_2 addition if the grapes have already had a treatment of KMS prior to shipment, which is usually done to help reduce microbial growth in transit.

It may be necessary to add 50 ppm of SO_2 again during the second or third racking of the young wines, especially if volatile acidity increases above .036 and .048 g/100 ml for white and red wines, respectively. Subsequent additions may be necessary to maintain a given level of free SO_2 during aging storage, or under prime conditions there may be no further need for SO_2 additions until bottling.

The novice winemaker may become distressed by the disparity between calculated and observed results in raising free SO_2 from, say, 32 ppm to 60 ppm. The unstable nature of potassium meta-bisulfite and other sources of sulfur dioxide regularly causes deficient results. For instance, it may be observed that the 28 ppm addition calculated in this example only realized a new free SO_2 analysis of 48 ppm. The remedy, of course, is to make another addition that is appropriate to fill the 60 ppm total needed.

One can make the following rationale for calculating SO_2 additions using potassium meta-bisulfite ($K_2S_2O_5$):

Atomic weights: $K = 39.10 \times 2 = 78.20$
$S = 32.06 \times 2 = 64.12$
$O = 16.00 \times 5 = \underline{80.00}$
Molecular weight of $K_2S_2O_5 \quad = 222.32$

Therefore, the SO_2 portion of $K_2S_2O_5$ is S_2O_4—a portional weight of 128.12. This, divided by the total molecular weight of 222.32, yields an active sulfur dioxide potential of about 57.6%.

Consider, then, an example where 1000 gal. of a table wine has a free SO_2 of 32 ppm and the winemaster desires a free SO_2 of 60 ppm:

Weight of dry table wine = 8.4 lb per gal. (see Brix table in Appendix B for wine weight information).

1000 gal. $\times$ 8.4 lb = 8400 lb

8400 lb $\times$ .000028 (the 28 ppm differential between 32 and 60 ppm) = .235 lb of SO_2 that needs to be added.

Using 100% sulfur dioxide gas as the SO_2 source, the winemaster has but to simply write the work order for a .235 lb SO_2 addition. If the gas is less than 100%, a simple factoring can be calculated. Sulfur dioxide gas at 91%, for instance, would be figured as .235 divided by .91 = .258 lb of SO_2 gas needed.

Using sulfurous acid solutions, the winemaster would need to factor the strength of his SO_2 source in calculating the correct addition. If the solution was 10% SO_2, then the proper treatment would be 2.35 lb (.235 lb $\div$ .10).

Using KMS the division of .235 lb by the .57 as determined above results in .412 lb of potassium meta-bisulfite required. Remember, though, that KMS is very unstable—the actual addition in this example may need to be increased from .412 lb up to .600 or .700 lb in order to achieve the desired 60 ppm level.

Sulfur dioxide instability is aggravated by humid or warm storage conditions, leaking containers, and other factors. The equilibrium reaction is portrayed in the following equation:

$$SO_2 + H_2O \rightleftharpoons H_2SO_3 \rightleftharpoons H^+ + HSO_3^- \rightleftharpoons H^+ + SO_3^=$$

Dry table wines made under clean conditions may be preserved with as little as 30 ppm of free SO_2. However, with any amount of fermentable sugar remaining in a wine, free SO_2 should not be allowed to decrease below a level of 60 ppm. Even these limits may be too low in many cellars. Much of the criteria for establishing SO_2 standards result from other operating conditions. The winemaster who enforces strict practices of quality control in terms of grape evaluation, cleaning of machinery and tankage, monitoring inventories and overall honest professionalism will encounter a minimum of serious problems. Lower levels of free SO_2 are required in operations where quality control is maintained, and lower levels of free sulfur dioxide will result in higher quality wines.

Sulfur dioxide is relatively inexpensive and does an adequate job in wine preservation, but it is a very pungent compound, and can mask and detract significantly from otherwise fine wines. Only the least amount of SO_2 necessary should be used.

Solutions of sulfur dioxide can be used to clean wine hoses, pipelines, pumps and other items of cellar equipment, including wooden cooperage. Normally 2 oz of KMS, with an equal amount of citric acid, may be dissolved in each 100 gal. of water to be used for cleaning. For heavy-duty work, the KMS may be doubled.

It is very important to distinguish carefully between the different sources of sulfur dioxide available when considering this chemical as a preservative and "sweetener" of clean empty wooden cooperage. Sulfurous acid solutions should not be used in order to avoid the formation of hydrogen sulfide. The use of potassium meta-bisulfite also is not advised, as some rinse water will almost surely remain in the cleaned tank, and this will dissolve the KMS, forming a sulfurous acid solution. Direct additions of sulfur dioxide gas, to expel air from the vessel and seal the SO_2 inside, are effective, but very dangerous and difficult to administer.

A few winemasters continue to order the burning of "sulfur sticks" in cleaned cooperage. Freshly-cleaned casks and vats are sealed except for the top bung. One sulfur stick (about ¼ in. × 1 in. × 12 in.) for each 1000 gal. tank capacity, is hooked at one end of a steel wire, while the other end is attached to the bottom of a bung used exclusively for this purpose. The wire length is gauged so that the burning sulfur sticks hang at the geometric center of the vessel when the bung is driven securely. This method utilizes most, or all, of the oxygen remaining in the tank, and the SO_2 gas given off is sealed inside as long as the vessel does not leak.

BLENDING

Many experts feel that any varietal wine is improved if it is blended with some other wine. Unfortunately, winemakers may blend wines in order to make up variances of specific analytical constituents such as alcohol and volatile acidity. The need for this arises, by and large, because of poor planning and negligent control rather than because of poor fruit.

The improvement of a wine by blending should take place in order to help attain designed standards of color, nose and taste. In some years, a particular vineyard may yield wines that are light and withdrawn. In other years, vines from the same plat may provide wines that are heavy and fragrant. Occasionally, a vintage may be recorded where the fruit was nearly ideal. Blending helps to maintain product consistency, in line with recognized demands from the marketplace.

Blending operations can increase or reduce color, brighten or subdue color hues, intensify or moderate aromas, distinguish or temper bouquet, stimulate or relax flavors and either nurture or discourage finite patterns of varietal taste values.

Some wines are the result of blending several individual wine constituents. However, simplicity, or relative simplicity, is usually the best winemaking philosophy. Too many wines being blended together can result in a potion that is neither distinctive nor exciting and, hence, is of little value.

Laboratory trials are necessary prior to the actual blending in the cellars. The wine industry has never known a winemaster whose anticipated results in blending were achieved without a little trial and error.

Normally samples will be gathered from those wine tanks that are being considered for the required blend. The winemaster may make several lab blends to test for the best possible product with regard to quality, economics and inventory control. After making a final lab blend from a new set of samples in order to double-check the blend formula, the winemaster should calculate the characteristics of his proposed blend on a *wine blend procedure* form similar to that illustrated in Fig. 5.2.

Bringing different wines together will necessitate the assignment of a new lot number so that a new bulk wine book inventory card can be made to reflect the new total gallonage of the blend (Fig. 5.3). As the blend is produced, the individual component wines used for the blend are subtracted from their respective bulk wine book inventory cards. When a constituent wine has been completely depleted this is noted on its card, which is then removed from the bulk wine book inventory and placed in the dead file. Most systems simply maintain a section at the rear of the file drawer for dead file storage.

Theoretical calculations of the laboratory analysis for the final blend should be reasonably close to the results actually observed. If not, one or more of the constituent wines either has been improperly sampled and/or has been analyzed yielding erroneous results. It is good practice, therefore,

WINE BLEND PROCEDURE

LOT **945 Seyval Blanc**   GALLONS _1,900_   DATE _3-18-80_

Tank	Lot	GALLONS			Alc.	Bal.	Ext.	T.A.	V.A.	SO₂		Remarks
		In Tank	To Use	As %						Total	Free	
S-31	943 Seyval	750	750	39.5	12.0	-1.8	2.2	.608	.033	—	56	✓
S-34	943 Seyval	750	750	39.5	12.1	-1.9	2.2	.615	.036	—	52	✓
1bbl.	943 Seyval	50	50	2.6	11.9	-1.8	2.2	.600	.042	—	74	✓
S-24	933 Ravat	1,050	350	18.4	12.3	-2.0	2.1	.623	.033	—	44	✓
Lab Blend Analysis		1,900	100.0		12.1	-1.9	2.2	.615	.039	—	54	
Analysis After Clarif. Lab Blend					11.9	-1.9	2.1	.608	.034	—	48	

SPECIAL INSTRUCTIONS: *Rack* 933 Ravat from TK # S-24 to fill TK # 37 — balance of 350 gallons to blend.

CLARIFICATION:
Blend and clarify in TK # BT-1
sample to lab — hold for
R-OK before racking to
cold cellar

Metafine: _4 lbs._
CaCO₃: _—_
Act. Char.: _—_
Sparkolloid: _3 lbs._
Bentonite: _3 lbs._
Other: _—_

% COMPOSITION				
Varietal Or Wine Type	As Type	As Old		NOT NYS
		1979	1978	
Seyval	81.6	—	—	—
Ravat	18.4	—	—	—
"Vintage 1979" label OK				
TOTAL	100.0			

FIG. 5.2.   WINE BLEND PROCEDURE FORM IN USE

Lot No. 945 | Type-Variety BLENDED SEYVAL | Class −14% | Color WHITE

Remarks: Blended 3-19-80 81.6% Seyval Blanc 18.4% Ravat Blanc

Date	Racked From Tank W.O.	Gallons	Treatment-Disposition	Racked To Tank W.O.	Gallons	PHYS INV	ALC	BALL	T.A.	V.A.	FSO$_2$
10-28 1979	F-6 79-198	1,670	2nd rack	S-15 79-198	1,602		12.2	−1.7	.645	.030	
10-31 1979					1,602	✓	12.1	−1.7	.645	.033	14
11-12 1979	S-15 79-207	1,602	3rd rack	3-31 79-207 / 5-34 (1 bbl)	750 / 750 / 50						
11-13 1979	(42 gals been destroyed − ATF approved − 11-13-79)										
11-30 1979						✓	12.2	−1.8	.630	.036	54
12-31 1979						✓	12.1	−1.75	.608	.033	40
1-31 1980						✓	12.2	−1.8	.615	.036	32
2-29 1980						✓	12.1	−1.8	.608	.039	60
3-19 1980	5-31 / 3-34 (1 bbl) 80-74	1,550	blended	BT-1 80-74	1,550						
3-19 1980	5-24 80-74	350	(RAVAT) blended	BT-1 80-74	1,900		11.8	−1.8	.615	.036	40

FIG. 5.3.   BULK WINE BOOK INVENTORY CARD IN USE FOR NEW WINE BLEND

to clarify the laboratory blend before writing the work order for the actual blend in the cellars. The addition of clarification agents can have a profound unanticipated effect upon color retention and various organoleptic factors. Another analysis of the proposed new wine, after clarification of the laboratory blend, should be made and posted in the laboratory analysis log and the wine blend procedure form.

The wine blend procedure form includes the provision for the winemaster to order further processing, including clarification and detartration, if necessary. This form is actually an extended work order, but a work order should also be written, attached and issued with the wine blend procedure form for continuity in the record system.

The wine blend procedure form should also describe the component and vintage percentages. Varietal-labeled wines must contain at least 75% of the variety stated upon the label, as stated in Title 27, CFR, Section 4.23. Section 4.39 rules that vintage-labeled wines may not be blended with more than 5% of wines from production of vintages other than those stated upon the label. The "% composition" feature of the wine blend procedure form (Fig. 5.2) illustrates such data.

CLARIFICATIONS

The clarification process is applied, as the name suggests, to clear new wines. This is an essential operation and small winery vintners may take this opportunity of handling the wine to perform other treatments that may be required.

Even after the strictest control, hardly any wine ever reaches this point in perfect chemical harmony. The total acidity may still be a bit high, iron and copper concentrations may be excessive, or color may be too intense. The most successful winemasters are quick to look for such inconsistencies and study materials and methods for correcting such conditions. The materials to be used are clarification agents, or "finings".

Young wines are not usually blended and clarified prior to the third racking, and at least three months of aging. White table wines are normally more difficult to clarify in early life than reds, as colloids have a tendency to remain in suspension following lower-temperature fermentations. The apprentice winemaker may tend to over-clarify, i.e., add too much fining in an effort to make the wine as bright as possible. However, most clarification agents usurp aroma, flavor and color constituents. Consequently, as with sulfur dioxide, finings should be used as sparingly as possible.

All blending operations should be completed prior to clarification and detartration. The blending of two stable wines can result in an unstable wine, due to the contribution of different thresholds of tolerance for color pigments, metals, tartrates and other compounds in solution.

Most of the clarification problems found in the small winery will involve acid, color and metal adjustment, as well as clearing the wine to a degree that will render filtration economical. Some small wineries never completely stabilize their wines, choosing to endure some precipitation of sediment later in the life of the wine rather than overprocess their products at the expense of quality. It is common to find statements to this effect on labels, assuring consumers that the potential sedimentation and/or haziness is not harmful.

Materials

All approved clarification agents and treatment compounds are listed by the ATF in Part 240, CFR, Subpart ZZ, entitled *Materials Authorized for Treatment of Wine*. It is recommended that the novice winemaker familiarize himself with ATF requirements, limits and methods, before beginning fining operations.

This text shall discuss the use of those authorized finings and treatments that are felt to be of the most value and importance in normal small winery clarifications. Such materials as Metafine, calcium carbonate, activated charcoal, Sparkolloid, bentonite and citric acid are considered to be essential compounds.

All clarification treatments should be made during one handling of the wine, if possible. This exposes the wine to minimal contamination and oxidation. Treatments for copper and iron normally will be made first; then reduction of total acidity, if necessary; then color adjustment, if needed; and finally addition of finings for clarity. This order is important to the clarification process, so that the proper reactions are achieved.

Copper and Iron: Metafine

Metal contamination is of great importance in any winery. Copper and iron can be absorbed from vineyard soils and some spray materials through vine uptake. More often, however, juice, must and wine gather the most significant concentrations of metals from contact with brass, copper and iron parts in winery equipment. Even stainless steel can impart some iron, but this is negligible as a rule.

Instability is manifested in wines by a *casse,* or haze, that usually develops later in the life of the product, often after bottling. Such formations can result from a concentration of only several parts per million of metal contaminant. For instance, copper can create a noticeable casse at a level of just 1 ppm, when the iron content in the same wine is in excess of 7 ppm.

On the other hand, a complete absence of copper may allow the formation of *mercaptans* (thioalcohols), which, in concentrations of only 1 ppm, can render a wine undrinkable! Consequently, the winemaster may choose to reduce only the iron content of his wine in clarification. This can be done with the use of a grain bran derivative called Afferin (magnesium phytate). This is a white powder that is readily soluble in wine, and normally takes several days of contact for full iron reduction. Afferin is calculated at the rate of 5 ppm for each 1 ppm of iron to be removed, up to a maximum of 5 ppm iron removal. The ATF limitation quoted from Subpart ZZ of Part 240, CFR, states:

No insoluble or soluble residue in excess of one part per million may remain in the finished wine, and the basic character of the wine may not be changed by such treatment, G.R.A.S. (generally regarded as safe).

Most winemasters set standards for copper and iron at 1 ppm and 5 ppm, respectively. Reduction of copper and iron content can be achieved safely with the proper use of readily available and approved compounds such as Metafine and Duvex, also known as "blue finings" (potassium ferrocyanide compounds). These compounds react rapidly and, in addition to the reduction of iron and copper, may also remove manganese, zinc, magnesium and other trace metals from wine. Some winemasters have noted that the use of blue finings may act as an antioxidant by slowing aging progress. Potassium ferrocyanide ($K_2Fe(CN)_6 \cdot 3H_2O$) compounds have also been found to reduce bacterial counts. These materials are generally manufactured so that 1 lb of blue finings will reduce the metals in 1000 gallons by 1 ppm. One must, of course, carefully read the label of any material being used every time before any calculations or additions are made. Should anything not be completely understood, the manufacturer of the additive should be consulted prior to its application.

The ATF limitation for Metafine (CUFEX) is quoted from Part 240, CFR, Subpart ZZ as follows:

No insoluble or soluble residue in excess of one part per million shall remain in the finished wine, and the basic character of the wine shall not be changed by such treatment, G.R.A.S.

Metals can be quickly and accurately analyzed with atomic absorption instrumentation, but such equipment is very costly. If such an instrument fits into the budget, however, it would be advisable to consider including supplemental equipment for analysis of other metals (potassium, sodium, calcium and magnesium). This will increase the functional capacity of the laboratory to include vineyard soil and leaf analyses, if desired.

Copper and iron testing performed chemically by separations and spectrophotometry is an arduous and lengthy task. Nevertheless, in order to avoid the capital investment of an atomic absorption unit, the winemaster may wish to make his tests by the methods described in Appendix A. It is advised when using these methods that as many samples as practicable be run at once in order to reduce the number of subsequent tests required.

Total Acidity: Citric Acid and Calcium Carbonate

The tartness of excessive total acidity is a quality that is more accepted in Europe than in America. It is important that the "bite" be subdued, even to "bland", for wines to be in line with demands in the United States.

Through the selection of grape cultivars which lend themselves to good year-in and year-out acid balance, many of the difficulties associated with acid adjustments in the cellars can be avoided. Good harvest control and proper amelioration levels can also help to reduce the need to treat wines for acidity imbalances. As has been discussed previously, blending is also a valuable tool in this regard. Nevertheless, every winemaster, at one time or another, is faced with the need for acid adjustments in his wines.

The analysis of total acidity is obviously necessary prior to any consideration of acid adjustment. Taste testing is also an important function, as the same wine type or varietal may vary in organoleptic factors from year to year so that total acidity levels may need to be readjusted. Both the analytical procedure for total acidity and the organoleptic evaluation of tartness are provided in Appendix A.

A standard for total acidity for a particular wine was exemplified earlier in this chapter as a range of .488 to .518 g/100 ml. Any result between these two acidity levels will be acceptable in a finished wine—as long as the organoleptic result concurs. A product one year may be more palatable nearer the .488 g/100 ml lower limit because the wine is rather heavy and full. In this instance the lesser total acidity creates a better balance because of the blandness. However, in other vintages, a total acidity nearer the .518 g/100 ml mark may be desirable because the wine is weak and needs some enhancement. These standards are quite narrow and a winemaster may choose to widen the range, for example to levels of .473 to .55 g/100 ml. The standards chosen should not have too great a tolerance, or else they will become meaningless and may quickly get out of line with market demands.

The apprentice winemaker should be cautioned about acidity additions prior to detartration, a function that should immediately follow the addition of clarification agents. The purpose of detartration is the formation and precipitation of the acid salt of tartaric acid, potassium bitartrate, which is also known as cream of tartar. Considerable acid reduction should occur during detartration, and in some cases it will be enough to bring an unbalanced wine into standard range.

As was explained in Chapter 4, an optimal pre-fermentation total acidity of about .770 g/100 ml could be expected to be reduced to approximately .510 g/100 ml by the time the wine is ready for bottling. The winemaster might expect about half of the .260 g/100 ml reduction to result from the effects of fermentation, and the other half in response to cold-temperature storage detartration procedures. Total acidity in the range of about .620 to .660 g/100 ml may well fall to within the standards of .488 to .518 g/100 ml mentioned earlier. Whatever standards are set, de-acidification should be considered only when absolutely necessary.

In some states, particularly in the west and south, excessive acidity is not normally a problem. On the contrary, additions of acid are commonly made to correct natural deficiencies. This is usually done with citric acid, although tartaric and fumaric acids are also used. The addition of citric acid, in particular, is usually made only after detartration and primary filtration. Not only does this practice aid in pinpointing the acidity level, but the citric acid contributes to ionic stability, which helps to stablize the wine after detartration.

Regarding the reduction adjustments of acidity, there are several modern agents which are currently approved for use. However, calcium carbonate remains a relatively inexpensive compound and a very reliable acid neutralizing agent. Subpart ZZ, Part 240, CFR, describes the use of calcium carbonate ($CaCO_3$) as:

to reduce the excess natural acids in high acid wine.

Further, this material is limited so that:

The natural or fixed acids shall not be reduced below five parts per 1,000.

In other words, deacidification with calcium carbonate cannot be used to reduce the total acidity of a wine below a level of .500 g/100 ml. If a winemaker desires further reduction of acidity, Sections 240.536 through 240.538, Part 240, CFR, must be complied with properly.

Calculation of calcium carbonate for a 2456 gal. lot of wine having a total acidity of .780 g/100 ml, and desired after detartration to be about .525 g/100 ml, may be made as follows:

.780 g/100 ml total acidity existing
−.117 g/100 ml typical amount estimated to be lost in detartration
.663 g/100 ml potential natural total acidity
−.525 g/100 ml desired remaining total acidity
.138 g/100 ml acidity to be neutralized with $CaCO_3$

.138 g/100 ml = 5.22 g/gal.
5.22 g/gal. × 2456 gal. = 12,820 g acidity to be neutralized with $CaCO_3$

$$\frac{12,820 \text{ g}}{454 \text{ g per lb}} = 28.25 \text{ lb of acidity to be neutralized with } CaCO_3$$

Atomic weights: Ca = 40.08 × 1 =   40.08
              C = 12.01 × 1 =   12.01
              O = 16.00 × 3 =   48.00
Molecular weight of $CaCO_3$ = 100.09

The calcium portion of the molecule is divalent, i.e., has the ability to neutralize, or tie up, two anionic charges indigenous to the acids being reduced. Therefore, 2 × the Ca atomic weight of 40.08 is divided by the molecular weight of 100.09 which results in about 80% effectiveness of the $CaCO_3$ to neutralize acids. The 28.25 lb of acid to be neutralized is then divided by .80 which results in approximately 35.3 lb of calcium carbonate necessary to realize the adjustment.

After this calculation is made and double-checked, the winemaster should confirm these results by making a mock-up in the laboratory. A sample of wine should be treated with the appropriate amount of calcium carbonate under simulated cellar conditions and studied carefully to be sure that the desired results are achieved.

Only food-grade calcium carbonate should be used, and it should be administered very slowly in a mixing tank that contains a small portion of the wine to be treated. The reaction of $CaCO_3$ with the wine acids is not immediate and gentle intermittent agitation is often required in the first hour or two, until all of the calcium carbonate has been sprinkled into the mixing tank. This is followed by a holding period of up to 24 h, after which the treated portion is returned to the bulk of the wine. Only a small portion of the wine is treated, because $CaCO_3$ causes a physical distortion of the natural acid profile. Re-blending this back with the bulk of the wine helps to dilute the effects of this distortion. Adding the calcium carbonate directly to the total volume of the wine would render the acidity profile of the entire lot distorted and perhaps unstable.

Oxidation and Color Intensity: Activated Charcoal

Oxygen may be the single worst enemy of the winemaster. The reader may recall from earlier discussions that volatile acidity buildup is primarily a result of the action of *Acetobacter* oxidizing ethanol into acetic acid. This reaction is portrayed in the following equation:

$$C_2H_5OH + \frac{1}{2}O_2 \longrightarrow CH_3CHO + H_2O$$

then:

$$CH_3CHO + \frac{1}{2}O_2 \longrightarrow CH_3COOH$$

Translated from chemical symbolism, this equation states that ethanol plus oxygen reacts to give acetaldehyde and water; then acetaldehyde and oxygen react to give acetic acid.

To achieve this reaction, *Acetobacter* require oxygen, as shown above, and are therefore aerobic.

While oxygen is required for yeast growth, the actual process of fermentation is best conducted in the absence of oxygen.

Oxygen can oxidize wine color pigments from delicate whites to heavy ambers, and from rich purple-reds to tawny-browns. Flavor constituents can also be adversely affected by oxidation. Many microorganisms detrimental to wine also require oxygen to live, such as the *Acetobacter* spoilage mentioned above.

White and rosé table wines are usually not in danger of oxidation when dissolved oxygen is present in concentrations of less than 5 ppm. Red wines, especially heavier-bodied types, can often tolerate higher levels of dissolved oxygen.

Excessive oxidation also results in the development of aldehydes, as was indicated in the previous chemical equation. This is readily apparent in the making of sherry (see Chapter 7).

The best small winery laboratories contain an oxygen meter, despite the cost of several hundred dollars. Figure 5.4 illustrates an oxygen meter that is considered satisfactory by many winemasters.

FIG. 5.4. OXYGEN METER

Many of the points made for the analysis of oxygen uptake in wines can be made for the monitoring of color intensity as an indicator of oxidation. In addition, colorimetry is a valuable tool for standardizing color intensities in wines.

Small spectrophotometers (colorimeters) are a welcome addition to the countertop of the small winery quality control program. The primary functions of this instrument are measurement of color intensity and analysis of iron and copper content by the separatory method. These methods are fully described in Appendix A.

Standards of color intensity should be developed for all wines to be produced. For example, a white wine that is of optimum color in the eyes of the vintner may read out with a light transmission of 87% at 425 μ wavelength. Therefore, the winemaster may decide that all white wines of that type should analyze between 85 and 89% light transmission at that wavelength in order to meet the standard. In like fashion, rosé standard transmittance may be found in the 475 μ range, while reds may be best analyzed on the spectrophotometer at 525 μ. The highest accuracy, as in almost any laboratory instrument, can be found in the middle portion of the scale.

One should bear in mind that the spectrophotometer does not measure color, but color intensity. The instrument cannot differentiate between color hues, such as red, green, brown or blue, but indicates how intense a color may be, whatever the hue.

Samples introduced into a spectrophotometer must be totally devoid of suspended solids—"bottle bright" or else the results from the instrument will be meaningless.

Figure 5.5 provides a visual description of a typical small winery spectrophotometer.

The use of activated charcoal is a last resort for salvaging white wines from browning, which is normally a result of excessive oxidation. It is, literally, a decolorizing agent and it is only rarely used in wineries that are properly operated.

However, there are times when excessively discolored wine, due to high levels of leucoanthocyanin pigmentation, requires bleaching. This occurs most often when winemasters allow extended skin contact in the white must before extracting the juice by pressing. White wines that have only slight browning due to oxidation may also be improved by the use of modest amounts of activated charcoal. However, because special permission must be obtained from the ATF before activated charcoal may be used, the winemaker is advised to be familiar with Part 240, CFR, Sections 240.527 and 240.527a.

Determination of proper levels of use of activated charcoal can be made in the laboratory in small experimental samples. One might divide a gallon

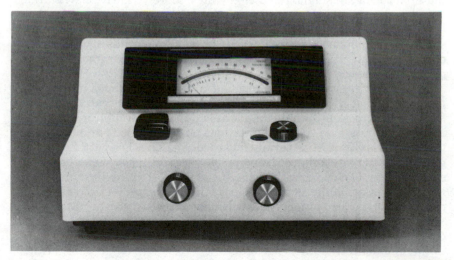

Courtesy of Bausch and Lomb, Rochester, N.Y.

FIG. 5.5.   SPECTROPHOTOMETER

sample of the wine into five equal portions, and treat four of these portions with amounts of activated charcoal equivalent to the following: ½ lb charcoal per 1000 gal. wine; 1 lb charcoal per 1000 gal.; 2 lb/1000 gal.; and 4 lb per 1000 gal. The fifth sample should be left undisturbed for reference as the control. After 24 h or so the samples can be carefully filtered, to determine the effects of the different levels of activated charcoal upon each individual portion. From this information the proper amount of activated charcoal can be determined and plugged into the calculation needed for the addition in the cellars. A treatment of more than 2 lb of activated carbon per 1000 gal. of wine is considered heavy, and in excess of 4 lb/1000 gal. is for salvage operations only.

The application of activated charcoal normally takes place along with the other clarification procedures, i.e. in a minor portion of the blend held in a mixing tank. Activated charcoal may be added directly in the mixing vessel, but caution is advised in handling this agent. Activated charcoal "flies" very readily and the dust can easily scatter. It makes sense to store and use activated charcoal in a confined area in the cellars.

Of course, the best manner to avoid the need for using charcoal in white wines is to minimize the conditions which promote oxidation. Cool fermentations, no air exposure and maintenance of good sulfur dioxide levels are the main practices to pursue.

Clarity: Sparkolloid

A century ago and before, winemasters clarified their young wines with materials such as isinglass and egg whites. More recently, clarifications with tannic acid and gelatin were common. The modern winemaster, however, uses materials such as Sparkolloid to help clear wines of suspended solids. This agent forms a gelatinous suspension which aids in precipitation of colloidal materials that remain in most untreated new wines, and especially white wines.

Sparkolloid is a commercial compound marketed in both hot mix and cold mix types. Most winemasters agree that the best results can be expected from the hot mix form, despite the additional time and trouble needed for its use. The ATF has no specific limitations on the use of Sparkolloid other than that it is G.R.A.S. (generally regarded as safe). A level of 2 lb per 1000 gal. of wine is commonly used, and up to twice this amount can be used in extreme cases of cloudiness. The specific amount can be determined in the laboratory in much the same manner as that described for activated charcoal in the previous section. An experiment of several concentrations (e.g. ½ lb, 1 lb, 2 lb and 4 lb per 1000 gal.) along with a control may be set up and observed over several days. The winemaster can then make the decision as to the minimum treatment necessary.

Some new wines, especially reds, may be so naturally clear that clarification with Sparkolloid may not be necessary. On the other hand, some new white wines may not clear entirely even after several months of cold room detartration storage. Sparkolloid treatment can only be expected to help in the clarification process—it is not a miracle fining. In most cases, the addition of bentonite, to be discussed in the next section, will greatly increase the effectiveness of Sparkolloid.

The application of Sparkolloid should be made in the form of a hot slurry, much the same way as gelatin is made in a kitchen. Only the amount of wine needed to properly suspend the Sparkolloid should be pumped into the mixing vessel (about 1 lb per 10 gal.). The wine should be heated and constantly agitated while the dry Sparkolloid powder is slowly added. After the suspension is completed, it should *immediately* be removed from the mixing vessel into the appropriate wine to be clarified, and gently mixed so as to be homogeneously distributed. If the suspension is left for any appreciable length of time in the mixing vessel, the Sparkolloid may set and solidify much like a gelatin preparation.

Protein Instability: Bentonite

Much has yet to be learned about protein solubility and instability in wines. It is known, as mentioned several times previously, that early racking procedures help to separate yeast cells from new wines, which reduces the potential for contamination by yeast cell autolysis and the release of yeast proteins.

This is a particularly difficult area to provide finite levels and thresholds of stability. In most cases, an application of bentonite, 2 lb per 1000 gal. of wine to be treated, may be considered empirically sufficient to remove protein haze potential. Individual experience, however, may suggest levels of 3 or even 4 lb per 1000 gal.

Bentonite is a montmorillonite clay that has been used for wine clarification and stabilization for several decades. Silica-alumina molecular "layers" in the crystal formation of this substance provide a tremendous negatively charged surface area with great adsorptive capacity. Type I (sodium bentonite) is found principally in South Dakota and Wyoming and is the grade used for wine clarification purposes. The proprietary name most common in the wine industry is KWK Volclay.

Bentonite is generally the last fining agent added. This very dense material contributes both chemical and physical characteristics necessary for the precipitation of suspended substances remaining in wines, and it comes closer than any other material to being a "cure-all" fining agent. It helps to remove protein and metallic hazes, and adsorbs many of the nutrients necessary for microbial growth as well as many enzymes. It is both adhesive and cohesive, and adds little or no flavor of its own, unless used in excessive amounts. The chief disadvantages of bentonite are a tendency to decolor the wine slightly (which can be an advantage in some cases); it removes amino acids and vitamins (which are in minimal proportions, usually); and it can form heavy lees sediment.

Levels of bentonite are calculated and administered in much the same manner as for Sparkolloid. A bentonite slurry is normally prepared for all wines, whether or not they appear clear and stable. Bentonite slurries must be made very carefully, using about twice the amount of wine needed to suspend Sparkolloid (about 1 lb per 20 gal. of wine in the mixing vessel). The bentonite slurry will clot unless the dry granules are added to the mixture very slowly. The ATF allows some water to be used in preparation of the bentonite slurry, as quoted from Subpart ZZ, Part 240, CFR:

Not more than 2 gallons of water shall be added to each pound of Bentonite used. The total quantity of water shall not exceed 1% of the volume of wine treated. G.R.A.S.

Some winemasters, however, dispense with this addition of water, choosing to use wine instead and accepting the negligible loss of alcohol due to the heating of the slurry portion.

Figure 5.6 illustrates a bulk wine book inventory card after a new blend of wine has been made, adjusted, and readied for detartration in a cold storage cellar.

Detartration

Most small wineries achieve stability in new wines by the proper use of clarifying agents together with the immediate precipitation of potassium bitartrate crystals in a chilled atmosphere. The alternative to this procedure would be the "softening", or *ion-exchange* process. By this technique, potassium ions from the unstable potassium bitartrate are partially exchanged with sodium ions from sodium chloride in order to form the more stable sodium bitartrate. This is an over-simplification of the process, but shall serve the purpose for this discussion. The technical and economic dynamics required for the successful operation of an ion-exchange column are usually out of reach for even medium-sized winery budgets and, consequently, details of ionic exchange technology and methodology are not included here.

The precipitation of unstable potassium bitartrate, or cream of tartar, is performed best in a controlled atmosphere of constant temperature. The normal duration is at least 4 weeks in a temperature range of from 25 to 28°F. In the case of Muscadine wines, this time period may need to be doubled in order to achieve stability. After this holding period, young wines may or may not have remaining haziness. Red wines usually clear rather quickly, but whites and rosés may not become brilliant before the primary filtration.

Fining progress can be monitored by holding samples of the wine being detartrated in hot and cold chambers in the laboratory. These facilities may be the same as used for post-bottling stability analyses that are described in Chapter 6. A refrigerator held at about 35°F and a small oven maintained at approximately 110°F are ideal. An observation period of from 2 to 3 weeks with no apparent precipitates in a sample can generally be interpreted as positive stability.

Lot No.	Type-Variety		Class	Color
946	BLENDED SEYVAL BLANC		-14%	WHITE

Remarks: Blended 3-19-80 81.6% Seyval Blanc 18.9% Ravat Blanc

Date	Racked From Tank W.O. Gallons	Treatment-Disposition	Racked To Tank W.O. Gallons	PHYS INV	ALC	BALL	T.A.	V.A.	FSO$_2$
3-20 1980	CLARIFIED: Yuba Metquire, 3 the Speedeloid	3 the Bentonite	80-76		11.7	-1.8	.608	.036	36
3-20 1980	BT-1 78 1,900	closed for detart.	CS-2 78 1,889		11.8	-1.8	.608	.036	32

FIG. 5.6.   BULK WINE BOOK INVENTORY CARD IN USE AFTER CLARIFICATION

Normally, once all clarification agents have been administered to a given lot of wine, and that wine has been appropriately analyzed to be sure that all adjustments have actually taken place, there is no need for further analysis of the wine in cold detartration storage. The temperature is such that almost no liability can be anticipated unless there is some obvious potential problem such as a tank that is unclean or unsound. Certainly it does no harm to gather a sample regularly from all lots of wine in detartration storage and perform an organoleptic analysis. This is especially so if there are tanks that are not completely full, in which case the head space can contribute to oxidation. However, this condition should be eliminated by planning ahead so as to insure full tankage. If this is impossible, the danger of oxidation can be greatly reduced by saturating the head space with nitrogen gas. Part 240, CFR, Subpart ZZ, restricts nitrogen gas only as follows:

The gas shall not remain in sparkling or still wine.

The use of carbon dioxide carries considerably more involvement with the ATF manual. The apprentice winemaker should be acquainted with Part 240, CFR, Sections 240. 531 through 240.535 before using carbon dioxide gas to purge tankage head space. Carbon dioxide gas will dissolve and later release from solution under the proper conditions, causing a wine to effervesce, or "sparkle". The production and taxation of sparkling and crackling (natural and artificial CO_2 gas, respectively) wine is very closely monitored by the ATF. To this end, carbon dioxide use should be such that no residual dissolved gases are present. Proper filtration procedures, however, will normally strip a wine of dissolved gases.

After the detartration period has been completed, the new wine can be prepared for the primary filtration out of cold storage.

PRIMARY FILTRATION

The primary filtration is generally considered the coarse filtration: normally it is done only to separate the newly stabilized wine from the gross elements of clarification lees and potassium bitartrate crystals. Additionally, any residual gases, such as the carbon dioxide mentioned in the previous section, are virtually removed. Filter media ranging from 2.0 to 5.0 microns porosity are commonly used for such a purpose. Wines with considerable haziness remaining may require even coarser media, perhaps to the 10.0 micron level. A medium that is less than 2.0 microns in porosity will almost certainly foul very quickly. This makes frequent changes necessary, which is a very costly situation.

The beginning winemaker should be cautioned to be sure that he or she has a proper storage place prepared for the filtered wine to be delivered. Usually such wines will first be filtered to a blending tank where citric acid may be added as a buffer (1 to 2 lb per 1000 gal.) in order to aid in maintaining ionic stability. If the wine is deficient in acidity in relation to the standards set, more than the token 1 to 2 lb per 1000 gal. will be calculated.

However, before any citric acid additions are made, the winemaster should be fully familiar with Part 240, CFR, Sections 240.364, 240.404 and 240.526.

The calculation procedure for citric acid additions are rather simple and straightforward, as indicated by the following example:

A wine measured to total 1955 gal. has just been filtered and analyzed revealing a total acidity of just .488 g/100 ml. The winemaster desires the total acidity to be .525 g/100 ml:

.525 g/100 ml total acidity desired
−.488 g/100 ml total acidity existing
.037 g/100 ml acid addition required

.037 g/100 ml = 1.40 g/gal. × 1955 gal. = 2737 g
2737 g = 6.03 lb of citric acid to add

Conversion factors used in this example can be found in Appendix C.

The holding period in the blending tank for the newly-filtered wine can also be used to make sulfur dioxide adjustments if needed, and allow the new wine to warm up. This temperature adjustment is necessary so that transfer to wooden storage vessels will not be such a temperature shock that it will cause leaking from the tank.

Filtration is often the process that permits microbial infection. The filter should be thoroughly cleaned with hot water before and after each run. New filter media should be treated with a cold water solution of citric acid and sulfur dioxide (approximately 2 oz per 100 gal. for *each* compound) to be sure that the filter medium is free from contamination before the wine filtration. Filter media should never be left overnight in the filter. Once used, the spent media should be properly removed from the cellars and the filter promptly and thoroughly cleaned with hot water.

A post-filtration analysis should include alcohol, Balling, total acidity, volatile acidity, free SO_2 and color intensity, as well as a rigorous and critical organoleptic evaluation. After posting the results routinely upon the laboratory analysis log, the analyst should be sure that the same data are recorded on the appropriate bulk wine book inventory card as shown in Fig. 5.7, so that the progress of the wine can continue to be monitored quickly and easily. Note the change in the third digit of the lot number as compared to Fig. 5.6.

AGING

There is perhaps no other single element of winemaking so controversial among winemasters as aging. White wines will age in stainless steel, glass-lined steel and other rather nonporous tankage, but will age much faster in

Lot No. 946			Type-Variety BLENDED SEYVAL BLANC				Class −14%		Color WHITE		
Remarks: Blended 3-19-80 81.6% Seyval Blanc 18.4% Vernot Blanc											
Date	Racked From Tank W.O.	Gallons	Treatment-Disposition	Racked To Tank W.O.	Gallons	PHYS INV	ALC	BALL	T.A.	V.A.	FSO₂
3-20 1980	CLARIFIGO:		4 bb. Metafine 3 bbs Sparkaloid 2 lbs Bentonite	80-76			11.7	−1.8	.608	.036	36
3-20 1980	80- BT-1 78	1,900	chilled for detect.	80- CS-2 78	1,889		11.8	−1.8	.608	.036	32
3-21 1980	(added nitrogen to headspace)			CS- 2 80- 80							
3-31 1980				CS-2	1,881	✓	11.8	−1.8	.555	.039	24
4-30 1980				CS-2	1,878	✓	11.7	−1.8	.525	.039	20
5-5 1980	80- CS-2 101	1,878	primary filtering	80- BT-2 101	1,867		11.5	−1.7	.510	.039	12
5-6 1980	80- 103		added 1.2 bbs Tricwine 134 lbs. KMS						.518		60
5-12 1980	80- BT-2 109	1,867	to give coverage	S-88 80- S-89 109	950 900		11.5 11.4	−1.6 −1.7	.518 .525	.042 .042	56 52
5-12 1980	total loss from processing = 50 gals.										

FIG. 5.7.   BULK WINE BOOK INVENTORY CARD IN USE AFTER PRIMARY FILTRATION

wooden vessels. While there is little exposure to oxygen in metal or glass tankage, some oxygen will remain from previous racking and filtration operations. This is usually enough to commence oxidation-reduction (aging) reactions.

More and more small wineries are turning to stainless steel aging tanks for white and rosé wine aging. These types of tanks are easier to maintain, provide more consistent results and generally provide a more antiseptic atmosphere. In particular, fine textured wines can be overpowered by the organoleptic effects of wooden cooperage. Unfortunately, stainless steel is very expensive, the type 304 somewhat less so than the 316 grade.

Wooden vessels, especially those made from cypress and redwood, are porous and allow microvolumes of air to be introduced into the wine for more rapid aging. Nevertheless, redwood continues to be very popular for the construction of white and rosé table wine storage tanks, especially in California, which is a source of this material. Redwood can offer some interesting constituents in the nature of "grassy" and "herbaceous" flavors.

In Germany, and in the best white wine regions of France, various species of white oak, principally *Limousin* and *Nevers*, are cherished for their aging effects which are often described as "vanilla" in character.

The length of time for aging after primary filtration can vary from a few weeks to a few years, depending upon the wine being produced. The size, type and condition of the cooperage used, the temperature of the aging cellar and the particular demands that varietal acceptance and tradition may require are all contributing factors. Normally, delicately flavored wines such as those from Chardonnay, Chenin Blanc, Pinot Blanc, Delaware, Dutchess, Cayuga, Seyval Blanc, Vidal Blanc and Villard Blanc are among the white wine cultivars not recommended for long periods of aging in wooden cooperage. While there are exceptions, the general rule is that white wines such the varietals just mentioned are most widely accepted as fresh, young and devoid of the effects of wood aging.

Other wines may be improved with the *tannins* that oaken cooperage contribute. Tannins are natural phenolic compounds that contribute bitterness and astringency to otherwise bland wines. The effects of tannins can be detected in tart wines as well. Tannins are found in the seeds, skins and stems of grapes, a prime reason that all, or most, of the stems are removed at the crusher-stemmer during the vintage season. Tannins also contribute to the flavor characteristics of many red wines, since red wines have considerably more exposure to skins and seeds during fermentation and, consequently, much higher tannin contents.

Perhaps the most common examples of matured white wines in wooden cooperage are those from Germany, notably those wines from Johannisberg Riesling grapes harvested at different stages of overripeness (spätlese, auslese, beerenauslese and trockenbeerenauslese). Others include the wines from Sauternes, Barsac and Graves in Bordeaux made from the cultivars Semillon and Sauvignon Blanc. Some of the great dry whites of Montrachet, Mersault and Pouilly-Fuisse in Burgundy, as well as those from Chablis, ripen well with rather extensive oaken aging. This is despite the fact that most of these wines are made from Chardonnay and Pinot Blanc, normally not recommended for long periods of aging in wooden cooperage.

Aging is the result of many highly complex reactions in the wine itself, along with the natural flavors contributed by the container in which the wine is stored. The smaller the vessel the faster the wine should age, all other factors being equal. For instance, a 50-gal. barrel has about twenty times the surface per unit volume as a 2500-gal. cask!

The oxidation-reduction reactions of aging are generally accelerated by higher temperatures. Free sulfur dioxide levels that are less than 50 ppm, especially in wooden containers, may allow the growth of *Acetobacter*. This problem is aggravated in containers where leaking and seeping occur so that oxygen enters the resulting head space. High concentrations of free SO_2 slow the aging process due to the anti-oxidative property of that compound. All of these factors should be kept in mind when designing an aging program.

The *ullage*, or head space, that develops in seeping wooden vessels should always be replaced by weekly cask fillings. The aging program should be monitored by the winemaster in the monthly analyses of alcohol, Balling, volatile acidity, total acidity, free SO_2, and the organoleptic evaluation made for each lot of bulk wine sampled during the bulk wine physical inventory.

Many vintners nowadays are turning more and more to light, fruity, even residually sweet red wines in response to the demands from the American marketplace. Many of these red wines are very similar to some white wines in both analytical and organoleptic characteristics. Aging for these types of red wines can be dealt with in much the same way as for white wines, with short aging periods and very little wood exposure.

American White Oak

This wood is famous in the manufacture of whiskey barrels which, in the case of Bourbon whiskey, may be used only once. Such used barrels can be a good selection for aging dessert wines such as sherry and port. However, used whiskey barrels should not be employed in aging programs for table wines. The whiskey penetrates deeply into the wood fiber, making it nearly impossible to completely neutralize. Some winemasters have experimented with dismantling barrels so that the inside "char" can be wire-brushed or shaved off, but even this relatively expensive process has been disappointing.

New barrels and casks made from American white oak can lend good values of tannins and vanillin in the aging programs for table wines— especially dry red table wines. Varietals made from Cabernet Sauvignon, Zinfandel, Cascade, Baco Noir, Maréchal Foch, Chelois, DeChaunac and Chancellor Noir can do very nicely in such containers, with careful aging programs.

Courtesy of St. James Winery
St. James, Missouri

FIG. 5.8.   STAINLESS STEEL STORAGE CONTAINERS

Courtesy of Sterling Vineyards
Calistoga, California

FIG. 5.9.   WOODEN AGING BARRELS

Limousin and Nevers Oak

Most wine experts agree that the flavor contributions from French and Yugoslavian oak of the *Limousin, Nevers* and related species, are the finest available for wine aging. Unfortunately, the European oak can be several times more expensive than the American resource, and the difference in quality is not nearly so great. To the red cultivars named in the previous section, one might add Pinot Noir, Syrah, Petite Syrah, Gamay and Rougeon to a list of varietals that may be aged very nicely in *Limousin* and *Nevers* oak vessels.

SUMMARY

Aging is defined in this text as the time that a wine spends recorded in the bulk wine book inventory. This particular inventory is a perpetual record designed to monitor all bulk wine movements and processing, providing finite quantities for good quality control functions.

Table wines, being less than 14% alcohol by volume and, therefore, subject to bacterial and other microbial spoilage, should be preserved with sulfur dioxide. This compound is the best known preservative for the small winery, and is also a good antioxidant for most wines other than the red varietals from *Vitis rotundifolia*.

Blending can help to correct problems in varietal wines, thus aiding in the maintenance of consistent products. Blending operations should be carefully planned and tested before the work order is written and the wine blend procedure is drawn for the actual blend to be made in the cellars. Blending is also a good occasion for clarifying and adjusting young wines in order to minimize handling.

Prior to adding any adjusting agents or finings, a winemaster must be familiar with ATF regulations dealing with the use of such compounds.

High metal contents can be reduced by the proper use of such materials as *Aferrin* and *Metafine*, while high total acidity may be reduced by the use of calcium carbonate. Activated charcoal may be used sparingly to help reduce the effects of oxidation and excessive color, but the vintner must keep in mind that this agent may significantly alter the aroma and flavor characteristics of the wines.

Oxygen is the worst enemy of wine quality, causing changes in pigmentation, aroma and flavor compounds. The winemaster must make every effort to maintain his wine vessels full at all times. The cellarmaster should conduct routine weekly cask-fillings to displace the head space with high free SO_2 ullage wine. In addition, monthly analyses should be performed in order to closely monitor quality control of wines in bulk storage.

Sparkolloid and bentonite are the most popular and effective clarification agents in use in the wine industry in America. Their use at levels of 1 to 2 lb per 1000 gal. of wine is usually sufficient to clear wines and stabilize proteins so that primary filtration can be economically effective after detartration.

Detartration is a function of cold storage that precipitates unstable tartaric acid salts such as potassium bitartrate, also known as cream of tartar and KHT. After a detartration period of about 1 month (up to 2 for Muscadine wines) in a $25-28°F$ cold cellar, wines may be primary filtered through a course-media filter to clear the wines prior to aging.

6

Quality Control During Bottling and Warehousing

The readiness of a wine for bottling is determined in most small wineries by both objective and subjective judgments. One year a white table wine may seem well rounded and mature in six months aging following the primary filtration. Yet another vintage may take nine months to a year or more, before qualifying as "bottle-ready".

It is during this time in the life of a wine that organoleptic analysis becomes an essential testing element in the laboratory.

If one were to follow the chemical analytical history of a wine from primary filtration to a year later, there may be only very slight changes that take place under normal circumstances. Figure 6.1 illustrates the analyses recorded over nine months on a typical bulk wine book inventory card kept for a dry red DeChaunac varietal table wine. Such records cannot be used to determine maturity and bottling time, although these numbers are invaluable in the maintenance of sound, spoilage-free inventories.

It is through our senses of sight, smell and taste, and to some extent, mouth feel—that we determine the maturity of a wine and its marketability. It is obvious that the primary value of any wine is the pleasure which it brings. This may, of course, reach much further than the eye, nose, and palate. The discovery and sharing of a superior inexpensive wine is often a source of personal gratification. Serving a quality wine to guests at table brings praise and pride. Connoisseurs delight in the possession of rare wines. Few wines are purchased because of their superior analytical values.

ORGANOLEPTIC ANALYSIS

As has already been shown, the laboratory analysis logs and other supplemental records have columns headed C, C, N, and T, which represent the primary parameters of organoleptic analysis: color, clarity, nose and taste.

Organoleptic analyses are inexpensive, take very little time and can be easily repeated in order to insure consistent results. The wine technician in the small winery laboratory should maintain the same approach to sensory

Lot No. 928 | Type-Variety DE CHAUNAC | Class −14% | Color RED

Remarks: Blended on 1-16-80 = 2,300 gal of De Chaunac (12%) + 500 gal Cascade (18%)

Date	Racked From Tank W.O. Gallons	Treatment-Disposition	Racked To Tank	W.O. Gallons	PHYS INV	ALC	BALL	T.A.	V.A.	FSO₂
3-31 1980	from previous card		5-27 80-84	2,775	✓	11.9	−1.4	.533	.051	48
4-30 1980			5-27	2,775	✓	11.9	−1.5	.525	.048	40
5-31 1980			5-27	2,775	✓	11.9	−1.4	.533	.048	36
6-30 1980			5-27	2,775	✓	12.0	−1.5	.525	.054	32
7-31 1980			5-27	2,775	✓	11.9	−1.4	.525	.051	52
8-31 1980			5-27	2,775	✓	11.8	−1.3	.533	.054	44
9-30 1980			5-27	2,775	✓	11.7	−1.4	.525	.051	36
10-31 1980			5-27	2,775	✓	11.8	−1.4	.518	.057	40
11-30 1980	aging seems slow		5-27	2,775	✓	11.6	−1.5	.525	.057	56

FIG. 6.1.   CHRONOLOGICAL ANALYTICAL HISTORY POSTED UPON A BULK WINE BOOK INVENTORY CARD

evaluation of new and developing wines as for older and more mature wines. However, allowances must be made for young wines, especially those that have not had the benefit of clarification, detartration and primary filtration. The novice winemaker should develop the insight to recognize promise and potential in younger wines.

Wine evaluation must be objective and scientific in order for results to be meaningful. The different moods and temperaments of the individuals involved should be considered. Tasting and decisions made while ill, stressed or weary should be avoided completely.

Wine evaluations are often based upon some variation of the 20-point scale originally developed in California. Wine judges will commonly agree that color and clarity are important to wine evaluation, but not as important as nose and taste. For example, some 20-point scales allow only 10% of the total points to be considered for color, and the same amount for clarity. This leaves 80% of the points for nose and taste. Figure 6.2 provides the *wine evaluation record* used by the American Wine Society.

Some winemasters and wine judges, primarily in the east, are formulating 100-point scales in the belief that the 20-point system is not broad enough to fully describe organoleptic evaluations. This writer is sympathetic to this belief and a sample 100-point scale is provided in Fig. 6.3.

Eight wines are considered by many to be the most that one can accurately evaluate at any one sitting. Water, bland cheese and unsalted crackers should be on hand to neutralize the palate between wines. After the round is finished, the taster should take a break for several hours so that the sensory organs can recover. Unless this rest is taken, wines may begin to seem quite similar to the senses. More definitive organoleptic subject matter can be found in Appendix A.

A format 20-point or 100-point scale is not necessary in the day-to-day routine organoleptic evaluation of samples that arrive in the laboratory. For such purposes the laboratory analysis log can be used directly in the posting of results. As long as everyone involved in the quality control program is aware of the scale being used, this can be adequate for communication and record-keeping. The winemaster may wish to devise a scale and post the parameters of this scale on a wall in the laboratory for ready reference. Individual standards pertinent to each wine type may supplement the scale.

It is common to use check marks to indicate the adequacy of each of the four routine organoleptic tests in the laboratory analysis log. These may be followed by comments in the "remarks" column. Each vintner, in establishing and maintaining organoleptic standards, will define the required characteristics for the wines, at each given point in aging, in order to earn the check mark as posted. This can be somewhat subjective, of course, as two knowledgeable palates may have two different opinions of the same wine. To reduce disparity of opinions, the standards should be realistic, simple and easily discernable. Organoleptic standards should also represent levels of quality which can be consistently achieved, and which closely resemble marketplace standards. The most basic use of sensory testing is for quality control.

WINE EVALUATION RECORD

Occasion: _dinner-party_ Place: _Horat_ Date: _6-7-80_

No. of Wine Tasted	No. 1	No. 2	No. 3	No. 4	No. 5	No. 6	No. 7	No. 8
Label Information	Blind Seyval Blanc	Blind Chard. #1	Blind Chard. #2	Blind Chablis	XYZ Pinot Noir	XYZ Cabernet Sauv		
Local Retail Price								
MAX. POINTS Clarity (2)	2	2	1.5	1.5	1.5	1.5		
Color (2)	2	1.5	2	1.5	2	2		
Aroma (4)	2.5	2	3	2.5	2.5	3		
Bouquet (2)	1.5	1.5	1.5	1.5	1.5	1.5		
Total Acidity (2)	2	1.5	2	1.5	2	2		
Tannin (2)	2	1.5	2	2	2	2		
Body (1)	1	1	1	1	0.5	1		
Sugar (1)	0.5	1	1	1	1	1		
General Flavor (2)	1.5	1.5	1.5	1.5	2	1.5		
Overall (2)	1	1	1.5	1	1	1.5		
Total Points	16.0	14.5	17.0	15.0	16.0	17.0		

FIG. 6.2.   WINE EVALUATION RECORD OF AMERICAN WINE SOCIETY IN USE

WINE EVALUATION ANALYSIS

EXAMINATION SEGMENT	WINE NO. 1	WINE NO. 2	WINE NO. 3	WINE NO. 4	WINE NO. 5	WINE NO. 6
COLOR (20)	Blind Seyval Blanc	Blind chardonnay #1	Blind chardonnay #2	Blind Chablis	X/2 Pinot Noir	X/2 Cabernet Sauvignon
Hue (5)	green 3	gold 3	sl. gold 4	sl. yellow 4	sl. tawny 4	great! 5
Intensity (5)	light 4	dark 3	4	4	white 4	heavy 5
RTV* (10)	8	7	8	7	8	9
CLARITY (20)		slight	fibers	haze	sl. haze	sl. haze
Suspension (10)	10	9	9	7	8	8
Precipitate (10)	specks 8	10	specks 8	sed. 7	sl. sed. 8	sl. sed. 8
NOSE (25)		?		slight	?	
Acescence** (10)	10	8	10	7	8	10
Bouquet (8)	5	5	6	5	5	6
RTV (7)	5	5	6	5	5	6
TASTE (25)		excess		excess		
Tartness (3)	3	2	3	2	3	3
Astringency (3)	3	excess 2	3	3	3	3
Sweetness (3)	slight 2	3	3	3	3	3
Balance (3)	2	2	3	2	3	3
Body (3)	3	3	heavy 3	3	thin 2	heavy 3
Flavor Taste (3)	2	2	3	3	3	off 2
Aftertaste (2)	1	1	? 1	sugar! 2	2	off 1
RTV (5)	4	3	4	4	4	3
IMPRESSION (10)	6	5	8	8	8	7
TOTAL (100)	79	73	86	76	81	85

*Representative of Type or Variety
**Acetic Acid Formation

FIG. 6.3.   FORMAT FOR 100-POINT WINE EVALUATION ANALYSIS IN USE

FINAL FILTRATION AND FINISHING

In large wineries, bottling time is often determined by factors of sales and inventory. In smaller concerns there are different indicators for the determination of bottling and these depend on the master's judgment.

Bottling is a common major contributor to oxidation, and this is often compounded by a final filtration just prior to bottling. Nevertheless, final filtration is highly recommended.

Final filtration should take place only after all preservatives and sweetening additions have been made and carefully mixed. A pre-filtration analysis should include alcohol, Balling, extract, total acidity, volatile acidity, free SO_2, total SO_2, color intensity, oxygen and a rigorous organoleptic evaluation. Each item of concern must, of course, come up to the standards and, most importantly, be within ATF and other regulatory limits. Figure 6.4 displays a comprehensive *daily bottling record* with the pre-final filtration analysis posted. This form will be followed throughout this chapter until the exemplified Seyval Blanc lot is bottled, packaged and fully accounted for.

Alcohol

Title 27, CFR, Section 4.26 fully explains the regulations concerning how alcoholic content should be detailed and displayed on wine labels.

ATF regulations provide for alcohol variance within 1.5% by volume of the stated alcohol content on the approved label to be used for the wine. Table wines that state 11% alcohol by volume on the label must actually contain an alcohol between 9.5% and 12.5% by volume.

For wines above 14% alcohol, dessert wines, the total must exhibit an alcohol statement, in percent by volume, that is within 1% by volume. Dessert wines labeled at 19% alcohol by volume must have an alcohol content between 18 and 20% by volume.

Finally, table wines with alcohol content of less than 14% need not indicate alcohol content at all, as long as the words, "table wine", or "light wine" are properly displayed on the label.

Dry and Sweet

Wines are generally considered dry if they contain less than about 4.0° extract, or a Balling of approximately 0.0° at 12.0% alcohol. At whatever level of sweetness of the finished wine, special preservation must be provided for wines that have any residual sugars that are fermentable.

Sweeter wines are usually made by two general methods: by fermenting juice that is significantly higher than the 22.4° Brix calculated for the Seyval Blanc lot (see Fig. 4.1), or by sweetening the wine during some later stage of production.

With some idea in mind as to what sweetness level the finished wine is to have, a value of Brix may be computed to generate the desired amount of alcohol. In other words, the first 22.4° Brix, as potential alcohol, added to say, 5.0° Brix desired for finished sweetness, gives a starting Brix of 27.4°

DAILY BOTTLING RECORD

Date 6-18-80	Lot Number 948	Gallons 1,850	Vintage 1979	Type-Variety SEYVAL BLANC		
From Tank(s) No. S-38 + S-39			To Bottling Tank(s) No.	Gallons	Received By—Date:	

| Pre-Final Filtration Analysis: | Alc. 11.5 | Ball. -1.7 | Ext. 2.2 | T.A. .518 | V.A. .045 | FSO$_2$ 40 | TSO$_2$ 104 | Color Int. 85 | O$_2$ <3 | C √ | C √ | N √ | T √ |

Final Adjustments: Sugar: Citric Acid: SO$_2$:   Other:

| Post-Final Filtration Analysis: | Alc. | Ball. | Ext. | T.A. | V.A. | FSO$_2$ | TSO$_2$ | Color Int. | O$_2$ | C | C | N | T |

Material and Supply Usage: Filter Pads: Bottles: Closure: Other:

| Post-Bottling Analysis: | Alc. | Ball. | Ext. | T.A. | V.A. | FSO$_2$ | TSO$_2$ | Color Int. | O$_2$ | C | C | N | T |

Production Data: Cases: Gallons: Serial Numbers: Received By—Date:

FIG. 6.4.   DAILY BOTTLING RECORD FORM IN USE AFTER PRE-FINAL FILTRATION ANALYSIS

prior to fermentation. Once the 22.4° Brix is fermented into 12.0% alcohol, a Balling of about 3.0° and an extract of approximately 7.0° will remain. There should be some 2.0° extract comprised of non-sugar solids (color pigments, glycerol, and other constituents) which, when subtracted from the total extract of 7.0°, results in the original 5.0° Brix (extract) intended. This analysis can be confirmed with some study and practice with the table wine nomograph found in Appendix A.

The winemaker may also wish to ferment to a slightly higher alcohol content than 12.0% when making residually sweet white wines in order to aid in preservation, and to help inhibit fermentation after the designated stopping point. A maximum of 13.5% alcohol is recommended in order to remain safely below the 14.0% ATF excise tax limit for table wines. Remember that the excise tax for wines under 14% alcohol by volume is levied at $.17 per gal. Wines of more than 14% alcohol, but less than 21.0% are rated at $.67 per gal. by the ATF. Wines that are labeled as less than 14% by volume which actually contain more than 14% are sure to attract disciplinary action by regulatory agencies.

Sweetness

A higher concentration of sugar for increased Brix prior to fermentation can be developed by allowing grapes to dehydrate on the vine, providing that the cultivar and the climate permit such a practice. Master winemakers find this practice very difficult and susceptible to many climatological and microbiological factors. More common is the addition of grape juice concentrate or cane sugar, as was explained in Chapter 4.

Stopping fermentation at the exact desired point can be an exasperating task, especially in small wineries that have only basic equipment. The usual method is to anticipate fermentation progress by preparing a graph from the data in the appropriate fermentation control record. When the desired Balling level is approached, prepare to adjust free SO_2 to about 200 ppm, and rapidly chill to about 25 to 28°F. This should shock yeast cells and hold fermentation progress until a coarse filtration can be applied. The resulting wine will be further reduced in free SO_2 than the novice winemaker may anticipate, as the carbon dioxide gas from fermentation will carry off much of the 200 ppm addition.

These parameters should be adjusted for each wine in each winery. The rate of fermentation, temperature, viability of the yeasts used, and many other factors can influence the way the winemaster goes about stopping a fermentation at a given point. In any case, the remaining cellar operations, after fermentation has been halted, may proceed as for any other table wine, except that free SO_2 levels may need to be maintained considerably higher throughout aging. A common range is 80 to 120 ppm free SO_2 which, again, as an antioxidant, can slow the aging progress to almost nil.

Many wineries, especially the larger ones, add cane sugar to dry wines just prior to bottling to achieve the desired sweetness levels. Dextrose is used, but it does not have the organoleptic sweetness value that cane sugar (sucrose) and fructose possess. The following calculation exemplifies an addition of cane sugar in a winery to raise 1675 gal. from 2.3° extract to 4.0°.

$$1675 \text{ gal.} \times .3390 \text{ lb/gal.} = 567.8 \text{ lb}$$
$$-1675 \text{ gal.} \times .1937 \text{ lb/gal.} = \underline{324.4 \text{ lb}}$$
$$\text{cane sugar addition} = 243.4 \text{ lb}$$

The lb/gal. factors of .3390 and .1937 can be found for 4.0° and 2.3° extract, respectively, in the Brix tables of Appendix B.

Top quality wine cellar production usually follows the method of adjusting for sweetness levels prior to fermentation. Wines made by this method are usually more stable than when sugar is added just before bottling. However, extreme care is necessary in stopping the fermentation at the predetermined point as well as in preserving the remaining sweetness throughout the aging operations.

The use of grape juice concentrate for pre-bottling sweetening is not considered because of the probability of gross instability in such a blend.

Volatile Acidity

Volatile acidity is limited to .140 g/100 ml by the ATF. This is officially explained in Part 240, CFR, Section 240.489. However, wines with more than half of this amount should be carefully examined for bottle worthiness. The winemaster should be well aware of any high volatile acidities in the wine inventory as bulk wines age. Any such wines should either be justified for bottling or separated from the inventory for use in a bulk outlet—some other winery, a vinegar producer, a sauce manufacturer, or some other processor. Rarely will a vintner choose to blend a wine with high volatile acidity with another that is low in order to obtain an acceptable level.

Sulfur Dioxide

Dry table wines, if filtered properly through a medium of less than 0.5 microns, may be bottled with free SO_2 at lower levels than if a higher porosity medium is used, providing bottling machinery and quality control can assure a very low contamination potential. Wines containing any amount of fermentable sugar solids must be adjusted to at least 100 ppm free SO_2, regardless of the filter medium porosity. Title 27, CFR, Part 4, Section 4.22, limits sulfur dioxide to 350 ppm.

Total SO_2 is a measurement of both combined and free SO_2. The combined form is usually in the form of sulfites that are ionically bonded and provide no preservation or antioxidation to wines. These bonds are made each time sulfur dioxide is added to wine and can build to significant levels with frequent additions. It is the combined, or total, sulfur dioxide content upon which the ATF limit is based. Obviously, if total SO_2 is less than 350 ppm, the free SO_2 must also be less than the 350 ppm limit.

Color Intensity

Skin contact and juice extraction pressures during fermentation, along with storage temperatures, oxidation levels and aging developments will heavily contribute to color values. As discussed during the clarification procedure in the previous chapter, wines can be bleached with activated charcoal so as to reduce color intensity, providing the winemaster is prepared to accept reduced aroma and flavor. There are also filter media available that contain activated charcoal. These are not recommended due to a high potential of instability that may be introduced. Stabilization problems occurring in color hue and turbidity due to metal imbalance and various minor precipitates are the most common. Most important is that this medium can significantly reduce aroma and flavor because of the activated charcoal properties.

Oxygen

If good cellar procedures have been followed, table wines should have endured a minimum of handling and movement in the winery. This should result in low levels of dissolved oxygen—certainly less than the 5 ppm normally considered a standard for table wines of distinction, especially whites and rosés. Browning of colors and the formation of aldehydes may be found in wines bottled with excessive dissolved oxygen.

Convinced that all items on the agenda for "checklisting" a wine prior to bottling have been properly examined and adjusted, the winemaster may perform an organoleptic evaluation and then proceed to a work order as exemplified in Fig. 6.5.

The Filter

The filter should be selected with the greatest of care. Wine contact parts that are made of stainless steel are best as they can be steam-cleaned and allow for durable long life. However, stainless steel is very expensive, especially when precision machining is necessary, as in the manufacture of a quality wine filter. The winemaster may opt for inert fibrous or plastic plates in the frame of a filter, providing he fully understands the limitations of such materials.

Final filtration should be made just prior to bottling, with the filtered wine delivered directly to an enclosed, vented bottling tank. The bottling tank should be ultra-clean, and even sterile, if possible. Vigorous brushing, steam and hot water should suffice.

Final filtration should be made through a cellulose-based medium, with porosity not exceeding 0.5 microns. With such a restrictive medium, this function must take place at a pressure usually not higher than about 26 pounds per square inch (psi). Such low porosity can render the filtered wine virtually sterile, which is optimal for reducing post-bottling microbial populations. Pressures exceeding 26 psi may force viable microorganisms

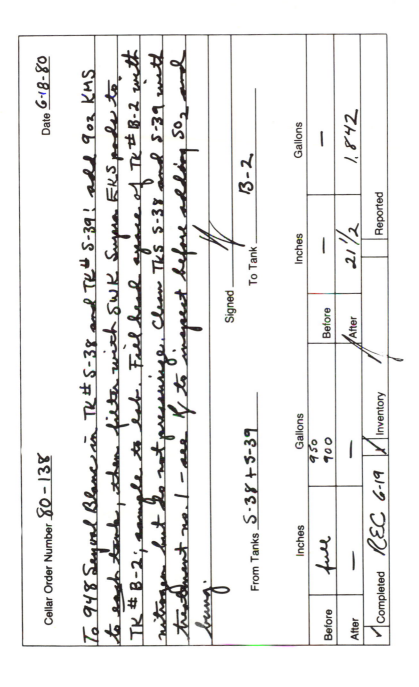

FIG. 6.5.   WORK ORDER FORM IN USE FOR WINE ADJUSTMENTS AND FINAL FILTRATION

through the filter pads into the finished wine, or the filter medium may rupture, allowing large numbers of bacteria, yeast and even larger impurities to enter the filtrate.

Some winemasters believe that filtration contributes to a reduction of fine wine quality, to the extent that these vintners rely only upon good clarification practices to brighten their wines. However, properly filtered wines are much brighter, and therefore, much cleaner. The virtual total removal of viable microorganisms is the best guarantee of reliable quality. The vast majority of America's finest vintners choose to final filter their wines. The evidence of *significantly* reduced wine quality due to a final filtration process is rare.

Cleanup of the filter should be such that all spent media are immediately disposed of, preferably at a remote site so that insects, rodents, and spoilage organisms have no chance to contaminate the winery. Each plate in the filter should be individually brushed and hot water rinsed, along with the filter frame. The filter may then be reassembled and completely re-rinsed with hot water or live steam, providing the filter is constructed of materials that will allow such treatment.

Figure 6.6 illustrates the daily bottling record now completed to the point where it may be considered the winemaster's bottling authorization.

BOTTLING AND PACKAGING

Bottles, closures and equipment, as well as the labor required to complete the bottling operation should be obtained, prepared and double-checked well in advance, with backup alternatives wherever possible. There should be sufficient material and labor to bottle the entire lot of wine prepared for bottling each day. Each individual day of bottling should be treated as a separate entity in terms of pre-filtration and post-filtration analyses.

Hard water may require ionization treatment prior to the bottle rinsing. The winemaster must be particularly careful when sulfur is present in natural water supplies, as "free" rhombic sulfur can form a hydrogen sulfide contamination. Bottle washers, complete with adjustments for hot and cold water at variable velocities, are simple yet adequate devices for the treatment of new wine bottles. Air-jet devices for the cleaning of wine bottles can perform inadequately, especially when the bottles are still cold from shipment, so that "case dust" is trapped inside the new bottles by condensed humidity.

The filling tubes of the bottling machine should be as long as is practical to reduce splashing and, in turn, oxidation, especially in white and rosé table wines. Nitrogen sparging is recommended as long as close attention is given to the manner and amount of nitrogen applied. Excessive nitrogen can "entrain" in the wine being bottled, emerging later as head pressure, possibly strong enough to push out the cork.

It is imperative that the bottling system include corking or capping operations immediately after filling. This will reduce exposure to oxygen, airborne microorganisms, dust and other contaminants.

DAILY BOTTLING RECORD

Date	Lot Number	Vintage	Type-Variety
6-18-80	948	1979	SEYVAL BLANC

From Tank(s) No.	Gallons	To Bottling Tank(s) No.	Gallons	Received By—Date:
5-38 + 5-39	1,850	B-2	1,842	TRL 6-19-80

Pre-Final Filtration Analysis:

Alc.	Ball.	Ext.	Sugar:	T.A.	V.A.	Citric Acid:	FSO2	TSO2	Color Int.	O2	C	C	N	T
11.5	-.7	2.2	—	.518	.045	—	40	104	85	<3	✓	✓	✓	✓

Final Adjustments: SO2: 1 lb. 2 oz.   Other: —

Post-Final Filtration Analysis:

Alc.	Ball.	Ext.	T.A.	V.A.	FSO2	TSO2	Color Int.	O2	C	C	N	T
11.3	-1.65	2.2	.510	.048	76	148	87	<3	✓	✓	✓	✓

Material and Supply Usage: Filter Pads: SWK Super EKS   Bottles: 750 Liter Burgundy   Closure: ENOPAK ST-11991 corks   Other: —

Post-Bottling Analysis:

Alc.	Ball.	Ext.	T.A.	V.A.	FSO2	TSO2	Color Int.	O2	C	C	N	T

Production Data: Cases: Gallons: Serial Numbers: Received By—Date:

FIG. 6.6.   DAILY BOTTLING RECORD FORM IN USE AFTER POST-FINAL FILTRATION ANALYSIS AS BOTTLING AUTHORIZATION

There are caps available with inert liners that can seal and protect wines adequately, although these will allow for greater head-space and, therefore, increase the possibility of oxidation. Caps may usually be applied directly to bottles if inspection proves that the closures have been shipped and stored in undamaged sealed containers, with no evidence of dust, dirt or other impurities.

Corks are marketed that have been pre-washed and sterilized to a dust-free condition. This extra treatment by cork manufacturers and importers is passed along in higher prices to the vintner. An advisable alternative is to purchase the extra-fine grades which have not been treated, providing they are soaked just before bottling.

A normal soaking procedure involves an initial soak in cold water for about 30 min, followed by another soaking in a 1 to 2% glycerine-water solution held at about 90°F for another 30 min. The small amount of glycerine will help to make the corks easier to drive in the corking machine and, more importantly, easier for the consumer to remove. Exact time schedules for the soaking operation will be difficult to maintain, as bottling is continuous while cork preparation must be batched. Most winemasters direct the bottling line foreman to have new batches of corks ready approximately every half hour in order to effectively supply needs.

New corks are currently available which have been manufactured from cork pieces glued together and formed under heat and pressure. These closures are recommended, as they are normally less expensive than natural corks. However, such agglomerated corks are very firm, and may require twice the time and amount of glycerine needed for standard corks. At this writing some new plastic straight corks are also entering the market that offer possible economic savings.

The new winemaker should be aware that labels on wine bottles must be fully approved by the ATF prior to usage. The ATF form that must be used for this purpose is ATF Form No. 1649: Application for Certificate of Label Approval under the Federal Alcohol Administration Act, which is fully discussed in Chapter 10.

A common practice is for capsules and labels to be applied during the bottling operation. However, should appropriate storage be available, one may stack bottles in *tirage* (bottles laid horizontally in tiers upon each other, with or without strips of lath between layers) for aging. The uncapsuled and unlabeled bottles in tirage can be inspected prior to packaging, culling out the "leakers" beforehand.

Corks, being a naturally porous substance are sometimes subject to closure failure. Consistently cool temperatures with low humidity and minimum light exposure are ideal conditions for tirage storage.

Bottling Line Quality Control

A relatively new method for the detection of viable yeasts in bottled wines now allows winemasters to collect data in less than one hour. With the potential for saving an indeterminate amount in bottling materials and labor, the modest cost of such testing is easily justifiable. This method is discussed in Chapter 9 and Appendix A.

It is recommended that a sample bottle be taken at random from the bottling line at regular intervals, to be sure that the free SO_2 level remains at the desired concentration, and that organoleptic characteristics remain acceptable. This sampling does not need to interrupt production progress and is a procedure normally made about every two hours. Acceleration of this sampling rate may be justified if there are special requirements or reasons for concern.

Newly bottled wines are almost always "bottle-sick" for several weeks or months, immediately following bottling. This phenomenon results from adjustments and handling of the wine in preparation for bottling. After the newly bottled wines have been rested, they will usually recover satisfactorily. Bottle-aging is therefore very important before the wines are released from the warehouse.

Post-Bottling Quality Control

At some point about midway in the bottling run, six bottles, fully packaged and inspected, should be delivered to the laboratory. A post-bottling analysis, as required by the daily bottling record (Fig. 6.7) should be performed on one of these bottles. Another of the samples should be carefully plated-out to examine for microbial life. Some winemasters feel justified in dispensing with the plating operation in the laboratory and using only the rapid viable yeast detection method. It should be noted, though, that the rapid method is for yeast only, whereas plating determines bacteria and mold as well.

The procedure for plating is discussed in Chapter 9 and Appendix A. Normally the results can be determined in several days. There are commercial devices available that do not save much time, but greatly simplify the detection of viable microbes. These are small plastic membrane filters that contain media already prepared and sterilized so that all the analyst need do is carefully filter an uncontaminated sample through the device and incubate it as directed. The use of these plastic membrane filters is described in detail in Appendix A.

Post-Bottling Stability Testing

There should be enough wine remaining from the two samples opened for post-bottling chemical analysis and plating to set up a stability test. For proper stability testing, the small winery laboratory should be equipped with a dependable refrigerator and oven held at about 35°F and 110°F, respectively. Uniformly sized and shaped flint bottles (clear and uncolored glass) with inert caps should be used. An example is the wide-mouth 4-oz bottle pictured in Fig. 6.8.

The stability test is made from three full bottles of wine. Prior to filling each test bottle, a little of the wine from the lot to be tested may be poured

DAILY BOTTLING RECORD

Date	Lot Number	Vintage	Type-Variety
6-18-80	948	1979	SEYVAL BLANC

From Tank(s) No.	Gallons	To Bottling Tank(s) No.	Gallons	Received By—Date:
S-38 + S-31	1,880	B-2	1,842	TRL 6-19-86

Pre-Final Filtration Analysis:

Alc.	Ball.	Ext.	T.A.	V.A.	FSO₂	TSO₂	Color Int.	O₂	C	N	T
11.5	-1.7	2.2	.518	.045	40	104	85	<3	✓	✓	✓

Final Adjustments:

Sugar:	Citric Acid:	SO₂:	Other:
—	—	1 lb. 2 oz.	—

Post-Final Filtration Analysis:

Alc.	Ball.	Ext.	T.A.	V.A.	FSO₂	TSO₂	Color Int.	O₂	C	N	T
11.3	-1.65	2.2	.510	.048	76	148	87	<3	✓	✓	✓

Material and Supply Usage:

Filter Pads:	Bottles:	Closure:	Other:
SWK Super EKS	.750 Ltr Burgundy	EUROPARK ST-11aql. corks	—

Post-Bottling Analysis:

Alc.	Ball.	Ext.	T.A.	V.A.	FSO₂	TSO₂	Color Int.	O₂	C	N	T
11.2	-1.7	2.2	.510	.045	72	140	86	<4	✓	✓	✓

Production Data:

Cases:	Gallons:	Serial Numbers:	Received By—Date:

FIG. 6.7.   DAILY BOTTLING RECORD FORM IN USE AFTER POST-BOTTLING ANALYSIS

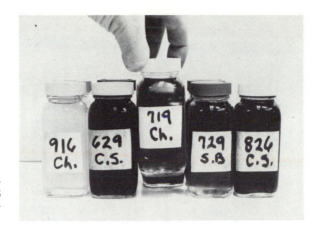

FIG. 6.8. FLINT BOT-
TLES WITH INERT CAPS
ARE USED FOR STABIL-
ITY TESTING

into the containers and used for rinsing. Even though the stability test bottles have been scrubbed clean and hot water rinsed several times, rinsing with a sample of the wine being tested will further insure that each stability bottle is free from contamination. Each bottle is then carefully identified with lot number, type or variety of wine, bottling date and any other data that may be necessary. One bottle is then placed in the refrigerator, another in the oven, and the third shelved at room temperature.

After an observation period of several weeks, the test results are recorded. If suspensions and/or precipitates form, appropriate action may need to be taken. The table provided in Fig. 6.9 may be helpful in the determination of positive stability test factors.

The four remaining samples should be logged into the sample library for later reference (Fig. 2.56 and 2.57). Some winemasters will take one of these samples at some point during the ensuing year and perform a complete analysis to insure shelf quality and stability. The three samples left in the library are held for reference.

CASED GOODS IDENTIFICATION

As each case is filled with bottles, that case should be stamped with a serial number or code, so it can be traced to the particular lot of wine and its position in the run. Often, winemasters will commence each year with case serial numbers starting with 1, followed by the appropriate calendar or fiscal year. For example, the case just described may have been numbered 589-80, meaning case number 589 bottled in calendar or fiscal 1980. There may also be a stamp applied to each case showing the type or variety of wine, along with the alcohol content. Small wineries often simply glue appropriate bottle labels on either end of each case in order to assure proper identification.

Designation of turbidity or separated material	CAUSE	PREVENTION	RECOVERY
Organic copper turbidity	Copper contact of beverages (Spray medium, copper piping, poorly silver plated or tinned fittings or equipment)	Avoidance of copper contact. Tinned copper pipes are often defective and should be replaced if possible with stainless steel.	About 90% of the existing copper is eliminated by fermentation. Addition of sulfurous acid is eventually able of reducing copper turbidity to a soluble stage (Total SO_2 must not be in unfermented beverages be above 80 mg/liter)
Calcium sulfate sediment	This turbidity occurs in brandy which has been diluted with calcium containing water.	Avoidance of calcium containing water for dilution purposes. Employment of softened water.	Cold storage and filtration, or ion-exchange treatment (IR 120)
Winestone sediment	Rapid cooling or higher tartaric acid content, or brief cask storage	Control of tartaric acid content before bottling. At 0°C about 1.5 g/liter tartar is soluble in the beverage (with 10 percent alcohol by volume)	Cold storage of wines and filtration or do tartaric acid determination and as far as possible on this basis carry out chemical deacidification.
Calcium tartrate sediment	During short observation period after chemical deacidification. Perhaps employment of calcium containing filter sheets.	Wash filter sheets in tartaric acid solution! Observation time before racking off from deacidification sludge should be at least 4 weeks.	Cold storage of beverages and filtration.
Copper sulfite turbidity	Copper containing beverages with simultaneous high total sulfurous acid content (over 120 mg/liter)	Prevent copper uptake! Remove copper piping and replace with stainless steel or hose.	Reduce copper concentration by blending; aeration of beverage to bring about disappearance of SO_2 (insoluble cuprous compounds thereby go over into soluble cupric compounds)

Type	Cause	Prevention	Recovery
Ferric phosphate turbidity	High iron content (more than 6–8 mg/l)	Prevent iron uptake! Lacquer bare iron parts.	Remove iron by Aferrin fining. Lower the iron-content by blending with iron-poor beverage Gelatin fining in connection with stronger dosing with sulfurous acid.
Pigment-tannin sediment	High pigment and tannin content.	By longer storage of harsh and highly colored beverages in casks, sedimentation in bottles can be largely avoided.	Gelatin fining with 20 g/100 liters and dosing with sulfurous acid up to a content of 25–30 mg/liter free SO_2 helps to remove an excess of tannin and leads to stabilized beverage.
Iron-tannin turbidity	High iron and tannin-content	Avoid iron uptake! Lacquer bare iron parts.	Gelatin fining for removal of excess tannin- Blend with iron and tannin poor beverages. Citric acid addition (50 g/100 liters in wine; 200 g/100 liters in fruit wine.)
Biological turbidity	Yeast and bacterial activity.	Sterile bottling as preventative measure. Sanitary operations!	Tight filtration (EK filter) and dosage with sulfurous acid; pasteurization.
Thermolabile protein turbidity	Unfermented beverages, especially grape juice, contain under certain circumstances thermolabile protein, which, after heating (pasteurization) and cooling lead to turbidity.	(a) Pasteurization done twice, filtration between times. (b) Bentonite treatment	Beverages with protein sediment are pasteurized, cooled and filtered. Repeat heating leads to no new turbidity.

Tanner, H., and Vetsch, U. 1956. How to characterize cloudiness in beverages. Am. J. Enol. 7, 142–149.

FIG. 6.9.   THE DETERMINATION, CAUSES AND RECOVERY OF POSITIVE INSTABILITY IN WINES

Full cases are normally stacked on pallets, ten cases to the tier, in order to simplify later inventory. Precise counts of each day's production are taken, including gains, losses and overall efficiency. The pre-bottling bulk wine gallonage is compared with the result obtained by multiplying the number of bottles filled by the contents of each bottle. These numbers should be appropriately recorded on the daily bottling record, the bulk wine book inventory card and the cased goods book inventory card, as shown in Fig. 6.10, 6.11, and 6.12, respectively. Note that this information serves as a receipt for the bulk wine from the cellars and as a receipt for the cased goods in the warehouse, making the transition and transfer complete.

WAREHOUSE QUALITY CONTROL

Packaged wine bottles should be placed upside down in the cases. The cases are then sealed and turned upside down, righting the bottles inside. Being left in this position will allow the corks to expand and form a proper seal. If the bottles were left upside down, exposing the wine to the cork, the wine could get between the cork and the neck of the bottle before the seal is made. Later, after drying and expansion, wine residue between cork and glass can make corks very difficult to remove.

After the cases have been left upside down (and bottles right side up) for two or three days, they should be returned to an upright position (bottles now upside down). This obviously will make the cases easier to identify, but more important, it keeps the moisture of the wine next to the expanded corks, helping to keep them supple and resilient.

Thus, with cases repositioned upon pallets, the production process is complete. The mature life of the wine, however, has just begun.

Cased goods physical inventory should be taken at the close of operations on the last working day of each month and reconciled with the cased goods book inventory in much the same manner as bulk wine control is administered. Figure 6.12 provides a format for an adequate cased goods book inventory record card. This should be kept in a 5 in. × 8 in. file drawer in the same way as is the bulk wine book inventory. Figure 6.13 illustrates how such a file is put to use.

Some winemasters choose to have separate forms for taking monthly physical inventories of bulk wines, cased goods and cellar materials, especially when there is a broad range of products or large quantities. Smaller wine cellars, however, often use forms similar to Fig. 6.14 and 6.15 for general physical inventory purposes.

The small winery should be designed with warehouse ceiling heights which will accommodate stacked pallets two high. The normal temperature range of warehouse storage is from 68 to 74°F, held as constant as is possible.

DAILY BOTTLING RECORD

Date	Lot Number	Vintage	Type-Variety
6-18-80	948	1979	SEYVAL BLANC

From Tank(s) No.	Gallons	To Bottling Tank(s) No.	Gallons	Received By—Date
S-38 + S-39	1,850	B-2	1,842	TRL 6-19-80

Pre-Final Filtration Analysis:

Alc.	Ball.	Ext.	T.A.	V.A.	FSO_2	TSO_2	Color Int.	O_2	C	C	N	T
11.5	-1.7	2.2	.518	.045	40	104	85	<3	✓	✓	✓	✓

Final Adjustments:

Sugar	Citric Acid	SO_2	Other
—	—	1 lb. 2 oz.	—

Post-Final Filtration Analysis:

Alc.	Ball.	Ext.	T.A.	V.A.	FSO_2	TSO_2	Color Int.	O_2	C	C	N	T
11.3	-1.65	2.2	.510	.048	76	148	87	<3	✓	✓	✓	✓

Material and Supply Usage:

Filter Pads:	Bottles:	Closure:	Other:
SWK Super EK5	750 Liter Burgundy	ENOPAK ST-11 agg/ corks	—

Post-Bottling Analysis:

Alc.	Ball.	Ext.	T.A.	V.A.	FSO_2	TSO_2	Color Int.	O_2	C	C	N	T	Received By—Date
11.2	-1.7	2.2	.510	.045	72	140	86	<4	✓	✓	✓	✓	YB 6-20-80

Production Data: Cases: 771 plus 5 bottles to TPR #1   Gallons: 1,834.4 + 2.0 gals to test = 13.6 gals   Serial Numbers: 12,446 - 13,216

FIG. 6.10. DAILY BOTTLING RECORD FORM IN USE AFTER BULK WINE RECONCILIATION AND TRANSFER OF CASED GOODS TO WAREHOUSE

Lot No. 949   Type-Variety BLENDED SEYVAL BLANC   Class —14%   Color WHITE

Remarks: Blended 3-19-80   81.6% Seyval Blanc   18.4% Pinot Blanc

Date	Racked From Tank W.O. Gallons	Treatment-Disposition	Racked To Tank W.O. Gallons	PHYS INV	ALC	BALL	T.A.	V.A.	FSO₂
5-80 1980		from previous card	3-38 80-750 3-39 107-900	✓	11.5	-1.7	.518	.045	40
6-19 1980	5-38 80-450 5-39 138-900	adjust SO₂ final filter	B-2 80-107 1,842		11.3	-1.65	.510	.048	76
6-20 1980	B-2 bottled:	1,834.4 gals in 750 liter oval and bottles 2.0 gals to lab for samples 13.6 gals loss	to dead file 6-20-80						

FIG. 6.11.   BULK WINE BOOK INVENTORY CARD CLOSED OUT FOLLOWING FINAL FILTRATION AND BOTTLING

Lot No. 949	Vintage 1979		Type-Variety SEYVAL BLANC		Size .750 LITER	
Class – 14%	Color WHITE		Minimum Inventory —		Closure ENOPAK ST-11 RED	Other —
Date	Cases In	Cases Out	Balance on Hand	Ref. No.	Inv.	Remarks
6-20-80	771	—	771	W.O. 80-107		SN 12,446-13,216 rec'd VB 4:45pm

FIG. 6.12. CASED GOODS BOOK INVENTORY CARD IN USE FOLLOWING RECEIPT OF CASED GOODS FROM BOTTLING LINE

FIG. 6.13. A TYPICAL FILE FOR CASED GOODS BOOK INVENTORY REC-ORD CARDS

SUMMARY

Wine maturity is indicated by organoleptic analysis, supplemented with chemical analyses. Sensory evaluation of color, clarity, nose and taste is individual and personal, but nevertheless should be scientific and objective.

The bottling and packaging of wine are critical procedures which must be prepared for well in advance. The wine to be bottled should be fully analyzed, then any adjustments can be made. Another analysis should subsequently be made in order to monitor the accuracy and effectiveness of these adjustments.

Sulfur dioxide addition and the measurement of free SO_2 are important considerations in wine preservation. Failure to monitor free SO_2 may result in post-bottling fermentation, which is a dangerous and expensive situation.

The small winery filter is a device that should be chosen carefully, as there is heavy expense involved. Final filtration of the wine to be bottled may be made through a washed medium that is not in excess of 0.5 microns porosity, at pressures that are held constant and below media limitations.

Bottling operations should be as gentle as possible, avoiding oxidation and the introduction of contaminants. Bottles and corks must be free of dust and prepared so as to provide a neutral, hermetically sealed environment for the wine.

Packaging materials should be tested before use so that potential difficulties can be corrected. Any descriptive literature on bottles must be approved by the ATF prior to use.

The winemaster should devise a quality control program for newly bottled wines that assures a minimum of "bottle-sickness". The rapid method

DATE _____   TAKEN _____   RECONCILED WITH _____   PAGE _____
COB _____   BY _____   BOOK INVENTORY BY _____   NO. _____ OF _____

TANK NO.	CAPACITY	LOT NO.	TYPE-VARIETY	−14% ALC	+14% ALC	SPARKLING	JUICE	OTHER
TOTAL								

REMARKS	SUGAR INVENTORY
	KG.

FIG. 6.14.   PHYSICAL INVENTORY—BULK IN LITERS

STORAGE AREA NO.	LOT NO.	TYPE-VARIETY	CASES @ 9.0 LITERS	BOTTLES @ .750 LITERS	BOTTLES @ .375 LITERS	OTHER	TOTAL LITERS
TOTAL							
REMARKS				TOTAL BULK LITERS FROM OPPOSITE PAGE			
				GRAND TOTAL LITERS			

FIG. 6.15.   PHYSICAL INVENTORY—CASED GOODS IN LITERS IN BOND (REVERSE SIDE OF PHYSICAL INVENTORY—BULK WINES)

now available for the detection of viable yeasts is recommended prior to each run. Spot checks for the analysis of free SO_2 and organoleptics should be made on a regular basis, e.g. every two hours, throughout the bottling run.

Post-bottling quality control should include plating the wine to examine for viable microbes. A stability test should also be conducted to determine suspension and/or precipitate formation.

Several bottles may remain in the library for future reference. One of these should be removed within a year and completely analyzed to check that the wine remains within specified parameters of quality and stability.

Cased goods should be properly numbered and identified so that each lot is easily traced. Pallets should be stacked so that corks remain dry for several days, allowing them to expand and fully seal the inner surface of the bottle neck. Cases may then be inverted and stacked on pallets in layers that are easily inventoried later.

Upon completion of the daily bottling record, the wine should be removed from the bulk wine book inventory and entered upon the cased goods book inventory. Entries for storage and maintenance in the cased goods book inventory are made in the same manner as for bulk wine, including reconciliation with the appropriate monthly cased goods physical inventory.

Construction of cased goods storage facilities should be such as to allow constant temperature maintenance at about 72°F.

Alexander Fleming
June 15 1948

Maria Pia de Savia
20-III-55

7

Dessert and Aperitif Wine Quality Control

The acceptance of dessert wines in the American marketplace has declined steadily during the past several decades, and table wines have become more popular. Aperitif wines have been more consistent, however, perhaps due primarily to the complementary nature that these wines have in the role of mixers and adjuncts to other more positively disposed products.

Dessert and aperitif wines are both made by wine spirits addition: grape brandy is added to raise the alcohol percentage. The major difference between these two types of wine is that aperitif wines result from a further addition of non-grape essences, flavorings, herbs and the like.

WINE SPIRITS ADDITION

Wine spirits addition is often referred to as "fortification" by both vintners and wine consumers. However, in this text, only the term wine spirits addition (WSA) will be used and it shall refer to the addition of wine spirits (brandy, usually at a high-proof of 184–192°, or about 92–96% ethanol) to natural wines in order to raise the alcohol content.

Sherry and port are typical dessert wine products from Spain and Portugal, respectively, but these names are utilized for similar products made in the United States, as well. Normally the range of alcohol found in dessert wines is from 16% to more than 20% by volume. Additions resulting in wines containing more than 21% alcohol are subject to a higher ATF excise tax rate. Wines are limited to 24% alcohol by volume, as beverages in excess of this level are considered spirits and carry different production regulations and tax liabilities.

Dry vermouth, the French type, and sweet vermouth, the Italian type, are good examples of aperitif wines, although there are many more. These wines are about the same as dessert wines in alcohol content.

The winemaster must be totally familiar with Part 240, CFR, Sections 240.374 through 240.386 for the production of dessert wines. Sections

Demonstration of the unusual skill of wielding a venencia to draw sherry from a cask
Courtesy of Sherry Institute of Spain

240.440 through 240.448 are necessary for aperitif wines. There is a great deal of control involved in the handling and formulation of wine spirits and non-grape flavor additives in the commercial winery.

BASE WINES

The experienced winemaker will know exactly what the end products of his grapes are expected to be each year, far in advance of the vintage season. The vintner may choose white grape cultivars which oxidize very easily for making sherry-type wines that require madeirization (see Glossary). Grapes that yield very dark, rich and heavy red color and body are ideal for port-type wines. The pale, water-white and delicate consistency of table wines is needed for conversion to dry vermouth. The initial table wines made from such grape cultivars may be properly listed in the bulk wine book inventory according to varietal status, as, say, 903 Niagara, 913 Ives, or 943 Elvira.

The WSA is always made prior to clarification and detartration so that the entire product, including brandy, concentrate or sugar, and other additions can be stabilized as a unit. As with other wine types, the blending of two stable wines, or of a stable wine with another ingredient (such as brandy), may create an unstable wine.

After the addition of wine spirits, the varietal grape names are usually replaced by more generic descriptions such as 904 shermat (sherry material), 915 port material, or 946 vermouth material. The winemaster can then, at a glance at his bulk wine book inventory cards, know the ultimate purpose of each wine treated with the WSA.

SWEETENING

Most dessert wines are sweet, with a few exceptions, such as dry cocktail-type sherries. In the east, dessert wines are most often sweetened with sugar under the ATF provisions, Part 240, CFR, Section 240.368. Aperitif wine sweetening rules are found in Section 240.430. Cane sugar (sucrose) is a more powerful sweetner than corn sugar (dextrose), but dextrose is becoming more popular, as it is considerably less expensive than sucrose.

In California and other areas where the addition of sugar is illegal, the sweetness of dessert and aperitif wines is derived either from the addition of grape juice concentrate or by arresting primary fermentation at some predetermined point by WSA. The increase in alcohol should be sufficient to stop any further action of the yeast, leaving the unfermented sugar to sweeten the resulting wine. All wines to be sweetened with grape juice concentrate are regulated by Section 240.403, Part 240, CFR. The regulations concerning WSA are found in Sections 240.374–240.383.

The procedure for calculating any form of sweetening is simple and straightforward, as long as the winemaker remembers that a subsequent brandy addition will dilute the extract and degree of sweetness. For example, 1000 gal. of 913 Ives inventoried at 13.4% alcohol and −1.2° Balling calculates to a *dry* (without sugar) extract of 3.3°. Recall from Chapter 4 that heavy-bodied red wines may have concentrations of non-sugar dissolved solids effecting higher dry extracts than the 2.0° or so anticipated from light-bodied white and rosé table wines. If the winemaster desires an extract of 10.0° *prior* to the WSA, then he can proceed as follows:

1000 gal. at 3.3° extract = 8.452 lb/gal.
1000 gal. × 8.452 lb/gal. = 8452 lb of wine

$$8452 \times \frac{(10.0 - 3.3)}{(100.0 - 10.0)} = 629.2 \text{ lb of cane sugar required}$$

The 8.452 factor can be found in the Brix Tables in Appendix B. The same tables can be used to double check this calculation:

1000 gal. at .27893 lb/solids/gal. =   278.93   lb solids
cane sugar addition   =   629.20   lb solids
   total   =   908.13   lb solids

$$\frac{908.13}{1046.56} \frac{\text{lb/solids}}{\text{gal.}} = .86781 \text{ lb solids/gal. after addition}$$

The dry Ives wine with 3.3° extract has a total solids content of 278.93 lb in 1000 gal. After the addition of 629.20 lb of cane sugar, the solids increase to a total of 908.13 lb. The sugar addition also increases the volume by 46.56 gal. (629.20 lb multiplied by .074 gal./lb) creating a new total volume of 1046.56 gal. The new total solids figure of 908.13 lb, divided by the new gallonage of 1046.56, results in .86781 lb of solids per gallon. This compares almost exactly to the .86786 lb solids/gal. listed for 10.0° Brix listed in the Appendix B Brix tables. Had dextrose sugar been chosen, a bit more would have been needed to raise the extract from 3.3° extract to 10.0°, as dextrose is slightly less dense than sucrose. More importantly, a wine sweetened with dextrose is much less "sweet" than one treated with the same level of sucrose—to the extent of about 30% or so in sensory response.

Observing the effects of dilution of alcohol we can calculate:

1000 gal. × 13.4% alcohol = 134 gal. of alcohol by volume

$$\frac{134}{1046.56} \frac{\text{gal. of alcohol}}{\text{gal. of wine}} = 12.8\% \text{ alcohol by volume}$$

The computation for adding grape juice concentrate for sweetening can be made in the same manner as described in Chapter 4, where concentrate additions to grape juice and must are discussed. In planning such additions to wine, however, the winemaster should be aware that sweetness additions of any type increase gallonage and decrease alcohol percentage. Conversely, brandy additions increase gallonage and alcohol percent, but decrease extract. The amount of wine spirits to be added will, of course, determine the amount of extract dilution expected.

WINE SPIRITS ADDITION CALCULATION

The same type of equation used to compute the sweetness addition can be used to determine the brandy requirement. Consider the following formula:

$$BW \times \frac{(AD - AE)}{(AS - AD)} = WSA$$

Where BW = base wine in gallons
AD = alcohol percentage desired
AE = alcohol percentage existing in base wine
AS = alcohol percentage of wine spirits
WSA = wine gallons of brandy required

Following the example considered in the previous section, we can assume that the winemaster desires the 1046.56 gal. of 913 Ives, at 12.8% alcohol, to be increased to 19.5% alcohol by volume:

$$1046.56 \times \frac{(19.5 - 12.8)}{(95.5 - 19.5)} = 92.26 \text{ wine gallons of } 191.0° \text{ proof brandy required}$$

This calculation can be double checked as follows:

1046.56 gal. × 12.8% alcohol   = 134 gal. of alcohol
brandy addition (92.26 × 95.5%) = 88 gal. of alcohol
   total   222 gal. of alcohol

$$\frac{222}{1138.82} \quad \frac{\text{gal. of alcohol}}{\text{gal. of wine}} = 19.49\% \text{ alcohol by volume}$$

This compares favorably with the previous calculation. Finally, we examine the dilution of extract:

1046.56 gal. × .86781 lb solids/gal. = 908.13 lb of solids

$$\frac{908.13}{1138.82} \quad \frac{\text{lb solids}}{\text{gal. of wine}} = .79743 \text{ lb solids/gal.}$$

.79743 lb solids/gal. = about 9.2° extract after WSA

The ATF computation methods for WSA additions may be found in Part 240, CFR, Sections 240.1015 through 240.1020.

None of the methods provided removes the need for "reciprocating" calculations. In other words, unless the winemaster has a standard recipe used each year in the formulation of each dessert wine type, it will be necessary to recalculate back and forth, making up for dilutions of alcohol due to sweetness additions, and then making up for dilutions of extract due to alcohol additions. This can be avoided by some rather complicated and cumbersome mathematical formulae. Instead, the table provided in Appendix B is designed to assist the winemaker in anticipating extract dilution.

Recall from our earlier example that a calculation for 10.0° extract was made prior to the addition of the high-proof brandy. The amount of brandy found necessary to raise the blend to 19.5% alcohol was 92.26 gal. or about 8% of the *resulting* product. Referring to the table in Appendix B, it can be found that an 8% addition of high-proof brandy to a wine with 10.0° extract reduces that extract level to about 9.2°. This can be compared with the actual results found in the post-WSA analysis listed on the *wine spirits addition record* in Fig. 7.1.

Before discussing the wine spirits addition record, note the 19.5% alcohol level, calculated in this example. In this case, the winemaster anticipates losing about .5% alcohol during the aging period of the port material. This is a normal occurrence and can be exemplified by tracing the production history of the Seyval Blanc lot in Chapters 5 and 6. It is necessary to anticipate and make allowance for some loss in alcohol, especially with dessert and aperitif wines, as they are regulated more closely than table wines. Specifics in this regard can be found in Title 27, CFR, Section 4.36.

WINE SPIRITS ADDITION RECORD

The wine spirits addition record is designed to provide a smooth transition from the under 14% alcohol bulk wine inventory category to the 14%−21% or 21%−24% alcohol categories. The wine spirits addition record should closely monitor and double check inputs, as well as provide a check list for critical analytical and regulatory requirements.

Wine spirits addition operations should be serialized by calendar, or fiscal year, and calculated so as to minimize the inventory of expensive high-proof wine spirits.

The ATF regulates the procurement, storage and addition of wine spirits very closely because of the increased tax liability generated when dessert and aperitif wines are made, not to mention the extreme liability of the brandy itself. These rules can be found in Part 240, CFR, Sections 240.160 through 240.163, 240.166 through 240.169, as well as those already mentioned (Sections 240.374 through 240.386 and 240.440 through 240.448).

The selection of good quality high-proof brandy, low in fusel oil content, is imperative for the manufacture of quality dessert and aperitif wines. The winemaster should work closely with a reputable supplier to select superior grades of brandy for WSA operations. The storage of flammable wine spirits is hazardous, and must be carefully monitored by both the winemaster and the cellarmaster.

The temperature of both base wine and wine spirits is very important in the calculation of WSA. Ethanol expands and contracts significantly with

WINE SPIRITS ADDITION RECORD

Date	WSBA Serial No.	Lot Number	Type-Variety
7-14-80	80-3	913	IVES

From Tank(s) No.	Gallons	To WSA Tank No.	Gallons
5-19	1,000	WSA-1	1,000

Pre-Sweetening Analysis:

Alc.	Ball.	Ext.	T.A.	V.A.	FSO₂	TSO₂	Color Int.	O₂	Wine Temp. C	C	WS Temp. N	T
13.4	-1.2	8.3	.660	.057	48	72	5/525	—	62°F ✓	✓	58°F ✓	✓

Final Adjustments:

Sugar:	Water:	Other:	ATF Sample:	Form 275
629.2 lbs	—	—	7-14-80 ✓	✓

Pre-WSA Analysis:

Alc.	Ball.	Ext.	T.A.	V.A.	FSO₂	TSO₂	Color Int.	O₂	Wine Temp. C	C	WS Temp. N	T
12.8	5.6	10.0	.630	.051	32	60	7/625	—	✓	✓	✓	✓

WSA Data:

WS Wine Gallons:	Proof:	Proof Gallons:	ATF Sample:	Form 2629
92.24	191.0°	176.22	7-17-80 ✓	✓

Post-WSA Analysis:

Alc.	Ball.	Ext.	T.A.	V.A.	FSO₂	TSO₂	Color Int.	O₂	Wine Temp. C	C	WS Temp. N	T
19.6	3.25	9.3	.585	.051	32	—	8/625	—	✓	✓	✓	✓

Production Data:

Gallons Result:	Loss:	Assimilation:	ATF Samples:	Form 2058
1,131 gals.	7.82 gals.	.7%	7-17-80 ✓	✓

Transfer Data:

To Tank(s) No.	New Lot Number	Type-Variety
BT-1	913	PORT MATERIAL

FIG. 7.1. WINE SPIRITS ADDITION RECORD FORM IN USE

temperature changes during transport and storage. A table is provided in Appendix B for factoring wine spirits at temperatures above and below the ATF 60° standard.

Note in the wine spirits addition record that three ATF samples should be taken. One 750 ml bottle sample should be prepared and reserved for the ATF gauger-inspector at each of the three processing stages: one of the base wine prior to sweetening; one of the base wine after sweetening (and any other pre-WSA adjustments); and one of the wine following the WSA. The ATF official may also require a sample of the high-proof bandy being used for the WSA.

Also note that there are three ATF forms involved in the WSA operations: forms 275, 2629 and 2058. Form 275 is used for reporting the addition of wine spirits. Form 2629 is executed on receipt of the high-proof brandy from the distiller, and form 2058 is required only when the wine produced is a special natural wine. These forms are fully described in Chapter 10.

Ethanol is a good desiccant and absorbs water readily. Consequently, during WSA operations, the addition of brandy may not yield the final volume anticipated. The differences observed from comparison of projected post-WSA gallonages to those actually measured often amounts to 1% or more. Acknowledgment of this assimilation is contained in the wine spirits addition record illustrated in Fig. 7.1.

Following the WSA and adequate mixing, the new dessert or aperitif wine material should be analyzed again. This time, however, the alcohol level will be high enough to require determination of alcohol content by the distillation method. As was stated in Chapter 4, the distillation procedure for alcohol will also be required following sweetening if the Balling was increased to a level exceeding about 0.0°. Methodology for testing alcohol by distillation is provided in Appendix A. A typical alcohol distilling apparatus for this analysis is pictured in Fig. 7.2.

The analytical rationale for alcohol, Balling, extract, total acidity, volatile acidity, free and total SO_2, color intensity and oxygen described for table wines also applies to dessert and aperitif wine quality control. The threat of bacterial infection is, however, reduced in wines with alcohol contents above 18% by volume. This, in turn, significantly lessens the need for frequent volatile acidity analyses. The methods for all of these quality control testing functions, along with a special nomograph for the calculation of dessert and aperitif wine extracts, are provided in Appendix A.

Many winemasters will bottle one or two samples of the newly created dessert or aperitif wine material, logging these laboratory-prepared samples into the retention library for future references (Figs. 2.56 and 2.57).

AGING

As defined in Chapter 5, aging is that period in the life of a wine when it is carried in the bulk wine book inventory—the time between fermenter and bottle. While the length of time for aging is usually much longer for dessert wines than for table wines, the postulates remain similar. The interim

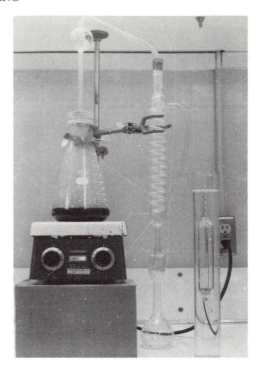

FIG. 7.2. ALCOHOL DISTILLING
APPARATUS

process of blending may be thought of as including WSA (and the generation of aldehydes in shermats), clarification, detartration, primary filtration and organoleptic development. In the case of aperitif wines, organoleptic qualities may be quickly attained by the addition of essences, spices,
herbs, and other non-grape infusions.

BULK WINE BOOK INVENTORY

The introduction, purpose and maintenance of the bulk wine book inventory was discussed in Chapter 5. The adoption of the same format can be
done for dessert and aperitif wines, although most winemasters prefer to
change the color of the card stock so as to quickly distinguish between
tax-classes of the wines in inventory. Figure 7.3 illustrates the bulk wine
book inventory card used following the creation of a new dessert wine
material by sweetening and WSA.

WINE DEVELOPMENT INSIGHT

There are instruments that perform chromatographic analyses which can
quantitatively pinpoint organic constituents affecting the aroma and taste

| Lot No. 644 | Type-Variety NIAGARA | | Class -14% | | Color WHITE |

Remarks:

Date	Racked From Tank W.O.	Gallons	Treatment-Disposition	Racked To Tank W.O.	Gallons	PHYS INV	ALC	BALL	T.A.	V.A.	FSO$_2$
4-18 1977	WSR 1 77-59	2,155	inventory + WSR	WSR 1 77-59	2,150		13.6	-1.7	.593	.042	84
4-19 1977	WSR 1 77-61	2,150	362 gal sugar	WSA 1 77-61	2,175		13.4	0.0	.585	.042	72
4-20 1977	WSR 1 77-62	2,175	174.6 WG-inventory	WSA 1 77-62	2,330		19.6	-2.0	.570	.036	60
	WSA # 77-3										
			Transferred to 644 Dry Sherry material Inventory Card								

FIG. 7.3.   BULK WINE BOOK INVENTORY CARD IN USE FOLLOWING WINE SPIRITS ADDITION

of wines. However, such technology is very expensive, and the data are difficult to interpret.

For the small winery quality control program, the case is made once more for the development of the human senses as essential analytical tools. The novice winemaker should make separate water and water/ethanol solutions of acetaldehyde, fusel oils, tannic acid and imitation vanilla, and learn to recognize the distinct nose and taste of each of these compounds. Once each can be identified "blind", samples should be diluted to half-strength and analyzed again. This dilution and blind recognition exercise should be repeated until each compound can no longer be detected. A "ringer" may be introduced, such as hydrogen sulfide, acetic acid, or a "blank" of plain water/ethanol, in order to expand the range of organoleptic components identified. Two or more compounds may be combined in the water and water/ethanol preparations, furthering sensory development. This should be followed by wines, such as sherries, ports or tawny ports, to see if these compounds can be distinguished. With hard work and dedication, plus comprehensive guidance from a reputable winemaster, the novice palate can develop quickly.

More related to the analytical parameters of organoleptic evaluation can be found in Chapter 4 and 6, as well as Appendix A.

BLENDING

The blending of young dessert and aperitif wines is not done in the same way as for table wines. The necessary blending operations for adjustments of color intensity and hue, and values of nose and taste, normally take place prior to the addition of the wine spirits. Few small wineries will need more than one or two separate shermats and port materials in a given production year. Consequently, the "recipe" for each dessert wine material and aperitif base wine is formulated by laboratory blending, and double-checked in the same manner as described for table wines in Chapter 5. Sweetening and WSA additions are then computed, and double checked by laboratory blends, in order to insure that the desired results are achieved in the cellars. This operational philosophy greatly reduces the necessity for any post-WSA blending adjustments among wines from the same vintage year.

One can find a few vintage-dated dessert wines in the market, particularly ports. These wines must comply with the vintage date ATF regulations as set forth in Title 27, CFR, Section 4.39. For the most part, however, dessert and aperitif wines are single production entities that are not vintage labeled. It is common in small wineries for some dessert wines to remain in bulk storage. These can be properly blended with new production lots of the same, or similar, types of wine, as long as such blends are followed by finings and detartration.

Some vintners apply for approval from the ATF (Part 240, CFR, Section 240.163) to utilize the specially designated WSA tanks for blending. This allows for the convenience of proceeding directly after the WSA with the addition of blend components, adjustments and finings, followed by detartration in cold storage. The administration of these processes is identical to that described in Chapter 5. The production of sherry-type wines, however, will require madeirization prior to clarification, detartration and primary filtration operations.

MADEIRIZATION

The making of sherry-type wines depends chiefly on the production of acetaldehyde from the oxidation of alcohol and acetic acid, or by the action of Flor yeasts (surface-growing film yeasts). This is known as madeirization, derived from the once-famous dessert wines from the island of Madeira. To a lesser extent, the formation of hydroxymethylfurfural from the dehydration of levulose sugar is also important in the organoleptic profiles of fine sherry types. Acetaldehyde has the characteristic nut-like nose and taste of sherries, while hydroxymethylfurfural is rather caramel and raisin-like. Both compounds result from the controlled heating of shermat wines in an atmosphere that can supply adequate oxygen. The basic chemical reaction in the formation of acetaldehyde from ethanol is as follows:

$$C_2H_5OH + \tfrac{1}{2}O_2 \xrightarrow{\text{Heat}} CH_3CHO + H_2O$$

Recall from the discussion in Chapter 5 that this reaction is the first step in the formation of acetic acid spoilage. However, the advanced alcohol levels provide a deterrent to the action of *Acetobacter* in dessert and aperitif wines.

Baking

The most popular method for making sherry-type wines in America is by *baking*. This process may range from the simple aging of shermat in barrels, where oxidation may take years to produce the desired effect, to very technically oriented programs of "Tressler" baking, where oxygen-diffusion and heat-applications are monitored in specially designed tanks. All baking processes are regulated by the ATF in Part 240, CFR, Sections 240.384 and 240.919.

The wines selected for shermat baking should be made from grape cultivars grown and vinified expressly for the purpose. Obviously, varieties that readily oxidize should be chosen. In the east, the cultivar Niagara is used extensively for the production of sherry wines, while Thompson seedless is a common shermat selection by California vintners.

The barrel aging of shermat can keep barrels "soaked-up" and preserved between the periods when they are required for aging table wines. Barrel-aging programs of shermats can be accelerated by placing the barrels outdoors so that the sun's heat can be utilized in baking. Such facilities must be approved by the ATF in accordance with Part 240, CFR, Sections 240.160 (b) and 240.521.

The baking of shermat can also take place in tanks located in cellars or rooms where temperature can be controlled. In some cases the baking shermat is continually, or intermittently, pumped from these tanks through a heat-exchanger in order to maintain or adjust temperature.

In most larger wineries primarily, and some smaller cellars, the Tressler system is generally used for the baking of shermat. Special tanks constructed of stainless steel, glass-lined steel, and wood are used for this purpose, each equipped with an exterior jacket of internal coils for the application of hot water or steam. An oxygen-injection device, usually a series of ceramic diffusers, is located inside the Tressler tank to efficiently distribute oxygen throughout the shermat during baking. Heat and oxygen inputs are monitored either manually or automatically with controls mounted locally or remotely.

Most Tressler tanks are also equipped with a reflux condenser. This collects significant amounts of distillate from the vapor that forms in the head space which is left to allow for expansion of the shermat. These distillates are returned to the shermat after baking has been completed, but before the finings and detartration operations take place. Losses of alcohol during the baking of shermat can be very significant when poor controls and equipment are employed. Some alcohol, however, is unavoidably lost, perhaps up to 1% by volume, and should, therefore, be added to other anticipated processing losses when WSA planning for shermat is made.

The phenomenon of increased alcohol concentration is often observed in some of the longer, naturally oxidized shermats aged in wooden vessels. This arises from the loss of water through the staves of wooden tanks by evaporation. The water molecule, H_2O, is much smaller than the ethanol molecule, C_2H_5OH, allowing for the dehydration of the shermat.

Quality control during the baking of shermat is highly dependent on organoleptic evaluation, even in larger wineries. The typical nose and taste of acetaldehyde and hydroxymethylfurfural may be difficult to distinguish from other aldehydes and esters that develop during the baking process. Nevertheless, the winemaster should strive to develop standards that are meaningful and repeatable for the production of undeveloped sherry types.

Color intensity can be measured during the baking process. The shermat should become noticeably darker to the eye, and significantly darker to the spectrophotometer. Remember that wines measured by the spectrophotometer must be absolutely brilliant or else the intensity measurement will be diffused and meaningless.

The most significant chemical test that can be included in the quality control program for baking shermat is the determination of acetaldehyde. Finer sherry types may have up to 600 ppm of acetaldehyde, while average is about half of that amount. Procedure for the testing of acetaldehyde is described in Appendix A.

Flor Yeast

Flor yeasts are surface-growing yeasts used primarily in Spain for the production of sherry. Flor yeasts, *Saccharomyces beticus*, must be carefully distinguished from other surface-growing film yeasts such as the genera *Candida, Kloeckera, Pichia,* and others, which are spoilage microorganisms.

In the Jerez de la Frontera area of Spain, where only sherries are made, the existence of Flor yeasts poses no threat to any other form of wine production. In America, however, where small wineries have several types of wine in close proximity, there is the potential for Flor infection. American Flor yeast products must therefore be made in total isolation. For this reason the baking of sherry-type wines is the preferred method of madeirization in America.

Flor shermat is made principally from Palomino and Pedro Ximenes (often referred to as simply "PX") grapes, both cultivars of the species *Vitis vinifera.* The Spanish sometimes prefer to "plaster" musts before pressing out the juice. Plastering is the addition of gypsum in order to convert potassium bitartrate into tartaric acid, increasing the total acidity. This lowers the pH, which helps to reduce the potential for spoilage due to lactic bacteria.

Spanish sherry material is made by the addition of wine spirits up to an alcohol level of about 15% by volume, after which the Flor yeasts are introduced. The surface-growth, or film, of Flor develops, creating acetaldehyde by the oxidation of ethanol, acetic acid and some forms of sugar.

In some Spanish wineries, however, there is no inoculation with Flor yeast cultures. Naturally occurring yeasts are allowed to develop the film and, in some cases, even the primary juice fermentation. In more advanced facilities the technology of submerged-growth Flor yeasts is utilized for more efficient fermentation and better quality control. Following fermentation more wine spirits are added to reach the desired final alcohol level. Flor yeast operations in American wineries are regulated by the ATF in Part 240, CFR, Section 240.385.

The parameters for the control of Flor yeast operations are similar to those already described for baked sherry-type products, except that the handling of samples must be closely controlled in order to avoid contamination and infection of other wines, both in the laboratory and the cellars. The making of Flor yeast products in any winery that makes other wines which might be susceptible to infection is not recommended, even under the most favorable conditions.

TYPES OF SHERRY

There are three basic sherry-type wines commonly produced in America. Dry, or cocktail, sherry is usually very pale amber in color with an extract of less than 6.0°, comparable to the Fino sherry from Spain. Very dark amber and sweet sherry types are labeled as cream sherry by most American vintners. These sherry types have an extract in excess of 10.0° as a rule, and are generally compared to the Oloroso sherry of Spanish production. Regular sherries of medium color and sweetness, are referred to as Amontillado in the Jerez de la Frontera.

In some wineries a special lot number system will be utilized to control the aging of dessert wines and aperitif wines. Most small wine cellars may follow production steps adequately with simple written notations on the bulk wine book inventory card in conjunction with the lot number system used for tables wines. Figure 7.4 is a continuation of the 644 dry sherry lot, after baking is completed and posted, as was noted on the bulk wine book inventory card following the WSA in Fig. 7.3.

SULFUR DIOXIDE

For the most part, sulfur dioxide does not play an important part in the production of dessert and aperitif wines. The higher level of alcohol acts as a natural preservative in most production situations. (A few exceptions exist which shall be discussed later.) Some winemasters choose total SO_2 levels in the $150-250$ ppm range in the belief that some increased stability is obtained. In the baked sherry types, medium-high levels of total SO_2 tend to "fix" aldehydes in solution, which is often considered desirable.

In the production of Flor sherry types, the winemaster should be careful in the use of SO_2 so as not to inhibit, or destroy, the Flor yeast during film development and fermentation. On the other hand, this medium can also support the growth of *Acetobacter* and wild yeasts, as it is low in alcohol (about 15% by volume) and there is ample oxygen in the head space of the fermenter. In some instances the conversion of ethanol and acetic acid to acetaldehyde by the Flor yeast is accompanied by the simultaneous conversion of acetaldehyde and ethanol to acetic acid by *Acetobacter*. In addition, some anaerobic yeasts can convert acetaldehyde to ethanol and acetic acid. Most important, however, is to prevent the formation of ethyl acetate from ethanol and acetic acid. The chemical equation of this reaction is as follows:

$$C_2H_5OH + CH_3COOH \longrightarrow CH_3COOC_2H_5 + H_2O$$

This reaction can be caused by acetic acid bacteria as well as other microorganisms. Ethyl acetate can also result from the action of lactic acid bacteria on glucose and gluconic acid. In any event, ethyl acetate is much more pungent than even acetic acid and, therefore, must be very carefully controlled.

In Michigan, dessert and aperitif wines that are sold in grocery stores and other private outlets must have an alcohol content not exceeding 16% by volume. (higher alcohol wines are marketed by the State Liquor Control Commission outlets and specially designated retail stores) Such low-alcohol wines must be preserved in much the same way as sweet table wines.

In recent years a trend has developed towards lower-alcohol aperitif wines. Dry and sweet vermouths can be found in the market that have less than 17% alcohol—the "pop wine" types with only 10% in some instances. These wines, too, must be preserved in the same manner as sweet table wines in order to prevent spoilage and/or secondary fermentation in the bottle.

Lot No. 644	Type-Variety DRY SHERRY MAT.		Class +14-21%		Color PALE AMBER				
Remarks: WSA #77-3 4-20-77									
Date	Racked From Tank W.O. Gallons	Treatment-Disposition	Racked To Tank W.O. Gallons	PHYS INV	ALC	BALL	T.A.	V.A.	FSO₂
4-25 1977	WSA 77-1/64 2330	for baking	BK-1 77-64 1,150		19.5	-1.9	.570	.036	56
			BK-2 77-64 1,150		19.4	-1.8	.563	.036	56
4-30 1977			BK-1 2,300	✓					
5-31 1977			BK-2 2,300	✓	19.4	-1.9	.563	.036	50
6-12 1977	BK-1 77-103 1,150	for fining	BT-2 77-103 1,140		19.1	-1.7	.556	.033	12
6-12 1977	BK-2 77-103 1,150	"	BT-2 77-103 2,290		18.9	-1.8	.555	.036	4
6-14 1977	BT-2 77-106 2,290	9 lbs R/CHAR, 2 lbs. SPARK. CS-4/108	CS-4 2,285		19.0	-1.8	.548	.036	4
		3 lbs. BENT.							

FIG. 7.4.   BULK WINE BOOK INVENTOY CARD IN USE FOLLOWING BAKING AND FININGS

CLARIFICATION, DETARTRATION AND PRIMARY FILTRATION

The procedure for finings and stabilization is basically the same for dessert and aperitif wines as for table wines. The most significant difference for dessert wines being that they do not need intense clarification treatments, since following detartration and primary filtration, long periods of aging stabilize the wine naturally. This is not the case with aperitif wines, however, as these are often very young wines when marketed.

The levels for copper and iron should be adjusted in the same manner as for table wines. A good standard is a range of ½ to 1 ppm of copper (some should remain to help avoid the formation of mercaptans) and less than 5 ppm for iron.

Total acidity may be slightly higher for sweet dessert and aperitif wines, but for dry cocktail sherry and dry vermouth, a popular range remains at .488 to .518 g/100 ml. Before adjusting acidity recall that considerable amounts may be lost in detartration due to the formation of acid salts.

Dry vermouth material is often bleached to a "water-white" colorless state by the use of activated charcoal. To a much lesser degree, baked dry sherry material is often bleached in order to achieve very pale amber shades of color. These practices require special permission from the ATF as set forth in Part 240, CFR, Sections 240.527 and 240.527a.

Under normal cellar conditions, young dessert wine materials, baked sherry-type materials and aperitif wine materials will react to the additions of Sparkolloid and bentonite much more quickly and effectively than table wines. This is primarily because the higher alcohol types are better solvents and, being usually a bit older than table wines, are naturally more clear to begin with.

The necessity for nitrogen-sparging head space in detartration tanks is significantly reduced for most dessert and aperitif wine materials. However, very pale sherry types and dry vermouth may need to be treated with nitrogen in order to avoid oxidation.

All other operations and quality control procedures for detartration and primary filtration of dessert and aperitif wines may be carried out in the same manner as discussed for table wines in Chapter 5. The careful review of that chapter is recommended before commencing with the analysis, calculation and application of finings, detartration and primary filtration.

APERITIF WINES: ESSENCES, HERBS AND SPICES

While sweet vermouth materials are sometimes aged following primary filtration, this is usually not the case. Dry vermouth materials are almost never aged, to avoid the possibility of color development. Vintners will often make only one vermouth material, separating the desired portions for dry and sweet into special tanks. The vermouth processing tank(s) must be approved by the ATF in accordance with Part 240, CFR, Section 240.164. Regulations for the addition of essences, herbs, spices and other ingredients are set forth in the same manual under Subpart S, Sections 240.440 through 240.448.

Dry vermouth types are usually flavored with very light-colored or colorless essences that are commercially available. Some winemakers purchase essences that are custom-made to an original exclusive blend or recipe. The direct infusion of herbs and spices is relatively rare today, with more uniform and economical results being achieved by the use of professionally prepared essences.

Sweet vermouth is flavored in much the same way as dry vermouth except that the essences are blended differently for taste. It is common to find flavors for dry vermouth to be rather citrus and herbal, while sweet vermouth reflects a heavy use of spices. Sweet vermouth is usually colored very dark, often with the use of caramel. As the name implies, sweet vermouth is very sweet, typically in excess of 15° extract. Dry vermouth may not be totally dry, but the most popular brands have an extract of less than 3°, as a rule.

Again, the novice winemaker is cautioned to be entirely familiar with the ATF regulations concerning the manufacture of aperitif wines, particularly the rules concerning required formula approvals. These statutes are included in Part 240, CFR, Sections 240.440 through 240.448, as stated earlier.

After vermouth and other special natural wines have been prepared according to individual color, sweetness and flavor standards, they are analyzed to confirm these product requirements, and then bottled without delay. Quality control for pre-final filtration, final adjustments and preservation, post-final filtration analyses, material and supply usage, post-bottling analyses and production data recording, can be monitored in the same manner as for table wines. The processing and recording of a sweet vermouth lot from the bulk wine book inventory to the daily bottling record and onto the cased goods book inventory is illustrated in Figs. 7.5, 7.6 and 7.7, respectively.

DESSERT WINE AGING

Most sherry and port-type wines can be effectively aged in American white oak. Several European oaks such as Limousin and Nevers, can be ideal for dessert wine aging. Redwood, however, has not gained acceptance with high-quality dessert wine aging programs among American vintners. Nonporous tankage of stainless steel, glass-lined steel and fiberglass severely restrict the oxidation reactions required for proper aging. These vessels do not have the capacity to impart tannins and other desirable organoleptic properties, as do oaken barrels and casks. However, some winemasters do use such containers for aging dessert wines, incorporating oak chips of various sizes and textures. Such programs should only be administered prior to finings and detartration, as stable dessert wines can be rendered unstable by the use of new oak chips.

Lot No. 949	Type-Variety VERMOUTH MAT.		Class +14-21%		Color VERY PALE AMBER

Remarks: WSA #80-1   3-17-80

Date	Racked From Tank W.O.	Gallons	Treatment-Disposition	Racked To Tank W.O.	Gallons	PHYS INV	ALC	BALL	T.A.	V.A.	FSO₂
5-6 1980	CS-6 80-104	2,445	prn. filt. Dry Verm.	VT-1 80-104	1,603	131.5 lbs sugar and 3.2 gals D Valencia	17.8	-3.5	.480	.036	28
5-7 1980	VT-1 80-105	1,603	3.2 lbs Citric, 3.0 lbs KMS	VT-1 80-105	1,605						
5-7 1980							17.5	-2.5	.495	.033	124
5-8 1980	VT-1 80-106	1,605	final filt. to bottling	B-2 80-106	1,601		17.4	-2.4	.503	.036	116
5-13 1980	CS-6 80-111	840	prn. filt. Swt Verm.	VT-1 80-111	830		17.9	-3.4	.480	.033	24
5-14 1980	VT-1 80-112	830	2.5 lbs Citric, 720 lbs sugar	VT-1 80-112	885	1.7 lbs KMS, 2 gals caramel, 9 gals SV-3 ea.	16.7	+6.8	.503	.030	128
5-14 1980				VT-1 80-112	885						
5-15 1980	VT-1 80-114	885	final filt. to bottling	B-2 80-114	880		16.8	+6.9	.495	.033	120
	to dead file			5-15-81							

FIG. 7.5. BULK WINE BOOK INVENTORY CARD IN USE FOLLOWING PREPARATION OF VERMOUTH MATERIAL FOR SWEET BOTTLING

DAILY BOTTLING RECORD

Date 5-15-80	Lot Number 949	Vintage —	Type-Variety SWEET VERM, (17% ALCOHOL)			
From Tank(s) No. VT-1	Gallons 885	To Bottling Tank(s) No. B-2	Gallons 880	Received By—Date: TRL 5-15-80		

	Alc.	Ball.	Ext.	T.A.	V.A.	FSO₂	TSO₂	Color Int.	O₂	C	N	T
Pre-Final Filtration Analysis:	17.9	-3.4	2.0	.480	.033	24	100	95/425	—	✓	✓	✓
Final Adjustments: Sugar: 720 lbs	Citric Acid: 2.5 lbs		SO₂: 1.7 lbs							✓	✓	✓
Other: 2 gals caramel 9 gals SV-3 essence												
Post-Final Filtration Analysis:	16.8	+6.9	12.1	.495	.033	12.0	248	7/525	—	✓	✓	✓
Material and Supply Usage: Filter Pads: Scott 8Y	Bottles: OI .75 liter demisot		Closure: alum. roll-on caps	Other: —								
Post-Bottling Analysis:	16.7	+6.9	12.1	.503	.033	112	232	8/525	—	✓	✓	✓
Production Data: Cases: 368 + 6 bot to lab. + 2 bot TPR-1	Gallons: 876.6 3.4 gals loss		Serial Numbers: 9,041 - 9,408	Received By—Date: VB 5-16-80								

FIG. 7.6. DAILY BOTTLING RECORD IN USE FOLLOWING THE BOTTLING OF SWEET VERMOUTH

Lot Number 949	Vintage —	Type-Variety SWEET VERMOUTH (17%)		Size .750 LITER	
Class +14-21%	Color DARK AMBER	Minimum Inventory 50 CASES	Bottle Type DESSERT	Closure ROLL-ON	Other —
Date	Cases In	Cases Out	Balance on Hand	Ref. No.	Inv.	Remarks
5-1 1980	1		66	—		from prev. card
5-9 1980	1	4	62	S.O. 80-271		
5-13 1980	1	5	57	S.O. 80-274		
5-14 1980	1	1	56	TPR #1		
5-16 1980	368	1	424	9,041-9,400		from bottling

FIG. 7.7.   CASED GOODS BOOK INVENTORY CARD IN USE FOR SWEET VERMOUTH RECEIVED FROM BOTTLING

The very high cost of oak barrels and casks, especially those manufactured in Europe, has opened the way for treating used whiskey barrels so as to be acceptable for dessert wine aging. The barrels are "shaved" to remove char from all inner surfaces, followed by several treatments of live steam, each for a duration of about three minutes so that the barrels do not get excessively hot. Finally, soaking for 24 h, filled with hot (160–190°F) water should render the barrels ready for the initial storage of wine. This first wine aged in a whiskey barrel should be an unbaked sherry-type material, which will serve to leach out most of any remaining whiskey flavors. This initial aging can range from 3 months to a year, and the barrels may then be used for the longer aging of finished sherry and port-type wines.

Apart from the more delicate dry, cocktail sherry types, most dessert wines can be aged in rather warm cellar areas, up to 80°F in some wineries. This will, of course, accelerate oxidation and aging and can help to reduce the number of years required to bring such wines to maturity. This, in turn, should help to reduce the inventory which accumulates when large volumes are aged for long periods.

SOLERA FRACTIONAL BLENDING SYSTEM

Earlier in this chapter the point was made that few blended dessert wines are made in most small winery operations, except perhaps the occasional blending of inventory remnants with new similar-type wines. It was also stated that blending stable wines may bring together elements that could render the blend unstable, so all blending operations should be completed prior to finings and detartration. The Solera fractional blending system is an exception.

The Spanish Solera fractional blending system contributes greatly towards making traditional sherries mellow and consistent, yet rich and distinctive. The operation of a Solera demands heavy investments in aging cooperage and wine inventories. It is common to find Solera systems in Spain where inventory is tenfold, or more, the amount of wine that is bottled from the process each year.

The term Solera actually relates to the final stage of aging, or *Criadera*, prior to bottling. Each Criadera, or row of barrels, is maintained at least half-full, and wine from the previous Criadera is added each year in order to fill the system. Wines undergo complex blending as they pass from Criadera to Criadera so that newer wines and older wines are mixed. Some of the wine from every year always remains—the average age of the blend becoming older and older with each vintage. Figure 7.8 outlines the operations of a typical Solera system in diagram form, and in operation in Fig. 7.9.

In addition to the application of the Solera for aging sherry-type wines, the system is also used for port wines made in America. In any case, the Solera is recommended in the small winery where dessert wines are to become a significant part of the product line.

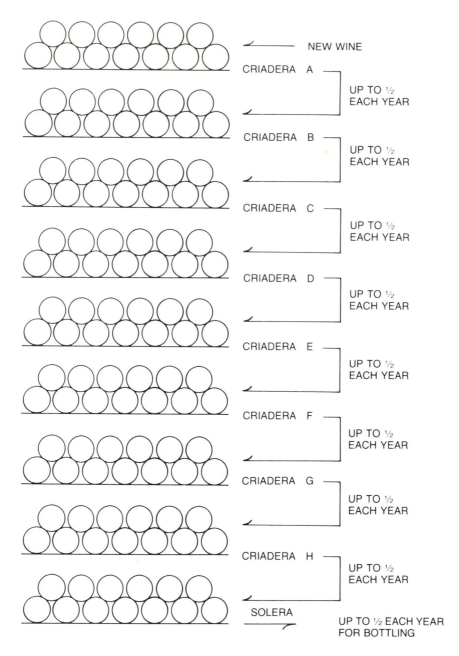

FIG. 7.8.   DIAGRAM OF SOLERA FRACTIONAL BLENDING SYSTEM

Courtesy: Sherry Institute of Spain

FIG. 7.9.   OPERATION OF A SOLERA SYSTEM

DESSERT WINE FINAL FILTRATION AND BOTTLING

The parameters for quality control during final filtration and bottling of dessert wines are similar to those for table wines as set forth in Chapter 5. When alcohol contents are 18% or more by volume, a case can be made for reducing preservation requirements. Some vintners, as stated previously in this chapter, prefer to bottle sherry-type wines with total SO_2 levels in the 150 to 250 ppm range. Filtration can be facilitated by the use of a filter medium that is slightly more coarse than that used for table wines. The "sterile" filtration concepts considered for table wines that are susceptible to microbial activity after bottling do not hold for the higher alcohol dessert wines.

As with the bottling of all other non-sparkling wine types, dessert wines should be analyzed prior to the calculation and addition of any final adjustments. Post-final filtration analysis, material and supply data, post-bottling analysis, and production data may be recorded in much the same way as described for aperitif and table wines. Figures 7.10, 7.11 and 7.12 show the transition of a dessert wine from the bulk wine book inventory to the daily bottling record and onto the cased goods book inventory.

WAREHOUSING

The storage of dessert and aperitif wine cased goods does not require the strict temperature control which table wines do. Dry, cocktail sherry types

Lot No. 649	Type-Variety DRY SHERRY		Class +14-21%	Color PALE AMBER

Remarks: WSA #77-3   4-20-77

Date	Racked From Tank W.O. Gallons	Treatment-Disposition	Racked To Tank W.O. Gallons	PHYS INV	ALC	BALL	T.A.	V.A.	FSO$_2$
2-29 1980	from previous card		42 bbls 2,100	✓	18.7	-1.8	.503	.039	4
3-31 1980			42 bbls 2,100	✓					
4-30 1980			42 bbls 2,100	✓					
5-31 1980			42 bbls 2,100	✓	18.8	-1.7	.495	.039	4
6-26 1980	42 bbls 80-141 2,100	to bottling	B-2 80-141 2,088		18.7	-1.7	.503	.042	4
	To dead file 6-26-80								

FIG. 7.10.   BULK WINE BOOK INVENTORY CARD IN USE FOR DESSERT WINE DELIVERY TO BOTTLING

DAILY BOTTLING RECORD

	Lot Number 649	Vintage —	Type-Variety Dry Sherry
Date 6-26-80			
From Tank(s) No. 42 bbls	Gallons 2,100	To Bottling Tank(s) No. B-2	Gallons 2,088 — Received By—Date: TRL 6-26-80

Pre-Final Filtration Analysis:

Alc.	Ball.	Ext.	T.A.	V.A.	FSO₂	TSO₂	Color Int.	O₂	C	N	T
18.7	-1.7	4.2	.503	.042	4	116	81/425	—	✓	✓	✓

Final Adjustments: Sugar: — | Citric Acid: — | SO₂: 1 lb. 28 | Other: —

Post-Final Filtration Analysis:

Alc.	Ball.	Ext.	T.A.	V.A.	FSO₂	TSO₂	Color Int.	O₂	C	N	T
18.6	-1.8	4.1	.503	.039	8	144	83/425	—	✓	✓	✓

Material and Supply Usage: Filter Pads: Scott SX | Bottles: .750 liter Dessert | Closure: Wood-top cork | Other: Dessert Wine Neck Tag | C ✓ N ✓ T ✓

Post-Bottling Analysis:

Alc.	Ball.	Ext.	T.A.	V.A.	FSO₂	TSO₂	Color Int.	O₂	C	N	T
18.7	-1.8	4.2	.495	.036	4	152	82/425	—	✓	✓	✓

Production Data: Cases: 874 + 6 bot to 1 Rb + 1 bot 7 bbl | Gallons: 2079.4, 8.6 gals loss | Serial Numbers: 14,107–14,977 | Received By—Date: VB 6-27-80

FIG. 7.11. DAILY BOTTLING RECORD IN USE FOR DESSERT WINE BOTTLING

Lot Number 649	Vintage —		Type-Variety DRY SHERRY			Size .750 liter	
Class +14-21%	Color PALE AMBER		Minimum Inventory 200 CASES	Bottle Type DESSERT		Closure WOOD-TOP	Other NECK TAG
Date	Cases In	Cases Out	Balance on Hand	Ref. No.	Inv.	Remarks	
6-12-80	1	1	314	—		from previous card	
6-16-80	1	11	303	S.O. 80- 310+311			
6-17-80	1	9	294	S.O. 80- 313,314+316			
6-20-80	1	15	279	S.O. 80- 320,321+323			
6-24-80	1	17	262	S.O. 80- 327,328,329,331,334+335			
6-27-80	874	1	1,136	14,104- 14,977			
6-27-80	1	11	1,125	S.O. 80- 338,340,342+344			
6-30-80	1	1	1,125		✓	F/Y END	
6-30-80	1	1	1,124		✓	phy inventory (see about)	

FIG. 7.12.   CASED GOODS BOOK INVENTORY CARD IN USE FOLLOWING RECEIPT OF DRY SHERRY BOTTLING

and dry vermouth wines are notable exceptions, and should be stored at fairly constant temperatures not exceeding 72°.

Most small wineries bottle dessert and aperitif wines only once or twice each year. This lack of frequency may require constantly accessible warehouse space so that such wines are not buried by more rapidly flowing products.

SUMMARY

The market share of dessert wines is diminishing in America, while the consumption of aperitif wines remains more level.

All dessert and aperitif wines result from the addition of high-proof wine spirits, or brandy, to increase the natural alcohol of 10–14% by volume to 16–24% by volume.

Dessert wines are distinguished from aperitif wines in that artificial flavoring and/or colorings may be added to the aperitif types, which are commonly called special natural wines. Popular aperitif wines are dry and sweet vermouth, while sherry and port typify dessert wine items. Special production regulations found in Part 240, CFR, Sections 240.374 through 240.386 pertain to dessert wines. Aperitif wines are governed by Sections 240.440 through 240.448.

The base wines intended for eventual dessert and aperitif wine production, should be made from grapes that are grown for this express purpose.

Calculations for sweetening must be within Federal and State limitations. Federal statutes for sweetening dessert wines can be found in Part 240, CFR, Section 240.368; for aperitif wines, in Section 240.430. Sugar may not be used in California; rather, concentrated juice is utilized for sweetening. Some wineries use high-proof brandy to arrest primary fermentations in order to conserve the remaining natural grape sugar as the sweetening source.

Computations for wine spirits additions are regulated in Part 240, CFR, Sections 240.1015 through 240.1020. The winemaker should remember that some alcohol will be lost in further processing. This should be compensated for in the level of alcohol calculated so as to insure that finished dessert and aperitif wines are comfortably within the 1% by volume tolerance limits.

The wine spirits addition record provides a format for recording calculations and results of sweetening and WSA operations by serial number within each fiscal or calendar year. Laboratory blends should be made and double-checked prior to blending, sweetening and WSA operations in the cellars. Samples of a base wine prior to sweetening, prior to WSA, and after WSA may be required by the ATF inspector-gauger. Winemasters may wish to hold samples of post-WSA wines, and the high-proof used in the WSA operations, in the reference library.

The ability to distinguish organoleptic characteristics of dessert wines must be learned to aid in quality control. The method for educating novice palates is much the same as described for table wines, except that different aroma and flavor compounds are considered.

Courtesy of The Sherry Institute of Spain

SHERRY "CATHEDRALS OF WINE"

While port-type materials and vermouth base wines may be immediately blended, adjusted, clarified and detartrated, the production of sherry types must undergo madeirization. This process, whether by baking or by the action of Flor yeasts, converts ethanol into acetaldehyde, the nut-like flavor compound common to all sherry-type wines. A minimum of 200 ppm of acetaldehyde is recommended, while a maximum of about 600 ppm may be obtained using optimum Tressler baking conditions and superior Flor yeast submerged-growth processes. Flor fermentations require a very comprehensive quality control program in order to avoid acetic acid and ethyl acetate spoilage reactions from *Acetobacter* infections.

Aperitif wines are usually processed with essences and other approved ingredients immediately following primary filtration. Most winemakers use prepared essences, for quality, consistency, and low cost. Specially approved tanks for the treatment of aperitif wines are required by the ATF in accordance with Part 240, CFR, Section 240.164.

Oak barrels are generally considered the best cooperage for aging dessert wines. Limousin and Nevers oak from Europe is much more expensive than American white oak, and the difference in price is perhaps not commensurate with the results obtained. Used oak barrels may also be used, provided they are properly treated so as to remove all residual whiskey effects in the wood. Redwood is not normally used for aging programs of high-quality dessert wines. Stainless steel, fiberglass and other nonporous materials will not allow proper dessert wine development, although these types of vessels can be used with oak chips prior to finings and detartration.

The highest quality dessert wines usually result from the Spanish Solera fractional blending system. This aging method requires large inventories stored in Criaderas that are maintained at least half-full, so that newer wines can be blended with those from the previous year. This insures that some wine from every vintage remains, and that the entire Solera becomes older each year.

Quality control of final filtration and bottling operations can be monitored in generally the same manner as for table wines. The proper use of the bulk wine book inventory, daily bottling record and cased goods book inventory provide a comprehensive program of check listing, control and record-keeping.

Dessert wines can be stored in the warehouse under somewhat warmer temperatures than table wines, except dry, cocktail sherry types. Dry vermouth and other very pale aperitif wines should also be stored in prime temperature-controlled areas.

8

Sparkling Wine Quality Control

The most famous of all sparkling wines is unquestionably champagne—named for the French Province where it is produced.

The ATF permits sparkling wines made in the United States to be labeled as "champagne" as long as certain criteria of production and packaging are met. Part 240, CFR, Subpart W, Sections 240.510 through 240.513 contains explicit regulations for the commercial production of sparkling wines.

The Federal authority requires separate areas in the winery for the production of sparkling wines. The vintner is directed to submit a formula that describes in detail the materials, containers, time durations and other significant aspects of each sparkling wine product to be made. The novice sparkling wine producer is advised to contact the Regional Administrator of the ATF prior to commencing any sparkling wine production, even if formulae are approved, so that an inspector may visit the premises and confirm that all requirements are fully met and understood. ATF form 2057, *record of effervescent wine,* is maintained on a monthly basis in accordance with Part 240, CFR, Section 240.909 (see Chapter 10). The definitions of different types of carbonated wines are set forth in Title 27, CFR, Section 4.21.

Sparkling wines obtain their effervescence from dissolved carbon dioxide gas which is released in the form of bubbles. This CO_2 gas must be the product of fermentation in order to qualify as "sparkling" wine. Still wines with CO_2 gas injected into them are known as "crackling" wines. The ATF excise tax rate for sparkling wines is $3.40 per gal., while the crackling wines are levied $2.40 per gal.

The production of sparkling wines commences with the *cuvée*, literally translated from the French as "tubful". The cuvée is simply the table wine from which the sparkling wine will be made, by means of a second fermentation. A cuvée may be a varietal wine or a blend. The term cuvée is sometimes used to describe the free-run juice from a press, but most often cuvée refers to wines that will become sparkling.

There are several methods of producing sparkling wines in America. The *Charmat,* or bulk, process uses specially designed tanks where cuvée wines are fermented up to about 6 volumes, of CO_2 (approximately 80 psi at 65°F), adjusted for sweetness and preservatives, counter-pressure filtered and bottled in a pressurized filling machine. Another method of sparkling wine manufacture is the *transfer*, where sparkling wine cuvée is fermented in glass bottles that are designed and constructed for high pressure. After fermentation "in the bottle" the wines may be aged and then finished in a manner similar to that just described for the Charmat process. Both of these methods require technically advanced equipment that is very expensive.

This text will be concerned with only the *méthode champenoise*: the bottle-fermentation and hand-finishing method that is required for sparkling wine manufacture in the province of Champagne in France. This process can be distinguished from the one used most often in American wineries by the statement "fermented in this bottle". While the méthode champenoise requires the smallest capital investment of the recognized methods for sparkling wine manufacture, it nevertheless, is expensive.

The demand for sparkling wine in America is primarily as a beverage for special occasions. The expense of Champagne has traditionally prohibited it from everyday consumption. While the increase in consumption of sparkling wines is greater than that of dessert or aperitif wines, sparkling types remain far behind table wines in U.S. consumption.

Before any discussion of sparkling wine quality control, the point should be made that extra care in cleanliness should be maintained throughout the sparkling wine process. "Hot-water" clean has been stressed in this text already, but in the case of the méthode champenoise, cleanliness is of utmost importance. Superior sparkling wines are difficult to make, at best, but with unclean equipment and materials, the task can be complicated and unrewarding.

CUVÉE WINES

The cuvée, whether white, rosé or red, should be produced from grapes that are chosen and processed for the express purpose of making sparkling wine. Cuvée wines should be fermented to about 10.5% alcohol by volume and maintained free of preservatives that can interfere with the secondary fermentation. Such wines should also be dry, with good acid balance, and finished with a complete clarification, detartration, and primary filtration. In some cases, the winemaster will choose to age his cuvée wines in wooden cooperage. Mature cuvée wines are, ideally, low-alcohol table wines finished and ready for bottling with no preservatives left to interfere. Unless specific wines are made for secondary fermentation, it will obviously be quite difficult to select suitable cuvée constituents from general table wine inventories.

Sparkling wines may be considered as either *fruity* or *vinous*. The classics of such wines are white and vinous—often described as having the "taste of the soil upon which the grape is grown". The best examples are products from Champagne, where the wines may be distinctively "chalky", sometimes attributed to the heavy concentrations of calcium carbonate that are found in the subsoil of the Champagne district. Few cultivars exhibit this trait, the most notable being Chardonnay and Pinot Blanc among white *V. vinifera* varieties.

Champagne from France may be made from Chardonnay, and also from the white juice of Pinot Noir and Pinot Meunier, both red cultivars of *V. vinifera*. Most sparkling wines made in America, certainly those made in the east, are of the fruity type. The various flavors derive from such cultivars as Delaware and Niagara, both listed among the better selections of *V. labrusca* descent. Nearly all Pink Champagne and Sparkling Burgundy types are fruity, perhaps owing to their American development. Until recently, sparkling red and rosé wines were not made in France to any significant degree. Champagne from the famous French province remains traditionally an entirely white product.

PREPARATION

The low alcohol content of 10.5% will allow viable yeast cells to ferment the properly prepared cuvée up to 11.5% or so, while generating the necessary carbon dioxide gas.

Dry cuvée wines are required prior to making a batch for secondary fermentation. Any sugars remaining may well ferment along with those added from the cuvée calculation, creating a fermentation that is overly productive and which can explode the bottles.

Acid balance is important in that finished sparkling wines are usually somewhat more acidic tasting than the same wines without bubbles. It is common to find sparkling wines with total acidity levels ranging from .550 to .650 g/100 ml. The acidity must, of course, be adjusted while the cuvée wine is in aging process.

The stability and brightness of the cuvée must be maintained to prevent the formation of potassium bitartrate crystals which can act as focal points for the release of CO_2 gas. This can cause excessive frothing and difficulty in handling during finishing processes. Hazy, or cloudy, wines can take years to clear and finish while the same wines brilliantly finished can take just months or weeks.

Secondary fermentations require several critical ingredients in any formula: a fermentable sugar, a nitrogen source, vitamins, and yeast. Prior to these being calculated, many winemasters will save about 1% of the cuvée wine for later use as a base for making finishing *dosage*, which will be discussed later in this chapter.

After an accurate measurement of the cuvée wine has been made and double-checked, the winemaker may make the necessary calculations for his batch.

Sugar

Cane sugar (sucrose) is preferred to dextrose, as dextrose can be up to 10% unfermentable and can be difficult to ferment in the sparkling wines process. Grape juice concentrate can have considerable solids and unstable tartrates that could prove troublesome in the later clearing operation.

The winemaster may consider that about 5 psi will develop for each 0.1° extract added to the cuvée wine. In other words, if one starts with an extract of 2.1° and this is raised to 4.1° extract with cane sugar (an addition of 2.0° extract), then about 100 psi should be generated in the secondary fermentation (20 × 5 psi). This is fairly typical of the pressures observed in commercial manufacture of sparkling wine by the méthode champenoise. Note that this pressure is quite high. Special care must be exercised in calculating the sugar additions, as well as when the sugar is actually measured and added in the cellars. Adding insufficient sugar may provide too little pressure for later disgorging operations to be carried out properly. Too much sugar will certainly overload the sparkling wine with dissolved CO_2 gas, making it impossible to handle during the finishing processes.

Nitrogen and Vitamins

Primary fermentations usually use all, or nearly all, naturally available nitrogen, an essential element for yeast growth and reproduction. The most common nitrogen sources used by winemasters today are specially designed products that also contain vitamins required by yeast cells. Two such materials are *Actiferm* and *Yeastex* as approved and listed in Subpart ZZ, Part 240, CFR, by the ATF. However, the limit for the use of these nitrogen sources is 2 lb per 1000 gal., which may often provide insufficient nitrogen for some secondary fermentations. The same is true for urea, another good source of nitrogen. Despite these shortcomings, some yeasts can produce enzymes for *deamination* reactions that can provide sufficient nitrogen to complete fermentations, although at a much slower rate. A nitrogen supplier should be consulted to determine the most economical approach to this problem.

Cuvée batches should be analyzed after all ingredients are thoroughly mixed according to calculations. The batch should continue to be gently mixed until the laboratory analyses confirm that specifications are met.

Yeast

The completed batch, temperature-adjusted to about 65°F, should be inoculated with a rapidly growing pure yeast culture. The yeast culture chosen should be one of superior viability, capable of withstanding the shock of being introduced to a medium that is about 10.5% alcohol. Several strains are produced and marketed commercially in the dehydrated form. Most experts agree that Champagne and Fermivin strains are superior choices.

A good fresh yeast source should be added at a rate of about 2 lb per 1000 gal. This is done by taking about 15 gal. of the cuvée batch, heating it to approximately 95°F and slowing adding the 2 lb of dehydrated yeast while agitating vigorously. This forms the yeast stock suspension which is added to the batch under gentle agitation. When adequate mixing has taken place, a sample should be removed, analyzed again to double-check batch specifications, and a fermentation tube test set up as described in Chapter 4. Once fermentation is apparent, it is time to bottle the entire batch.

The transfer of cuvée wine from aging cooperage to the batch is recorded in the bulk wine book inventory and the *tirage record*, as shown in Figs. 8.1 and 8.2, respectively.

FILL

The bottling and capping of the cuvée batch is called the "fill". Preparations for the bottling of still wines (Chapter 6) apply also to the planning done prior to the fill of the cuvée batch. This includes hot water cleaning and re-cleaning of bottling tanks and filling machines. Adequate supplies of new champagne bottles and crown caps, as well as sufficient manpower, should be anticipated, to assure that the entire batch can be completely filled in one working day.

The pressure-glass bottles used must be hot water clean, and free of chips, scratches and other imperfections which might weaken them. Once the cuvée has begun to ferment in the special bottling, or "fill", tank, the bottles may be filled. The bottling tank(s) used for still wines should not be used for the fill in order to avoid any possibility of yeast contamination when returned to still wine usage.

The cuvée wine must be continually agitated throughout the fill in order to keep yeast cells in homogeneous suspension. Bottles should be filled to normal fill-height, 750 ml, and immediately crown-capped before any dust or other airborne contaminants can be introduced.

Top quality crown caps are necessary to provide a secure seal that is inert to the cuvée wine acids and other organic constituents. Poor quality crown caps can rust and yield to pressure. Equally important is that the crowns be properly applied with a machine that is designed and built for the purpose.

TIRAGE

In France the term, *tirage*, refers to the addition of sugar, nitrogen, vitamins and yeast for secondary fermentation. In America the *tirage* also includes the laying of filled bottles on their sides in rows, and stacking them, row upon row. A pile of tirage is pictured in Figure 8.3.

Tirage is not only an attractive means of storage, but also practical. The "layering" is a more efficient use of space than case-storage would be. Plus, there are always some bottles that will explode if fermentation proceeds as planned. In tirage, such explosions are usually relatively insignificant (0.2% or so may be expected). In case storage, however, such explosions can ruin the entire case, due to glass splinters and wetness. If several such explosions should occur at the bottom of a pallet full of fermenting cuvée, the entire pallet may fall.

Whether the fill is stacked in tirage or in cases, the inventory should be kept on the cased goods book inventory, despite the fact that the wine is not

| Lot No. | 819 | Type-Variety | WHITE CUVÉE | Class | -14% | Color | WHITE |

Remarks: Blended 7-11-79   50% Chardonnay   50% Seyval Blanc

Date	Racked From Tank W.O. Gallons	Treatment-Disposition	Racked To Tank W.O. Gallons	PHYS INV	ALC	BALL	T.A.	V.A.	FSO₂
10-31 1979	from previous card		S-38 79-268	950	✓	10.6	-1.8	.630	.036	8
11-30 1979			S-38	950	✓	10.5	-1.7	.630	.023	8
12-12 1979	S-38 79-319 950	to file	FT-1 79-319	945					
	to dead file 12-12-79								

FIG. 8.1.   BULK WINE BOOK INVENTORY FORM IN USE FOR AGING AND TRANSFER OF CUVÉE WINE TO A BATCH

TIRAGE RECORD											
CUVÉE											

Date 12-12-79	Lot Number 819		Vintage 1978			Type-Variety BLANC DE BLANCS					
From Tank(s) No. S-38	Gallons 950		To Filling Tank(s) No. FT-1			Gallons 945			Received By—Date: TRL 12-12-79		

Pre-Final Filtration Analysis:	Alc. 10.5	Ball. -1.8	Ext. 1.8	T.A. .623	V.A. .033	FSO₂ 4	TSO₂ 64	Color Int. 80/425	O₂ 4	C ✓	C ✓	N ✓	T ✓

Secondary Fermentation Ingredients:	Sugar: 160 lbs.	Nitrogen: 15 oz.	Vitamins: YEASTEX 61	Other: —

Post-Final Filtration Analysis:	Alc. 10.4	Ball. 0.2	Ext. 3.8	T.A. .615	V.A. .030	FSO₂ 4	TSO₂ 60	Color Int. 89/425	O₂ 4	C ✓	C ✓	N ✓	T ✓

Filling Data:	Yeast: CHAMPAGNE 1 lb. 15 oz.	Vault No. 2	Temperature: 64° F	Other: —

Post-Filling Analysis:	Alc. 10.4	Ball. 0.2	Ext. 3.8	T.A. .615	V.A. .030	FSO₂ 4	TSO₂ 60	Color Int. 80/425	O₂ 4	C ✓	C ✓	N ✓	T ✓

Bottles Filled: 4,789	Size: .750 LITER	Gallons: 948.9	Tirage Pile(s) No. 16, 17 & 18	Received By—Date: VB 12-13-79

TIRAGE							
	Press.	Temp.	Vols.	Press.	Temp.	Vols.	Remarks:
1st Week:							
2nd Week:							
3rd Week:							
4th Week:							
5th Week:							
6th Week:							
7th Week:							
8th Week:							

Post Fermentation Analysis:	Alc.	Ball.	Ext.	T.A.	V.A.	FSO₂	TSO₂	Color Int.	O₂	C	C	N	T

Riddling Data:	Date to Tables:	Table(s) No.	Broken:	Received By—Date:

FIG. 8.2.   TIRAGE RECORD FORM IN USE FOR RECORDING A CUVÉE BATCH

FIG. 8.3.   WINE BOTTLES STACKED IN TIRAGE

finished or marketable. The rationale is that the fill is *not* still in bulk storage but is in bottles—the bottles being the unit of inventory control. Figure 8.4 illustrates the cased goods book inventory in use following the transfer of the fill from the batch exemplified in Fig. 8.2.

In tirage, as fermentation nears completion, the yeast cells will *autolyze*—physically break down, causing cell walls to crack and protoplasm from inside the cells to diffuse into the wine. This accounts for the fresh yeasty flavor that better examples of méthode champenoise sparkling wines have. When case storage is used, the bottles are on end rather than horizontal so that precipitated yeast cells are confined to a smaller area and there is less exposure of the wine to yeast cell protoplasm.

The length of time needed for secondary fermentation can vary widely— from under two weeks to much longer. This phenomenon is due to yeast viability in combination with temperature and other factors. Many winemasters choose tirage temperatures ranging from about 60–65°F. Whatever temperature is chosen, it should be as constant as possible.

The tirage record should be updated at least twice weekly in the partial analysis of cuvée wines undergoing secondary fermentation. Progress can be effectively monitored by the determination of carbon dioxide gas *volumes*, an industry term for *atmospheres* of CO_2 pressure. An atmosphere is equal to about 14.69 psi at standard temperature. As the secondary fermen-

Lot Number 819	Vintage 1978	Type-Variety BLANC DE BLANCS			Size .750 LITER
Class SPARKLING	Color WHITE	Minimum Inventory —	Bottle Type CHAMPAGNE	Closure CROWN	Other —
Date	Cases In	Cases Out	Balance on Hand	Ref. No.	Inv.	Remarks
12-13-79	399	—	399	VAULT #2		piles 16,17+18

FIG. 8.4.   CASED GOODS BOOK INVENTORY CARD IN USE FOLLOWING TRANSFER OF FILL TO TIRAGE

tation progresses in tirage, pressure will vary and, to a much lesser extent, so will temperature. The use of a piercing device, that can measure both of these functions simultaneously, can provide the information necessary to determine CO_2 volumes.

Zahm and Nagel Piercing Device

Despite its simplicity, the Zahm and Nagel piercing device, as shown in Fig. 8.5, is a rather expensive tool for the small winery laboratory. The sample bottle from tirage is placed directly below the hollow needle that extrudes on the underside of the crossbar. With the bottle exactly centered, the crossbar is driven down firmly so that the hollow needle pierces completely through the crown cap. It is held in that position by the binding clips on either end of the crossbar. The device and bottle are picked up and inverted, perhaps a dozen times, in order to establish an equilibrium between the carbon dioxide in solution and the gas in the head space. The bottle is then set upon the laboratory countertop, and the pressure gauge read and recorded. The thermometer is then pushed downward through the needle into the wine so that an immediate temperature can be recorded. These two results are posted routinely in the laboratory analysis log, and are used to determine the volumes of CO_2 gas from the table provided in Appendix B. All results should also be posted in the tirage record for each wine being analyzed.

The piercing device may be relieved of its pressure, after the thermometer is raised from within the sample bottle, by turning the small valve at the rear of the crossbar. Most analysts attach a short piece of tubing to the outlet valve spout so that it can be drained into the laboratory sink with minimal spatter. The piercing device is then thoroughly cleaned and rinsed with hot water several times before securing. The sample bottle of tirage should be destroyed, as much of its carbon dioxide gas will have been lost in the testing procedure. An example of progress in secondary fermentation, as recorded on the tirage record, is shown in Fig. 8.6.

There should be no attempt to finish sparkling wines in tirage until fermentation has been complete for at least three months. This will provide ample time for the CO_2 to dissolve and help to make handling much more efficient. Winemasters often age their tirage for several years prior to *riddling* and *disgorging* in order to maximize "yeastiness". Such care and expense are normally reserved only for the classic white vinous types of champagne. The more fruity types of sparkling wines may well suffer from excessive exposure to the products of yeast autolysis.

RIDDLING

The dangerous task of shaking the yeast sediment from the sides of the bottles (which have been laid horizontal in tirage) down into the bottle neck is called the *remuage* in France, and *riddling* in America. This is also called the clearing process.

FIG. 8.5. THE ZAHM AND NAGEL PIERC-
ING DEVICE

For safety during riddling, a mask that fully covers the face and throat is
an absolute necessity. Failure to wear such a device after appropriate
instruction and warning should result in dismissal of a riddler. This matter
should be constantly stressed when the winemaster conducts periodic meet-
ings with the cellarworkers. Along with the mask, the riddler should wear a
high collar, long sleeves, and long pants. Thick soled, non-slip footwear
should be worn as well. All of this is, of course, to prevent serious injury from
broken glass.

In "A" tables similar to those pictured in Fig. 8.7, the riddler, several
times daily, raises each bottle of sparkling wine an inch or two above its
resting place, gives it a quarter-turn, and then firmly replaces the bottle
into the receptacle. This action jolts and spirals the sediment down into the
bottle neck, usually within a month or two.

There is no simple analytical method to make sure that this step is
properly progressing, other than by the judgment of the trained eye. Such
testing is usually performed by selecting several random samples for very
close visual examination in the laboratory using a candle or unfrosted bulb.

TIRAGE RECORD

CUVÉE

Date 12-12-79	Lot Number 819	Vintage 1978	Type-Variety BLANC DE BLANCS				

From Tank(s) No. S-38	Gallons 950	To Filling Tank(s) No. FT-1		Gallons 945	Received By—Date: TRL 12-12-79		

Pre-Final Filtration Analysis:	Alc.	Ball.	Ext.	T.A.	V.A.	FSO₂	TSO₂	Color Int.	O₂	C	C	N	T
	10.5	-1.8	1.8	.623	.033	4	64	88/425	4	✓	✓	✓	✓

Secondary Fermentation Ingredients:	Sugar: 160 lbs.	Nitrogen: 15 oz.	Vitamins: YEASTEX 61	Other: —

Post-Final Filtration Analysis:	Alc.	Ball.	Ext.	T.A.	V.A.	FSO₂	TSO₂	Color Int.	O₂	C	C	N	T
	10.4	0.2	3.8	.615	.030	4	60	89/425	4	✓	✓	✓	✓

Filling Data:	Yeast: CHAMPAGNE 1 lb. 15 oz.	Vault No. 2	Temperature: 64°F	Other: —

Post-Filling Analysis:	Alc.	Ball.	Ext.	T.A.	V.A.	FSO₂	TSO₂	Color Int.	O₂	C	C	N	T
	10.4	0.2	3.8	.615	.030	4	60	88/425	4	✓	✓	✓	✓

Bottles Filled: 4,789	Size: .750 LITER	Gallons: 948.9	Tirage Pile(s) No. 16, 17 + 18	Received By—Date: VB 12-13-79

TIRAGE

	Press.	Temp.	Vols.	Press.	Temp.	Vols.	Remarks:
1st Week:	—	65°	1.0	—	64°	1.0	
2nd Week:	4	65°	1.2	8	64°	1.4	fermentation is apparent
3rd Week:	15	66°	1.85	23	65°	2.6	
4th Week:	39	65°	3.35	54	64°	4.2	
5th Week:	70	64°	5.2	78	65°	5.7	
6th Week:	86	66°	6.3	94	66°	6.9	
7th Week:	101	65°	7.3	99	65°	7.1	
8th Week:	98	66°	7.0	100	64°	7.2	

Post Fermentation Analysis:	Alc.	Ball.	Ext.	T.A.	V.A.	FSO₂	TSO₂	Color Int.	O₂	C	C	N	T
	11.4	-1.8	2.1	.608	.036	4	48	89/425	—	✓	✓	✓	✓

Riddling Data:	Date to Tables: 5-6-80	Table(s) No. 25-51	Broken: 41	Received By—Date: DEC 8-19-80

FIG. 8.6.   TIRAGE RECORD FORM MAINTAINED THROUGH SECONDARY FERMENTATION PROGRESS BY MEASURING CO_2 VOLUMES

FIG. 8.7. "A" TABLES FOR RIDDLING

DISGORGING

Once the winemaster has determined that the sparkling wine batch is ready for finishing and the riddling is complete, a representative sample should be completely analyzed. As usual, the analytical results are posted in the laboratory analysis log and the appropriate tirage record. Note that, apart from determinations of CO_2, the tirage and riddling processes are not monitored by the usual methods of analysis such as alcohol, Balling, total acidity, volatile acidity, etc. Once secondary fermentation commences, most, or all, of the remaining oxygen in the cuvée will be utilized by the yeast, resulting in an anaerobic atmosphere that is quite stable inside the bottle. Consequently, until the riddling process is completed, there is no reason for routine analysis of the fermented cuvée batch.

Disgorging is the process of chilling the sparkling wine bottles down to approximately 25–30°F and freezing an ice-plug in the neck of each bottle next to the crown cap. This traps the sedimentation in the ice-plug. The disgorger then removes the crown cap in an "alcove" device (usually a barrel partially cut away) so as to reduce the spatter of the ice plug report. The disgorged bottle is immediately transferred to the finishing *dosage* machine. This machine will measure and add a specific amount of dosage (sweetener and preservative) in equilibrium with the closed system. The volume of dosage is designed to replace the volume lost by removal of the ice plug. The sparkling wine is then corked and wire-hooded so that the cork cannot be blown out by the remaining CO_2 gas pressure.

It is not necessary for the entire batch of cuvée to be finished from riddling in the same day. For the sake of consistency, however, the winemaster should insure that significant lapses of time do not take place between disgorging runs of the same batch.

During the initial stage of the finishing process, the chilling is easily carried out in cases with the bottles upside down, so that the sediment remains in the neck next to the crown cap. The temperature should be monitored carefully, as excesses over 30°F may render the sparkling wine "wild" and difficult to finish. This very cold temperature reduces the pressure (but not the volumes of CO_2), so that loss and waste are minimized.

Freezing the ice-plug is performed many different ways, from ice and salt to propylene glycol-refrigerated tables. An experienced winemaster can determine when an ice-plug is ready for disgorging. The actual disgorging process takes a bit of practice and skill.

DOSAGE

There are as many different formulae for making *dosage* as there are kinds of sparkling wines. Each winemaster will have his own method and formulation.

The normal dosage volume added to each bottle of disgorged sparkling wine is about 50 ml. This volume should contain the proper amount of preservative and sweetness necessary for the entire volume of 750 ml in each bottle of sparkling wine. Consider the following dosage formula:

Cuvée wine   85% (saved from the original cuvée)
High-proof brandy   15%
    Total   100%

Add cane sugar so as to raise the extract to 40°, and SO_2 to maintain a level of about 800 ppm.

A 50 ml per bottle addition of this dosage should do the following:

(a) Raise alcohol from 11.5% by volume to about 12.0%.
(b) Raise extract from approximately 1.8° to about 4.7°.
(c) Raise free SO_2 from 0 to about 50 ppm.

These figures are just examples and will have to be adjusted in actual practice. Each winemaster must calculate the optimum dosage for the finished sparkling wine product. All calculations are made from weighted averages, with the extract being determined by referring to the solids per gallon table provided in Appendix B.

The administration of dosage is prescribed in a work order, along with quality control analyses for alcohol by the distillation method, extract, and free SO_2. These methods are detailed in Appendix A.

Dosage must be added with a machine or device that is designed expressly for that purpose. If one were to try adding dosage to a sparkling wine by hand, equilibrium would be violently disrupted, causing excessive frothing that will nearly empty the bottle. The dosage machine provides a closed system to contain the reaction, allowing equilibrium to develop before the cork is driven.

CORKING

The cost of natural cork has risen so much recently that many vintners have chosen to use the plastic stoppers now available. An additional advantage of these stoppers is that they are often driven only by means of a small mallet, thus dispensing with the expensive corking machine. Plastic stoppers do not add much in the way of aesthetic appeal, however, and are sometimes thought to represent poor quality sparkling wines.

Natural corks require careful selection. The best are laminated to avoid imperfections. Some have a thin layer of paraffin applied around the "belt" of the cork in order to improve sealing capabilities. Others are made of agglomerated cork, and are available in a number of slightly different shapes and styles. Some suppliers even offer closures made from both cork and plastic. The winemaster is encouraged to try different types before placing a large order.

A sparkling wine must be finished with a closure operation that eliminates the possibility of leakage, which would result in a lifeless, unmarketable product.

Some vintners soak their natural corks in a solution of 85°F clean water with 2−4% glycerin, for about 1½ h. Corks soaked for shorter periods with less glycerin may be difficult to remove, even after several months. Soaking in the solution for longer periods with increased glycerin may result in a high percentage of "leakers" and difficulty in keeping the cork in the bottle long enough to apply the wire hood. The winemaster and finishing foreman may wish to experiment to determine the optimal combination of temperature, glycerin and time. In any event, soaked corks should be well drained before using, in order to avoid "cork-water" being squeezed into the sparkling wine by the corking machine.

WIRE HOODING

There are a number of different types of wire hoods available. It may be worthwhile to try several that are reasonably priced and choose the one that seems to work with the least difficulty, both in application and removal. Some of the very inexpensive types will not endure the "reverse-twist" necessary for removal—the wire may break under the stress. The wire hood may perform properly when applied by one machine and fail when utilized in another device because of different stresses and strains. Again, the best method is trial and error, accomplished with the cooperation of a reputable supplier.

Generally, two samples are taken to the laboratory during each disgorging and finishing day, one in early morning and the other in early afternoon. Each of these samples should be crown-capped, rather than corked and wire-hooded, in order that the CO_2 volumes may be measured accurately with the Zahm and Nagel piercing device. The complete analysis for the *daily disgorging record* is a composite of both these separate analyses. In addition, most winemasters will require a microbial and stability examination, similar to that described for table wines in Chapter 6.

The administration of disgorging and finishing operations should be recorded on the daily disgorging record (Fig. 8.8).

FINISHED SPARKLING WINES

In some small wineries the winemaster may require that finished sparkling wines be stored in the cellars in piles or bins for several months prior to any further packaging. The reason for this is to allow the wine to rest, and to watch for the phenomenon of bottle-sickness described in Chapter 6. Intermittent checking for leaking corks is done by simple visual examination, and the analyst should also scan the bottles for signs of tertiary fermentation and haze or sediment development.

However, today most sparkling wine cellars dispense with piling or binning. The finished wines are placed in cases, in the same manner as table wines, directly from the bottling line but without packing materials applied. The cased goods book inventory account should be entitled "finished unlabeled". The newly finished wines are observed for leakage, etc., as described in the previous paragraph. After a desired period of storage, perhaps three months or more, the wine is packaged with labels and capsules, and transferred to the "finished labeled" account in the cased goods book inventory. It is then ready for warehousing and shipment. Four samples are taken for storage in the reference library in the same manner as was described in Chapter 6.

WAREHOUSING

Sparkling wines should have priority for the most consistently cool warehouse space. Cool temperatures and lack of movement help to minimize pressure stresses on the cork. Other storage and inventory practices are the same for sparkling wines as for table wines (see Chapter 6).

SUMMARY

The making of sparkling wines is a highly demanding and costly operation. Apart from the dangers of working with glass under high pressure, the tax liability is 20 times that of table wines. The vintner is advised to carefully consider his potential and the problems involved before applying for ATF permission and formula approval.

While several methods of sparkling wine manufacture exist for larger-scale wineries, most small wineries choose the traditional French méthode champenoise in order to simplify handling operations and reduce capital outlay for equipment. Most experts also agree that this method produces the finest sparkling wines.

DAILY DISGORGING RECORD

Date 8-19-80	**Lot Number** 819	**Vintage** 1978	**Type-Variety** BLANC DE BLANCS
From Riddling Table(s) No. 25-32	**Bottles:** 1,044	**Gallons:** 206.9	**To Cold Room No.** @ Temp. C5-1 @ 27°F	**Received By—Date:** DEC 8-19-80

Pre Disgorging Analysis:

Alc.	Ball.	Ext.	T.A.	V.A.	FSO₂	TSO₂	Color Int.	O₂	C	C	N	T
11.5	-1.9	2.0	.608	.036	4	48	89/425	—	✓	✓	✓	✓

Disgorging Data:

Pressure: 99 | **Temperature:** 64°F | **Volumes CO_2:** 7.05 | **Remarks:** —

Dosage Data:

Amount per Bottle 50 ml | **Lot Number Used:** D80-3 | Alc. 14.1 | Ext. 40.6° | FSO₂ 788 | **Remarks:** —

Post Disgorging Analysis:

Alc.	Ball.	Ext.	T.A.	V.A.	FSO₂	TSO₂	Color Int.	O₂	C	C	N	T
11.9	1.8	4.7°	.615	.039	48	92	88/425	—	✓	✓	✓	✓

Material and Supply Usage:

Corks: DODGE lemmetzel | **Wire Hoods:** Buller + west. | **Cases:** OI rectingman | **Other:**

Production Data:

Cases: 86 2 bot. to lab. | **Gallons:** 204.9 | **Serial Numbers:** 16,707 – 16,792 | **Received By—Date:** VB 8-20-80

FIG. 8.8.   DAILY DISGORGING RECORD FORM IN USE

THE "DOSAGE"

Secondary fermentation calculations and analyses should be recorded in a well designed tirage record so that exact control of production variables is maintained.

There are two specific points to be made regarding the processing of sparkling wines. The entire operation requires absolute cleanliness, in order to avoid the complications of interfering microbial growth. The other point is that of safety. Specially designed masks that cover the face and throat should be worn by riddlers and disgorgers, along with other appropriate clothing, to eliminate the hazard of broken glass.

Riddling and disgorging operations require semi-skilled labor that should be trained to handle these duties efficiently. Dosage, corking, and wirehooding materials are critical in the production of a sparkling wine product that will be successfully marketed.

A sampling should be made twice daily during the disgorging operation in order to insure proper levels of preservative, sweetness, volumes of CO_2 gas, and other pertinent factors. The entire disgorging and finishing of sparkling wines should be administered and recorded in a daily disgorging record.

Finished sparkling wines should be placed unlabeled back in piles or bins (or cases) so that the possibility of leaking corks and tertiary fermentation can be monitored. This also provides a rest period for the dosage to properly blend with the sparkling wine.

After several months in the "finished unlabeled" account, the wines may be packaged, sampled and transferred to the "finished labeled" account in the cased goods book inventory, ready for shipment.

General warehousing and inventory techniques are much the same for sparkling wines as for table wines.

Courtesy of Schieffelin & Co.
New York, N.Y.

STATUE OF DOM PERIGNON

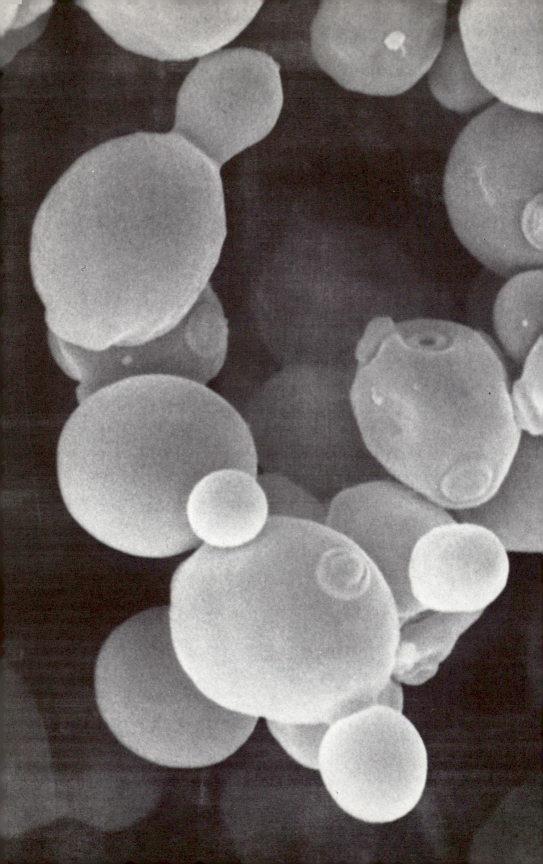

9

Microbiological Analysis in the Small Winery Laboratory

More often than not, the difference between ordinary and good, or great, wines can be measured by the winemaster's knowledge and control of microorganisms in the cellars. A wine may be totally destroyed by ignoring the overwhelming potential of microbial action.

Apart from the investment required for a good microscope, the capital outlay for microbiological analysis is minimal.

The scope of this chapter shall be to acquaint the reader with some of the common bacteria, molds and yeasts that are found in winemaking, and to establish a framework upon which the winemaster may build further knowledge.

GRAM STAIN

Although there are millions of different types of microorganisms, there is one procedure that can be used to help identify certain groups. *Gram* staining was discovered accidentally by Christian Gram in 1884. Organisms that turn violet-black are called *gram-positive*, while those that stain a pink or red color are *gram-negative*. The age and condition of microorganisms may influence the results. No procedure or technique can yield finite results every time unless highly technical equipment is employed. In any case, gram staining remains a very important tool in the small winery laboratory. For instance, *Acetobacter aceti*, the vinegar producing bacteria, will stain gram negative, while *Lactobacillus brevis*, a bacteria that ferments malic acid to lactic acid, is gram positive. Both are bacilli (rod-shaped bacteria) and may be difficult to distinguish without this staining procedure. Obviously, the winemaster will want to take immediate action to inhibit the vinegar bacteria, while he may opt to encourage growth of the malo-lactic bacteria. The procedure for gram-staining can be found in Appendix A.

Scanning electronmicrograph of Saccharomyces ellipsoideus
Courtesy: Michael Sullivan, SEM Mississippi State University

DIFFERENTIAL STAIN

The primary purpose of *differential staining* is to determine the viability of yeast cells in a culture. This rather simple procedure, described in Appendix A, distinguishes dead cells from live ones, in that dead cells will stain with methylene blue, and viable cells will not.

This is used to determine whether or not stored yeast cultures are sufficiently active to carry out normal fermentation. Lagging fermentations and other abnormal reactions may also be examined for percent viability of yeast cells. For example, several samples may exhibit microscopic counts averaging 75 viable cells and 25 stained, or dead, cells per 100 yeast cells counted. Such a culture is considered 75% viable. Cultures exhibiting less than about 50% viability should be destroyed and replaced with fresh yeast from reputable suppliers.

PLATING

Many winemasters "plate-out" samples of juice, wine, or other materials in order to observe the different kinds of microorganisms that may be present. A precisely controlled amount of sample is placed on a sterile medium, usually agar, and incubated. This procedure is especially helpful in testing the sterility of post-bottling samples as described in Chapter 6. The full technique for plating is provided in Appendix A.

The colonies formed by bacteria will be very different in size, color and texture from those of molds or yeasts. The size of the sample inoculated will determine the number of colonies that develop. Dilutions of 1/1000, 1/10,000 and 1/100,000 are commonly utilized for inoculations in order to effectively separate the colonies across the medium in the petri dish (a glass or plastic covered dish used for plating). Plating is used both quantitatively and qualitatively to ascertain the numbers and types of viable microbe(s) in the sample. For example, if 100 colonies developed from the 1/1000 dilution, and 10 colonies developed from the 1/10,000 dilution, then we could conclude that our sample was comprised of approximately 100,000 cells per ml. The winemaster may need to continue dilution as it is common to find more than 3 million yeast cells per ml in the growth phase of a healthy culture.

Developing colonies also allows for separation of different microorganisms. A sample from any given colony in a petri dish may be "picked" with a sterile loop and inoculated free of contamination into another sterile medium. This medium can then serve in the propagation of pure cultures for immediate use, or the pure culture can be stored in agar slants (test tubes partially filled with agar media, inoculated, cultured, and then refrigerated; see Appendix A).

MODERN QUALITY CONTROL METHODOLOGY

The winemaster can find a number of commercial systems available nowadays which simplify qualitative and quantitative microbiological analysis. This is especially helpful in the determination of contaminants in bottled wines. Neradt and Kunkee have developed the *Rapid Method for Detection of Viable Yeast in Bottled Wines*, which is described in Appendix A. In general, microorganisms are collected on a membrane, stained with two separate stains and microscopically examined after each staining. The dead cells are differentiated from the total number of yeast cells so that the viable cell count can be determined by difference. This method is especially useful in that it can determine minute insterility of wine in less than one hour; hence, bottling operations can be stopped so that a minimum of time and materials are wasted.

A more refined membrane apparatus is available from the Millipore Corporation. However, this system requires several days of incubation to determine the microbiological state of a wine, which is its chief disadvantage. Its advantages are in the form of quantitative determinations, and in its ability to trace and locate sources of infection. Procedure for this analytical system in presented in Appendix A.

THE MICROSCOPE

As has already been discussed, a good microscope in the laboratory is indispensable for effective analysis of microbial activity.

Modern compound light microscopes, such as the one pictured in Fig. 9.1, can magnify images between 100 and 1000 times (100 × and 1000 ×). All the microorganisms that are important to practical winemaking quality control can be seen in this range. Electron microscopes can reach magnifications in excess of 100,000 ×, however, their cost is prohibitive. Figure 9.2 provides a scanning electronmicrograph of a yeast cell, *Saccharomyces ellipsoideus*, var. Champagne, magnified 8,000 ×.

It goes without saying that even larger wineries cannot justify the full-time use of a scanning electron microscope. A simple, modern compound light microscope is much more practical, but even so, will be an expensive instrument to buy, and must be operated with the utmost care. Appendix A describes the proper use and care of the microscope.

YEAST

Yeasts are lower-form plants which have no leaves, stems, roots or chlorophyll. They are unicellular and have no locomotion of their own. Rather, they are mobilized by natural forces which may exist in water and wine. The manipulation of wineyard equipment by man is a mechanical source of yeast mobility.

Wild yeasts may exist as airborne cells or may be found in the soil and on plants in a vineyard. One of the most common places to find large concentrations of wild yeasts is on ripening grapes, as part of the *bloom* (waxy coat—cutin) on the outermost surface of each berry. Once the berry is broken so that the yeast is exposed to the interior sugars, fermentation begins. For that reason, wine is often referred to as a natural beverage. Nowadays, however, the existence of such wines is, fortunately, very rare.

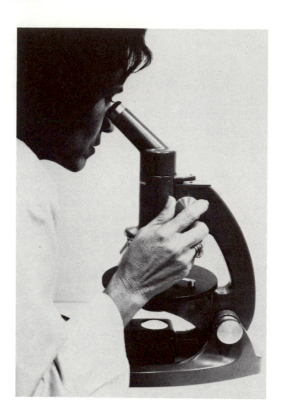

FIG. 9.1. A MODERN COM-
POUND LIGHT MICROSCOPE

The term yeast may be derived from the ancient Greek *Zestos*, meaning to "boil without fire". Wine yeasts, as mentioned previously, are plants, being thallophytes and belonging to the subphylum *Fungi*. Further taxonomic classification follows:

Class:	Eumycetes
SubClass:	Ascomycetes
Order:	Endomycetales
Family:	Saccharomycetaxeae
Subfamily:	Saccharomycoideae
Genera:	*Saccharomyces*
	Dekkera
	Candida
	Kloeckera
	Pichia

Wine production is carried out with strains of the genus, *Saccharomyces*. *Saccharomyces cerevisiae* is grown for juice and must fermentations, as well

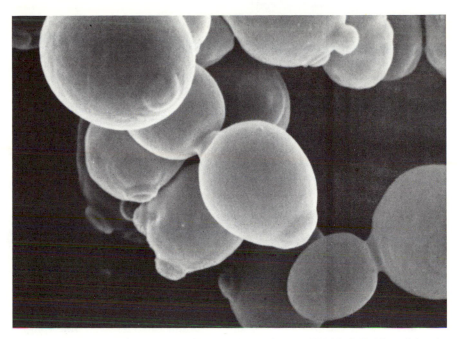

Courtesy: Michael Sullivan, SEM Mississippi State University

FIG. 9.2.   SCANNING ELECTRONMICROGRAPH OF *SACCHAROMYCES ELLIPSOI-DEUS*

as for secondary sparkling wine fermentations. The genus *Dekkera* resembles *Saccharomyces cerevisiae* in several ways, but is considered a spoilage yeast.

Saccharomyces beticus is a surface-growing yeast, also described as a "film", or "Flor" yeast. It is used principally for the production of Sherry-type wines which require high levels of acetaldehyde, a product of such surface-growth yeasts (see Chapter 7). It is essential that the winemaker handle Flor yeasts with the utmost of care. Cellar infections of such yeasts may convert all table wines into sherry-like products, or destroy the wines.

The genera *Candida, Kloeckera,* and *Pichia,* among others, are also surface-growing yeasts, but are spoilage yeasts. They are sometimes referred to as wild, or *apiculate*, yeasts. The winemaker must be able to recognize and destroy these in order to avoid ruinous consequences.

Grapes may carry many different species of yeasts and other microbes upon arrival at the winery. Most winemasters treat the newly arrived fruit with 50−75 ppm of free SO_2 just prior to crushing (see Chapter 3). This treatment normally inhibits the wild yeasts and other microorganisms present on the bloom. After crushing and any other immediate processing, inoculation with cultured yeasts can take place. Most cultured strains are somewhat acclimatized to the effects of free sulfur dioxide in lower concentrations. This allows the cultured yeast to dominate the fermentation and overgrow any competing wild yeasts that may remain viable. Fermentation may then be anticipated to proceed in a predictable fashion. Fermentation arising from wild yeasts will require stopping the undesired fermentation with gross amounts of sulfur dioxide and/or pasteurization, and inoculating the remains with an abnormally heavy population of cultured yeast cells. Such operations almost always result in an inferior product. In addition, the time and effort required to repair such fermentations is much more costly than simply looking after the sulfur dioxide requirement at the crusher.

WINE YEASTS

Two of the most common strains of wine yeasts commercially available today are the Pasteur strain of *Saccharomyces cerevisiae*, var. bayanus, and a cultivar by the proprietary name of Fermivin. Both are available in dehydrated form with approximately 10 trillion cells per pound. An inoculum of 2 lb per 1000 gal. results in approximately 5 million cells per ml, which is usually sufficient for wine fermentation.

Bayanus is more commonly known as "Champagne" yeast, which does not imply that it is limited to secondary Champagne purposes. Champagne yeast has proven to be a fine strain for fermenting both white and red wines, as well as for secondary fermentation of cuvées. The dehydrated form of this yeast should first be suspended in an appropriate amount of warm (100–105°F) water for a few minutes, and then added to a rather small amount of the juice, must or cuvée to be fermented. Once about half of the sugar solids in this small volume have been fermented, this starter culture is ready to be added to the entire lot. Reputable active dry yeast suppliers will provide more detailed instructions for handling and usage.

Fermivin is a newer development, thought to be somewhat superior to Champagne yeast by about as many winemasters as those who think the opposite. Fermivin is added directly to the juice, must or cuvée, eliminating the need for warm water suspension and starter activation. On the other hand, many feel that Champagne yeast is more susceptible to sulfur dioxide and, therefore, easier to control when SO_2 is used as a preservative just before bottling.

Figure 9.3 illustrates the morphology of a wine yeast cell.

Wild Yeasts

As mentioned previously, wild, or apiculate, yeasts are potentially devastating and must be controlled. These microorganisms may arrive at the winery in concentrations exceeding a million cells per ml! Inhibiting them is absolutely necessary to prevent them from dominating the culture. At a reasonable rate of reproduction, after only 2 h of unchecked activity in juice or must, there may be a population in excess of 10 million cells per

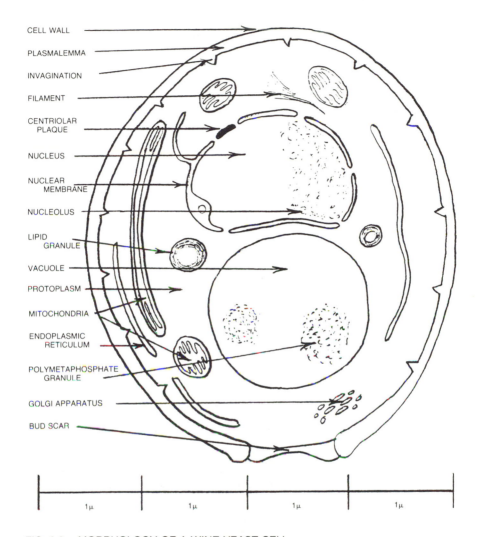

CELL WALL

PLASMALEMMA

INVAGINATION

FILAMENT

CENTRIOLAR
PLAQUE

NUCLEUS

NUCLEAR
MEMBRANE

NUCLEOLUS

LIPID
GRANULE

VACUOLE

PROTOPLASM

MITOCHONDRIA

ENDOPLASMIC
RETICULUM

POLYMETAPHOSPHATE
GRANULE

GOLGI APPARATUS

BUD SCAR

1μ   1μ   1μ   1μ

FIG. 9.3.   MORPHOLOGY OF A WINE YEAST CELL

ml—more than twice the number of cells in an inoculum of 2 lb of cultured yeast per 1000 gal.

Wild yeasts are easily distinguished from wine yeasts under magnification. Wild yeasts are characterized by a variety of irregular shapes with apiculate points, while wine yeasts are egg-shaped, and of consistent form. Some wild yeasts reproduce by fission, while wine yeasts can reproduce either sexually or by asexual budding.

VINEGAR BACTERIA

The most well known bacteria in wine technology are wine spoilage microorganisms from the family *Pseudomonadaceae*, and the genus *Acetobacter* (there are several species—the most common being *aceti*). These bacteria can turn a fine wine into vinegar overnight. Figure 9.4 pictures a microphotograph of a typical *Acetobacter* infection during the growth phase.

Acetobacter are rod-shaped cells that are approximately ½ micron wide and 1 micron long—about half the size, or less, of a common wine yeast cell. They may appear as single rods, in pairs, in short chains, or in long chains. Depending upon the species, they can be either motile by polar flagella or non-motile, and are Gram-negative.

Acetobacter are often called Mycoderma by some winemakers, but that is a misnomer. Pasteur used these bacteria to demonstrate the production of vinegar, the chemical formula for which is provided in Chapter 5. While ethyl alcohol (normally in concentrations less than 16% by volume) is the preferred substrate for the generation of acetic acid, some vinegar bacteria may utilize lactic acid, and even sugars. The cellar that is kept clean and maintained with good programs of quality control does not usually harbor significant problems.

Acetobacter can be introduced from many sources, one of the most common being from contact with fruit flies and other insects. Vinegar bacteria do not form spores, but can remain viable in wooden aging barrels and other

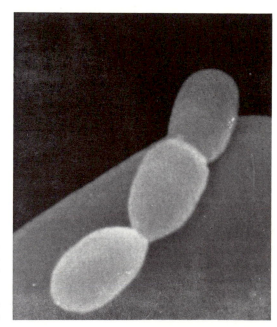

FIG. 9.4. SCANNING ELEC-
TRON MICROSCOPE MICRO-
PHOTOGRAPH OF *ACETOBAC-
TER* INFECTION

Courtesy: Michael Sullivan, SEM
Mississippi State University

areas of the wine cellar for years, waiting for the proper conditions of growth. Fortunately, most bacteria are highly sensitive to free SO_2. Monthly inventory sampling should include a volatile acidity test of each bulk table wine vessel in the cellars. This testing will indicate any activity over the previous month by keeping records on the bulk wine book inventory card, as depicted in Figs. 5.1 and 6.1. The method for volatile acid testing is provided in Appendix A.

LACTIC ACID BACTERIA

Many winemasters mistakenly discount lactic acid bacteria as spoilage organisms. These microbes are acid-tolerant and are readily found in most places. While certain of these bacteria may be desired in quality winemaking, many are deleterious.

Lactobacilli are normally observed as rods in chains, are non-spore formers and non-motile. They are Gram-positive and may be either anaerobic or microaerophilic.

Genus *Pediococci* are spherical and often occur in tetrads. They are non-spore formers and are non-motile. They are Gram-positive and may be either aerobic or facultative.

Leuconostoc genera are spherical cells, occurring in chains, and are non-spore formers. They are non-motile, aerobic and Gram-positive.

Lactobacilli may be either homofermentative, such as the species *casei* and *plantarum*, or heterofermentative, such as the species *brevis, buchneri* and *fermenti*.

Lactobacillus brevis can cause a wine spoilage known as "Tourne", a disorder characterized by an unusual depletion of tartaric acid and the formation of lactic acid, acetic acid and carbon dioxide. The resulting wine, if not totally spoiled and undrinkable, will be insipid and/or sour.

Lactobacillus buchneri, among other species, can effect a spoilage known as *Piqure Lactique*. Pentose sugars (5-carbon sugars not fermentable by wine yeasts) are utilized by these heterofermentative bacteria to form lactic acid, acetic acid, and carbon dioxide. Homofermentative types ferment residual glucose and/or fructose to lactic acid and manitol. The result is a pungent and sour wine.

Pediococcus cerevisiae is a homofermentative spoilage bacterium that can produce histamine from the amino acid histadine. This is more commonly found in red wines. Even in minute quantities, histamine can stimulate blood circulation in the human body. For this reason some wines can be positively affected by the formation of histamine, but it remains classified as a spoilage reaction since it is detrimental to wine flavor.

Leuconostoc mesenteroides is a particularly harmful heterofermentative cocci. It can cause a wine spoilage called *Amertume*, in which the naturally occurring glycerol is broken down into acrolein. The acrolein is very bitter and irritating to nasal membranes. This same bacteria can also cause "ropy" wine, known as *Vins Filant* or *Graisse* in France. The wine becomes rather "oily" and very heavy-bodied as sucrose is polymerized into strands of dextran. Similarly, *Lactobacillus brevis* and *Lactobacillus pastorianus* can form glucan from the dextrose sugar which may remain in some wines.

MALO-LACTIC FERMENTATION

Natural malo-lactic fermentations may occur in virtually all viticultural areas, although cultured malo-lactic bacteria are frequently introduced by winemasters.

The most common strains cultured for malo-lactic fermentation are *Leuconostoc oenos*—ML-34 and PSU-1. Both of these are heterofermentative cocci.

The chemical reaction for the malo-lactic fermentation is as follows*:

$$\begin{array}{c} CH_2COOH \\ | \\ CHOHCOOH \end{array} \xrightarrow[-CO_2 - H_2]{NAD,\ Mn^{++},\ malic\ dehydrase} \begin{array}{c} CH_3 \\ | \\ COCOOH \end{array}$$

$$\xrightarrow[+ H_2]{NADH_2,\ lactic\ dehydrase} \begin{array}{c} CH_3 \\ | \\ CHOHCOOH \end{array}$$

This is a stoichiometric reaction in which very little energy is released.

Whether natural or inoculated, the malo-lactic fermentation is not easily predicted. The reaction may commence immediately or lag for months. Superior growth conditions, such as low levels of free SO_2, pH higher than 3.3, temperature in the range of $50-85°F$ and lower concentrations of alcohol, all contribute toward more dynamic reactions.

Some winemasters choose to leave the lees sediment in contact with the wine so as to obtain nutrients from the autolyzed yeast cells. This, however, can also result in enzymatic reactions which may deteriorate, or destroy, the wine. Aeration is a deterrent to malo-lactic fermentation, despite the fact that the *Leuconostoc* cultured strains may be either aerobic or micro-aerophilic. The winemaster may expect a higher incidence of natural malo-lactic fermentations in wines made from free-run than those made from pressed juice, or must, in that there will be more nutrients available and a higher pH in the free-run product.

The most common benefits derived from malo-lactic fermentation are reductions in total acidity, an increase in pH and a mellowing of tartness. On the other hand, volatile acidity can be expected to increase. Excessive volatile acid buildup may necessitate the arrest of the malo-lactic fermentation with a sulfur dioxide addition and/or the immediate chilling to 30°F, or lower.

Malo-lactic bacteria may be cultured and tested on tryptone-glucose-fructose-yeast extract (TGYE medium) and tomato juice. Inoculations are usually made when the wine is about 5° Balling during primary fermentation.

---

* Amerine, M.A., Berg, H.W. and Cruess, W.V. 1972. The Technology of Wine Making, 3rd edition. AVI Publishing Co., Westport, Conn.

This will insure that sufficient nutrients remain to support the bacterial growth.

A reliable qualitative method for monitoring malo-lactic fermentation progress is paper chromatography. This is a rather simple, inexpensive technique whereby the sample components will separate into "spots" at different distances from a "base line" at the bottom of a special paper when suspended over a solvent/indicator solution. The distance of the spot from the base line is measured and calculated as a ratio known as the "R_f". In turn, the spots can be interpreted as tartaric, citric, malic, lactic, succinic or fumaric acid. Full methodology for this paper chromatographic technique is provided in Appendix A.

MOLDS

Winemasters are sometimes confused by the morphological similarity between some yeasts (particularly some strains of *Saccharomyces cerevisiae* in dormancy) and molds. Certain yeasts have elongated pseudomycelia when dormant. However, culturing these yeasts on nutrient-rich media should quickly restore them to the ovoid shape characteristic during reproductive growth. Culture of the mold should result in true mycelia—branched filaments with single limbs separated by cross-walls rather than elongated cells formed by budding. Consequently, molds are distinguished by the formation of true mycelia.

Grapes may arrive at the winery already infected with a number of different molds, particularly during vintage seasons that have frequent rainfall. Among the most common of these molds are brown rot, black rot, downy mildew and powdery mildew. All of these infections can be controlled in the vineyard with proper use of fungicides. Failure to identify and control these and other molds will most likely result in fruit that is discolored and dehydrated. In addition, some molds create unpleasant odors and flavors. The economics of such conditions necessitates rejection of the fruit by the winemaster.

One favorable species of mold is the *Botrytis cinerea*, often called the "noble mold". Under proper conditions this microorganism can improve the quality of grapes. *Botrytis* usually develops during the ripening season, especially when there is wet weather. The mycelium penetrates the bloom and the skin of the grape berry, causing dehydration which results in fruit of higher sugar concentration. The most noteworthy wines made from Botrytised grapes are the Sauternes from Bordeaux, France, and the *Beerenauslese* and *Trockenbeerenauslese* products from Germany.

Botrytis will also attack tartaric and malic acids in grapes, which aids in mellowing the final wine products. However, the winemaster must be cautioned against allowing the development of the noble mold because of one major disadvantage. The permeation of the berry skin can also allow the entrance of wild yeasts, bacteria and other molds that may deteriorate, or destroy, the fruit. To the naked eye, *Botrytis cinerea* appears on grapes as a fuzzy gray growth.

Molds do not grow in, or on, wines that are properly cellared and preserved. These microorganisms are principally aerobic and, therefore, cause most problems inside wine tanks and in other areas that are not properly maintained. *Penicillium* is a mold that can endure some rather extremely cold temperatures, surviving as a white growth which later may sporulate into a bluish-green powdery appearance.

In the vineyards, *Penicillium* may attack grapes through skin cracks, or permeations, leaving the fruit worthless. This mold may survive on almost any moist surface in the winery, with corks and wooden wine vessels being prime targets. *Penicillium* renders a very unpleasant taste which makes only slight infections sufficient to ruin a wine.

Figure 9.5 provides a table for trouble-shooting the causes of most microbial spoilage occurring in the small winery.

SUMMARY

This book cannot provide a full course in the application of microbiological principles to top-quality winemaking. The winemaker should consider this chapter only as a framework upon which to build a personal expertise in the recognition and handling of microorganisms in the small winery laboratory and cellars.

Different types of staining and propagation procedures are helpful in distinguishing between separate genera and species of microorganisms and, in some cases, determine viability.

The most important tool for the winemaster, however, is the microscope. It cannot function well without proper use and care. A number of common terms are used in the language of microscopy with which the winemaster should be familiar in order to efficiently operate the instrument and interpret the results.

There are a number of microorganisms that are either essential, or helpful, in producing quality wines. Primary fermentations of juice and must, as well as secondary fermentations of sparkling wines, require cultured yeasts of the genus, *Saccharomyces cerevisiae*, primarily the strains Champagne and Fermivin. Yeasts of the genus *Saccharomyces beticus*, or Flor yeasts, may be used under controlled conditions for the production of Sherry-type wines and other madeirized wine products. A culture of *Leuconostoc oenos* bacteria may help to mellow some wines by the malo-lactic fermentation. The mold, *Botrytis cinerea*, can concentrate sugars and mellow acids in certain grapes when weather conditions permit and adequate controls are maintained in the vineyards.

Many other microorganisms, such as *Acetobacter*, can be very destructive. The prepared individual should be able to recognize such microbes as apiculate yeasts, lactic acid bacteria, the acetic bacteria just mentioned, and certain molds. This is absolutely necessary for consistently high-quality winemaking.

	Yeast	Bacteria		
		Acetic Acid Bacteria	Bacilli	Lactic Acid Bacteria
Visual appearance	Fine haze or precipitate or film; wine may be gassy.	Gray film on surface.	Fine haze or precipitate or silky, streaming cloud when shaken. Wine may be dull.	
Odor	Not characteristic.	Vinegary (acetic acid and ethyl acetate).	—	Wine may be slightly gassy. Sauerkraut or diacetyl character.
Microscopic appearance	Greater than 4–5 mμ.	Ellipsoidal or rods (involutionary forms may be present).	Less than 4–5 mμ. Rods.	Less than 4–5 mμ. Rods or cocci (like tangled mass of hair=*Lactobacillus trichodes*).
Growth on Basic medium[1]	+	+	+	+ (Ferments malate to lactate= malo-lactic bacteria.)
Basic medium + cycloheximide[2]	0 or +[3]	0 or +	0 or +	+
Catalase test[4]	—	+	+	—
Spore formation	0 or +	—	+	—
Calcium carbonate plates[5]		Clearing.	No clearing.	No clearing.

[1]Basic medium: The basic culture medium contains 2.0 g/100 ml of Bacto tryptone, 0.5 of Bacto peptone, 0.5 of Bacto yeast extract, 0.3 of glucose, 0.2 of lactose, 0.1 of liver extract (Wilson), 0.1 ml of 5% aqueous Tween 80, 100 ml of diluted and filtered tomato juice. To prepare tomato juice, dilute 4-fold with distilled water, filter through Whatman No. 1 paper using a Büchner funnel and Super Cel filter aid. To prepare the basic medium dissolve solid ingredients in diluted tomato juice by heating (avoid scorching by frequent agitation of the flask). When cool, adjust pH to 5.5 with concentrated hydrochloric acid, add 2 g agar/100 ml, and autoclave 15 min at 15 psig.

[2]Acti-dione: 1 ml containing 10 mg of cycloheximide is added to each 100 ml of medium.

[3]Growth with cycloheximide is presumptive evidence of *Brettanomyces* or *Dekkera* yeast (see Chap. 4 for confirmatory evidence).

[4]Basic medium with cycloheximide: Add a drop of 3% of fresh hydrogen peroxide to colonies. If gas is evolved the organism is catalase positive.

[5]Basic medium with cycloheximide minus the glucose and lactose and containing 2% calcium carbonate and 3% ethanol (the ethanol is added after the autoclaving): If acetic bacteria grow they will cause clearing of this cloudy medium around the colonies.

Source of data: Tanner and Vetsch (1956) and personal observations of authors.

(From Amerine et al. 1980. Technology of Winemaking. AVI Publishing Company, Westport, Conn.)

FIG. 9.5.   DETERMINATION OF CAUSES OF MICROBIAL SPOILAGE

ATF and Supplemental Record-keeping

Good records and record-keeping are as important as accuracy of analysis in the cellars. The records in a small winery laboratory are the first line of data that the ATF and other authorities will demand in the course of inspection. These records are important internally, as well, providing the basis for processing, production and marketing decisions. Data recording is imperative for administrative reports to monitor economic efficiency. A winery cannot function effectively without a sound system of records.

Superior records do not necessarily mean large quantities of data. The best systems are the simplest to maintain and demand the least amount of time. However, complete basic information is required. Official records are essential to the continuance of an ATF permit, and supplemental records provide the material for keeping official records. The obligations and responsibilities are clear.

SUPPLEMENTAL RECORDS (FIG. 10.1)

We shall discuss supplemental records first because they have already been mentioned in the earlier chapters in this text. Throughout this chapter, it is hoped that each record format can be justified, and illustrations are provided to show how these fit into the total administrative picture. The full development of this material becomes the foundation support for the ATF forms that are required.

Laboratory Analysis Log

As was discussed in Chapter 2, the *laboratory analysis log* is the backbone of all daily records kept for the winery. It bears repeating that every sample that is brought into the laboratory, whether tested or not, must be recorded in the laboratory analysis log. Every analysis, whether casual, formal, routine or specific, is also entered into the log. The policy set by the winemaster should allow no exceptions. Figure 2.49, p. 000 shows a common format that meets with acceptance in most small winery laboratories.

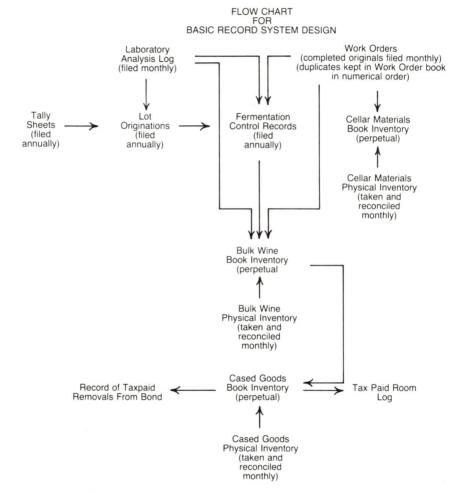

FIG. 10.1.   SUPPLEMENTAL RECORD-KEEPING FLOW CHART

At the end of each month, the laboratory analysis log pages should be stapled together and filed in the appropriate file of monthly operations.

Annual File or Vintage File

The *annual file*, or as some winemasters prefer, the *vintage file*, is used primarily for storing tally sheets, lot origination forms and fermentation control records. These should be filed for each lot of wine made during the vintage season. The serial numbers used should be representative of individual lots, and all three records should be stapled together so that wines may be readily traced back to the source vineyard(s). Convenient reference can be made if these serial numbered lots are kept in numerical order in the file. Each annual file should be kept for a minimum of seven years prior to being discarded. Even then, it is advisable to inform the ATF of plans for destroying these files.

Tally Sheet

The *tally sheet* serves the very important function of recording grape source and net weight data as grapes are received at the winery during the vintage season.

Each specific source area should carry a separate digit in the lot number assigned to fruit from that particular place. The tally sheet should display this source information clearly so that the link from vineyard to winery is properly recorded. The net weight is the starting point for computing winery production and for accurate preparation of lot origination. To that end, the tally sheet provides the basis for calculating volume measurements in amelioration and fermentation. It should be attached to the appropriate lot origination form and fermentation control record and filed in the annual file when completed. Figure 3.7, p. 109 displays a tally sheet format that may be used.

Lot Origination

The *lot origination* form is often called the "birth certificate" of a wine. With this form and the proper information provided by the laboratory analysis log (Fig. 3.8, p. 114) and the tally sheet, the winemaster can accurately calculate amelioration materials necessary for white wine fermentations and closely estimate those for red must fermentations. These efforts were explained in Chapter 4.

The lot origination is an annual supplemental record which is important for tracing the history of a wine. This information is necessary for ATF inspections.

Most supplemental records are prepared as originals, with no copies. The lot origination may, in the case of larger wineries, or under special management circumstances, be copied and submitted as a reporting form. The fully completed format contains a great deal of accounting and management information.

Completed lot origination forms should be stapled to the appropriate tally sheet and fermentation control record for filing in the annual or vintage file. A format for lot origination is depicted in Fig. 4.1, p. 119.

Fermentation Control Record

The primary purpose of the *fermentation control record* is to monitor fermentations of juice and must. This form provides a daily history of the new wine, during its most vulnerable stages. These data, in the hands of a competent winemaker, provide the basis for decision-making and execution of winery policy.

After the juice or must is deemed ready to be called wine, the data should be transferred from the fermentation control record to the bulk wine book inventory. This was presented in detail in Chapter 4.

All of the three stapled annual forms—tally sheet, lot origination and fermentation control record—are then placed together for each lot of wine in the annual file.

Figure 4.5, p. 126 serves to illustrate a format for the fermentation control record.

File of Monthly Operations

This is a self-descriptive manner of record storage. In the file drawer that contains current production records, the winemaster should set up twelve file folders at the beginning of each annual or fiscal year. Each may be tabbed with consecutive months as follows:

January Monthly Operations—1981
February Monthly Operations—1981
March Monthly Operations—1981
    etc.

Some winemakers prefer to title the monthly operations file folders with the year expressed first:

1981—January Operations
1981—February Operations
1981—March Operations
    etc.

Whatever the title system, the basic premise is to establish and maintain file folders for consecutive months and years with tabs that are easily identifiable in a logical sequence.

Along with the pages from each month of the laboratory analysis log, the file of monthly operations will contain the completed originals of work orders, reconciled bulk wine physical inventory and cased goods physical inventory, plus a number of other supplemental and official ATF records.

This monthly file should be kept intact for at least seven years, after which the winemaster may request, as with the annual file, destruction permission from the ATF.

Work Order

The *work order form*, introduced in Chapter 4, was described as a vital communications link between the winery laboratory and cellars. Completed work order forms returned from the cellars and appropriately posted are kept chronologically. Each month these forms are stapled together in daily progression from top to bottom. Copies of work orders are kept in serial number order in the work order "prescription" book. Figure 4.4, p. 125 displays a general work order format sufficient for most small winery purposes.

Bulk Wine Book Inventory

The purpose of this perpetual record is to maintain an up-to-date profile of every unbottled wine in inventory. This includes location, volume, state of production and analytical history. The *bulk wine book inventory* is the check or proof of the monthly bulk wine physical inventory.

When a card is removed from the bulk wine book inventory, it is considered "dead" and should be appropriately posted and filed in the bulk wine book inventory dead file.

Figure 4.13, p. 141 illustrates a format for the bulk wine book inventory.

Daily Bottling Record

This format, as discussed in Chapter 6, acknowledges the receipt of bulk wine deliveries from the cellars to the bottling line. It also records pre-bottling and post-bottling analyses, accounts for bottling and packaging materials used and provides delivery data for cased goods from the bottling line to the warehouse.

The *daily bottling record* is maintained daily as bottling operations are performed. The completed and posted daily bottling records are kept chronologically in the file of monthly operations.

Figure 6.4, p. 191 displays a daily bottling record format.

Cased Goods Book Inventory

This card file is the sequel to the bulk wine book inventory. It is necessary for maintaining an accurate current account of all cased goods in the winery warehouse. The *cased goods book inventory* should provide the warehouse number, date(s) of transaction(s), and other pertinent data such as lot numbers, taxpaid removal numbers, case serial numbers (or bottling dates), volume of cases, vintage year, etc. Complete information should be logged and readily available.

As lots become depleted a "dead" card will be written for the cased goods book inventory. This card will be posted and filed in the cased goods book inventory dead file.

Figure 6.12, p. 207 illustrates a card format for the cased goods book inventory.

Cellar Materials Book Inventory

As work order forms are issued and carried out, the materials used should be accounted for. Of these, sugar must be inventoried and carefully controlled as prescribed in Part 240, CFR, Section 240.914. Most winemasters, while taking inventory, will take the opportunity to fill out a *cellar materials book inventory*. This includes other expensive and pertinent materials besides sugar that should be closely controlled in the winery. This form is usually printed on a card the same size as used for the bulk wine book inventory and the cased goods book inventory so that it can be filed with these other inventories.

The cellar materials book inventory can be as comprehensive as the vintner wishes. In larger wineries everything is inventoried, not only in the cellars, but also in the bottling room and warehouse. The cellar materials book inventory is effective only to the extent that it is maintained and utilized. Generally, the small winery should make a monthly inventory of such items as activated charcoal, bentonite, calcium carbonate, citric acid, filter pads, metal-reducing agents, potassium metabisulfite, Sparkolloid and, of course, sugar. In wineries that make dessert wines, high-proof brandy must be inventoried and controlled in the same manner as the bulk wine book inventory. Corks, capsules and labels may be also counted monthly, but this is more often done quarterly, semiannually, or as necessary to maintain adequate supplies.

Physical Inventory: Bulk Wine, Cased Goods and Cellar Materials

The bulk wine book inventory is, like the cased goods book inventory and the cellar materials book inventory, a perpetual record that is reconciled at the end of each month with an appropriate *physical inventory* form as shown in Fig. 6.14, p. 209 and Fig. 6.15, p. 210. Note that these forms incorporate the physical inventory of bulk wine, cased goods and sugar. This can simplify record keeping for smaller wineries, but larger concerns may wish to keep each physical inventory on a separate form. The format chosen should be custom-designed to fit the breadth and scope of the inventories in the individual winery.

Wine Spirits Addition Record

The *wine spirits addition (WSA) record* is necessary when dessert wine or aperitif wine is made by the addition of high-proof brandy to table wines (see Chapter 7). The transition from the "under 14%" tax-class to the "14−21%" category in the bulk wine book inventory should be fully documented by such a record.

This record accounts for the selection and analysis of wines prior to sweetening and WSA, the materials used, the samples and forms required by the ATF, post-WSA analysis and production data. Upon completion, wine spirits addition records are filed in the appropriate file of monthly operations.

Figure 7.1, p. 218 is an example of the wine spirits addition record.

Tirage Record

A *tirage record* is necessary to account for the transition of cuvée wines from the bulk wine book inventory to sparkling wines in the cased goods book inventory. This process is termed the *méthode champenoise* and is fully described in Chapter 8.

The tirage record is divided into two major parts: the cuvée portion and the tirage section. The cuvée half of this record monitors the cuvée wine analysis, fermentation dosage ingredients and post-final filtration analysis. Filling data, post-filling analysis and production results should also be posted in this portion of the tirage record format. The tirage segment is primarily designed to analyze the progress of secondary fermentation by the determination of CO_2 gas volumes measured twice weekly. Once secondary fermentation is complete, post-fermentation analysis and riddling data are recorded, marking the transition of the tirage status to riddling storage. Discharged tirage record forms may be filed in the appropriate file of monthly operations. Where sparkling wine operations are significantly large, vintners may choose to maintain a separate file for sparkling wine operations.

Figure 8.2, p. 249 illustrates a comprehensive tirage record.

Daily Disgorging Record

The *daily disgorging record* is necessary to monitor sparkling wine production from riddling storage through the finishing processes.

Figure 8.8, p. 259, illustrates a format for the daily disgorging record wherein pre-disgorging analysis, disgorging and finishing dosage data, post-disgorging analysis, material and supply usage and finished production data are recorded.

As each day of disgorging activities is finished, the daily disgorging record is completed and attached to the appropriate tirage record. In the case of two or more lots of sparkling wines being finished in a day, a daily disgorging record should be discharged for each lot handled.

Taxpaid Room Log

Most small wineries will maintain a *taxpaid room* which, in accordance with Part 240, CFR, Section 240.145, may be used to store wines that are taxpaid for retail sales at the winery and wines that have been returned, rather than storing them in the bonded (not taxpaid) warehouse. Section 240.921 is definitive for the records necessary in this special area.

ATF Correspondence File

The ATF uses a coding system for correspondence reference that is usually rather foreign and meaningless to most vintners. Consequently, the winemaster, in his written communication with the ATF, should establish a system that accounts for both ATF and State incoming and outgoing correspondence.

The simplest system is comprised of single files tabbed appropriately by calendar or fiscal year. These are used to store letters, circulars and other such material as may be received, as well as copies of material sent. More commonly, however, there will be several files such as the following:

ATF Correspondence—1981
ATF Circulars and Notices—1981
ATF Official Documents—1981
State Correspondence—1981
State Circulars and Notices—1981
    etc.

An important thing to remember is that neatness and organization count. An ATF inspector will usually have much more confidence in an operation that is well documented and managed than one that is haphazardly run.

All written correspondence with the ATF and State should be serialized each calendar or fiscal year. In other words, if a fiscal year commences on July 1, the next letter sent to the ATF should carry the notation, "Serial No. 1-81/82", or whatever the year may be. This notation is usually made in the upper right corner of the letter or near the date. In this manner, the ATF (or State) and the vintner have a common reference in official correspondence.

Supplemental Files

In most small wineries the files for copies of purchase orders, invoices, shipping orders, bills of lading, etc., are kept in a separate office. These should be readily available for ATF inspection regardless of their location, especially the documents concerning purchases of grapes, juice, wine, sugar, cellar materials, wine spirits, as well as sales and marketing data. During the course of an ATF or State inspection, all of these records may be required.

Supplemental Record-keeping Flow Chart

Up to this point the discussion of records and record-keeping has been limited to supplemental materials. The flow chart provided in Fig. 10.1 may bring the supplemental records system into perspective. This chart will be expanded upon in the discussion of official ATF records and record-keeping.

OFFICIAL ATF RECORDS AND FORMS

During the past several decades, ATF regulation in the wine industry has been simplified. This pattern continues as, at this writing, significant changes in ATF forms and statutes are being considered and studied. While a good portion of required records have been removed from the ATF system, the March 1979 edition of Part 240, CFR, lists the forms that still remain effective in wine production reporting and record-keeping.

The latest edition of Part 240, Title 27, CFR, dated January, 1981 retires several forms from required record-keeping in the winery. Among these are forms ATF 2056, ATF 2057, ATF 2621, ATF 2058 and others. Many winemasters continue to use these formats anyway, despite the relaxed regulations, in that familiarity with this system of control is much easier than drafting custom formats for supplemental records to achieve the same goals.

Record of Still Wine—Form 2056

Part 240, CFR, Section 240.908 describes ATF *Form 2056* and its use. The ATF 2056 is maintained on a monthly basis for transactions that take place in the winery. Consequently, it is filed in the appropriate file of monthly operations when completed. Figure 10.2 provides a reduced-size reprint of a typical ATF Form 2056. This form is not supplied by ATF, although that bureau will usually provide a master Form 2056 upon request which the vintner can pass on to his printer. Most winemasters opt for having the pertinent winery information printed as the top of the form in order to save time and effort each month as a new Form 2056 is begun.

As new wine is removed from fermenters and placed into the bulk wine book inventory, it must also be logged in Form 2056 under column (c). The terms "produced by" and "used for" can create confusion until the new winemaker becomes familiar with this form. As an example, suppose that a vintner had 1000 gallons of wine that was to be sweetened with cane sugar just before bottling. In turn, say that 100 lb of sugar was used in the sweetening process. Recall from the lot origination form that each pound of sugar is equal to .074 gal. of volume. It is then calculated that 100 lb × .074 = 7.4 gal., and thus 1007.4 gal. is "produced" from sweetening, while 1000 gallons was "used" for sweetening.

After all operations for each month are completed in the cellars, Form 2056 is tallied along the "total" line across the bottom of the form and this figure is entered in the monthly report of wine cellar operations, which will be considered later.

Transfer of Wine in Bond: Form 703

Returning to the ATF Form 2056, make note of columns (h), "Received From Other Wine Cellars", and (j), "Transfers To Other Wine Cellars".

DEPARTMENT OF THE TREASURY
BUREAU OF ALCOHOL, TOBACCO AND FIREARMS
RECORD OF STILL WINE
SEE REVERSE FOR INSTRUCTIONS

FORM IS EXECUTED USING
☐ LITERS
☐ WINE GALLONS
☐ LITERS & WINE GALLONS

NAME

BONDED
☐ WINE CELLAR ☐ WINERY
NO.

ADDRESS

TAX CLASS[1]

MONTH AND YEAR

DATE (a)	EXPLANATION (b)	PRODUCED BY FERMENTATION (c)	PRODUCED BY SWEETENING (d)	PRODUCED BY ADDITION OF WINE SPIRITS (e)	PRODUCED BY BLENDING (f)	PRODUCED BY AMELIO-RATION (g)	RECEIVED FROM OTHER WINE CELLARS (h)	TAXABLE REMOVALS (i)	TRANSFERS TO OTHER WINE CELLARS (j)	REMOVED FOR DISTILLING MATERIAL (k)	REMOVED TO VINEGAR PLANT (l)	REMOVED FOR EXPORT (m)	OTHER TAX-FREE REMOVALS (n)	USED FOR SWEETENING (o)	WINE USED FOR ADDITION OF WINE SPIRITS (p)	USED FOR BLENDING (q)	USED FOR AMELIO-RATION (r)	USED FOR EFFERVESCENT WINE (s)
1.																		
2.																		
3.																		
4.																		
5.																		
6.																		
7.																		
8.																		
9.																		
10.																		
11.																		
12.																		
13.																		
14.																		
15.																		
16.	This is a sample form for use of commercial printers, and all such forms printed by them must conform therewith, unless changes are authorized by the Regional Director, Bureau of Alcohol, Tobacco and Firearms.																	
17.																		
18.																		
19.																		
20.																		
21.																		
22.																		
23.																		
24.																		
25.																		
26.																		
27.																		
28.																		
29.																		
30.																		
31.																		
32.																		
33.																		
34.																		
35.																		
36.																		
37.																		
38.																		
39.																		
40.																		
	TOTAL																	

ATF FORM 2056 (5120.12) (6-76) PREVIOUS EDITIONS ARE OBSOLETE

[1] 14%; 14-21%; 21-24%.

FIG. 10.2. ATF FORM 2056—RECORD OF STILL WINE

These transactions are monitored by the ATF inter-regionally by the use of ATF *Form 703*. Figure 10.3 is a reprint of the front side of the Transfer of Wine in Bond. When fully executed and posted, Form 703 may be filed in the appropriate file of monthly operations.

Record of Effervescent Wine: Form 2057

ATF *Form 2057* is required in wineries that produce sparkling or crackling wines. As was explained in Chapter 8, sparkling wines are produced from a secondary fermentation and carry an ATF tax rate of $3.40 per gal. Crackling wines are "artificially carbonated" by injection of CO_2 gas, and are assessed $2.40 per gal.

Form 2057 is similar in scope and execution to Form 2056. ATF Form 2057 is not provided by the bureau except for a master that can be reproduced.

At inventory, after the close of business each month, ATF Form 2057, where applicable, is tallied and held for reconciliation with other supplemental and ATF records, including the monthly report of wine cellar operations.

Figure 10.4 is a reduced reprint of the front side of Form 2057. Instructions for executing this record of effervescent wine are provided on its reverse wide, as with most ATF forms. Regulations regarding its use are defined in Part 240, CFR, Section 240.909.

Fully executed ATF Forms 2057 are filed in the appropriate file of monthly operations, or where larger volumes of sparkling wine are involved, a special file may be employed for sparkling wine documentation.

Record of Bottled Wine: Form 2621

The purpose and scope of ATF *Form 2621* are set forth in Part 240, CFR, Section 240.912. Form 2621 is procured, maintained, tallied, reconciled and filed in much the same manner as Form 2056, except that Form 2621 includes changes in the cased goods book inventory, while Form 2056 is concerned with the bulk wine book inventory.

Figure 10.5 is a reduced reprint of ATF Form 2621, record of bottled wine.

Wine Tax Return: Form 2050

As was discussed in earlier chapters, wine manufactured and sold in the United States is subject to an excise tax imposed and collected by the ATF. Under the provisions of Part 240, CFR, Section 240.901, the ATF requires that Federal excise taxes on wines be paid semimonthly along with the properly administered and submitted ATF *Form 2050*.

One should make note that ATF Form 2050 and all other ATF forms discussed subsequently in this chapter are required to be executed in gallons, whereas Forms 2056 and 2621 may be expressed in either gallons or liters. The author suggests that all records be maintained in gallons in

Department of the Treasury - Bureau of Alcohol, Tobacco and Firearms

Transfer of Wine in Bond

(See footnotes and instructions on back)

1. Serial No. (Begin with "1" each Jan. 1)

2. Shipped From

☐ Bonded Wine Cellar ☐ Bonded Winery

3. Number	4. Name of Proprietor

5. Address (Street, City, State, ZIP Code)

6. To ☐ Bonded Winery ☐ Vinegar Plant
☐ Bonded Wine Cellar ☐ Distilled Spirits Plant

7. Number	8. Name of Proprietor

9. Address (Street, City, State, ZIP Code)

10. Description of Wines and Containers

Number of Cases, Packages, or Tanks[1] (a)	Serial Numbers (b)	Gallons (c)	Percent Alcohol (d)	Kind[2] (e)	Produced at[3] (f)

11. Date Shipped	12. Signature of Shipper

13. Consignee's Receipt

Date (a)	Loss or Discrepancy, and Cause[4] (b)	Consignee's Signature (c)

ATF Form **703** (Rev. 8-72) *Dispose of all prior issues*

FIG. 10.3.   ATF FORM 703—TRANSFER OF WINE IN BOND

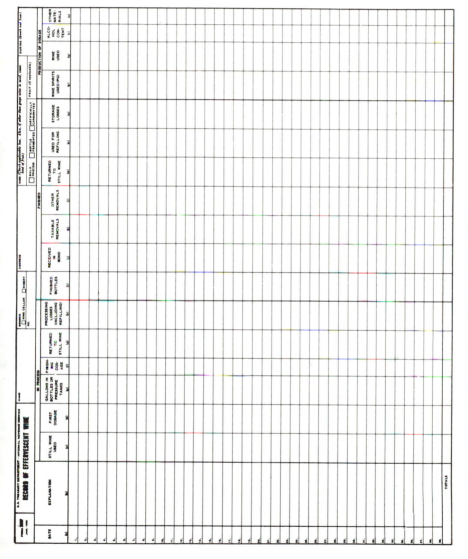

FIG. 10.4.   ATF FORM 2057— RECORD OF EFFERVESCENT WINE

SAMPLE COPY
FOR THE USE OF
COMMERCIAL PRINTERS

Department of the Treasury - Bureau of Alcohol, Tobacco and Firearms			MONTH AND YEAR	TAX CLASS (Check one)
Record of Bottled Wine				☐ NOT OVER 14% ALCOHOL
NAME	ADDRESS		REGISTRY NO.	☐ MORE THAN 14%, NOT OVER 21%
				☐ MORE THAN 21%, NOT OVER 24%
				☐ ARTIFICIALLY CARBONATED
				☐ SPARKLING

WINE GALLONS

DATE (a)	BOTTLED (b)	RECEIVED IN BOND (c)	TAXPAID WINE RETURNED TO BOND (d)	REMOVED TAXPAID (e)	TRANSFERRED IN BOND (f)	DUMPED TO BULK WINE OR DESTROYED (g)	USED FOR SAMPLES OR TASTING (h)	OTHER 1/ (i)
TOTALS								

1/ Identify each entry in column (i) as breakage, exported, removed for use of U.S., inventory overage, inventory shortage, etc., as the case may be. Show separate totals for debit items and for credit items.

INSTRUCTIONS

1. A daily record is required to be maintained by each proprietor of a bonded wine cellar, of wine filled into bottles, bottled wine received in bond, and bottled wine removed (26 CFR 240.912).

2. A separate Form 2621 will be prepared to cover each tax class of wine.

3. Quantities in wine gallons will be based on the number and size of bottles rather than the quantity used for bottling or the quantity ascertained after dumping from bottles.

4. Wine in process, such as sparkling wine still to be disgorged, will not be included in this record.

5. All columns will be totaled at the end of each month and the totals will be used in preparing the summary of bottled wine in bond, Part I, Section B, of Form 702.

ATF Form **2621** (Rev. 8-72) Revisions of 4-61 and 7-69 on hand at Bonded Wine Cellars may be used until supplies are exhausted.

FIG. 10.5. ATF FORM 2621—RECORD OF BOTTLED WINE

order to save confusion and needless effort. As the ATF progresses toward implementing the metric system, however, the vintner will need to make arrangements to convert the record-keeping system accordingly. The conversion tables in Appendix C may be of assistance.

Figures 10.6 and 10.7 are reduced-size reprints of the front and reverse sides, respectively, of ATF Form 2050.

The removal of wine from bond, where the total amount of unpaid tax is in excess of $100 may be authorized by the ATF in accordance with Part 240, CFR, Section 240.222 with the execution of Form ATF F 2053 as illustrated in Figure 10.8.

Gauge Report: Form 2629 (Revised as ATF 5110.26)

The ATF closely controls the handling of high-proof brandy in wineries. *Form 2629* is used to report the receipt of high-proof at the winery. In that such shipments are apt to be intermittent, the execution of this form is made commensurately, with a copy stored in the appropriate file of monthly operations.

Figures 10.9 and 10.10 are reduced-size reprints of the front and reverse sides, respectively, of ATF Form 2629.

Application for Release of Wine Spirits and Report of Addition to Wine: Form 275 (Revised as ATF 5120.28)

Dessert and aperitif wines that have spirits added are reported by ATF *Form 275*, recently revised as ATF *Form 5120.28* (see Figure 10.11).

Upon completion of WSA operations, the necessary Form(s) 275 are submitted, with a copy placed in the appropriate file of monthly operations. In cellars where WSA operations are frequent, ATF Forms 2629 and 275 may be filed in a system that is more suited to the purpose.

Special Natural Wine Production Record: Form 2058

As was discussed in Chapter 7, special natural wines are flavored and/or spiced wines such as vermouth and other high-alcohol aperitif types. Part 240, CFR, Section 240.910 is explicit in the use of ATF *Form 2058*, which administers the record of special natural wine production. Form 2058 is maintained daily as such operations take place and can be generally compared to ATF Form 2056, except that Form 2058 is designed only for vermouth and other special natural wines. The ATF provides only a master Form 2058 from which the vintner may have a supply custom-printed.

At the end of each month, Form 2058 is tallied along the bottom of the document and held for reconciliation with other ATF and supplemental records. Finally, the tally line is posted on the monthly report of wine cellar operations and then filed in the appropriate file of monthly operations. Again, should a winery have a larger production profile for special natural wines, a separate filing system may be employed to store such records.

DEPARTMENT OF THE TREASURY – BUREAU OF ALCOHOL, TOBACCO AND FIREARMS **WINE TAX RETURN** *(Prepare in quadruplicate - See instructions on back)*	1. SERIAL NUMBER *(Begin with "1" Jan. 1 each year)*

2. PERIOD COVERS *(Inclusive dates)*		3. FORM OF PAYMENT
From	To	☐ CHECK ☐ MONEY ORDER ☐ OTHER *(Specify)*

4. REGISTRY NUMBER OF BONDED WINERY OR WINE CELLAR	5. AMOUNT OF PAYMENT	**NOTE:**
6. EMPLOYER IDENTIFICATION NUMBER		PLEASE MAKE CHECKS OR MONEY ORDERS PAYABLE TO THE INTERNAL REVENUE SERVICE (Show Employer Identification Number on all checks or money orders)
7. NAME AND ADDRESS *(Begin one space below dots)*		Internal Revenue Service Use Only

TAX	$
PENALTY	
INTEREST	
TOTAL	$

RECAPITULATION OF TAX *(Before making entries 1 through 5, fill in applicable schedules)*

1. TOTAL TAX *(From line 11)*	$
2. ADJUSTMENT INCREASING TAX *(From line 17)*	
3. GROSS TAX *(Total of lines 1 and 2)*	
4. ADJUSTMENTS DECREASING TAX *(From line 23)*	
5. TAX TO BE PAID WITH THIS RETURN	$
(Line 3 minus line 4) (Includes $ net interest paid)[1]	

Under the penalties of perjury, I declare that I have examined this return (including any accompanying explanations, statements, schedules, forms, transactions, and tax liabilities required by law or regulations) and to the best of my knowledge and belief it is true, correct, and complete.

SIGNATURE	TITLE	DATE

SCHEDULE A - WINE TAX

(a) TAX CLASS	(b) GALLONS	(c) TAX RATE	(d) AMOUNT OF TAX
6. NOT OVER 14% ALCOHOL *(From line 40, column (c))*		$0.17	$
7. MORE THAN 14% BUT NOT OVER 21% ALCOHOL *(From line 40, col. (d))*		$0.67	
8. MORE THAN 21% BUT NOT OVER 24% ALCOHOL *(From line 40, col. (e))*		$2.25	
9. ARTIFICALLY CARBONATED *(From line 40, col. (f))*		$2.40	
10. SPARKLING *(From line 40, column (g))*		$3.40	
11. TOTAL AMOUNT OF TAX FOR PERIOD			$

SCHEDULE B – ADJUSTMENTS INCREASING AMOUNT DUE *(See instructions)*

EXPLANATION OF INDIVIDUAL ERRORS OR TRANSACTIONS (a)	AMOUNT OF ADJUSTMENTS	
	(b) TAX	(c) INTEREST
12.	$	$
13.		
14.		
15.		
16. TOTAL AMOUNT OF ADJUSTMENTS	$	$
17. TOTAL TAX AND INTEREST ADJUSTMENTS *(Col. (b) plus col. (c))*		$

SCHEDULE C – ADJUSTMENTS DECREASING AMOUNT DUE *(See instructions)*

EXPLANATION OF INDIVIDUAL ERRORS OR TRANSACTIONS (a)	AMOUNT OF ADJUSTMENTS	
	(b) TAX	(c) INTEREST
18.	$	$
19.		
20.		
21.		
22. TOTAL AMOUNT OF ADJUSTMENTS	$	$
23. TOTAL TAX AND INTEREST ADJUSTMENTS *(Col. (b) plus col. (c))*		$

[1]Subtract interest in line 22, column (c), from interest in line 16, column (c), and enter net interest paid with this return, if any.

ATF F 2050 (5120.27) (6-77) EDITION OF 7/75 MAY BE USED

FIG. 10.6.   ATF FORM 2050—WINE TAX RETURN (FRONT SIDE)

		SCHEDULE D – DAILY REMOVALS				
DATE *(Account for each day of period)* *(a)*	**SERIAL NO. OF ATF FORM 2052** *(To be completed for prepayments)* *(b)*	**STILL WINE**			**ARTIFICIALLY CARBONATED** *(Gallons)* *(f)*	**SPARKLING** *(Gallons)* *(g)*
		NOT OVER 14% ALCOHOL *(Gallons)* *(c)*	MORE THAN 14% BUT NOT OVER 21% ALCOHOL *(Gallons)* *(d)*	MORE THAN 21% BUT NOT OVER 24% ALCOHOL *(Gals)* *(e)*		
24.						
25.						
26.						
27.						
28.						
29.						
30.						
31.						
32.						
33.						
34.						
35.						
36.						
37.						
38.						
39.						
40. TOTALS (Enter also in Schedule A)						

INSTRUCTIONS

1. ATF Form 2050 shall be prepared in quadruplicate to cover the tax on wine for each semimonthly period as provided in 27 CFR Part 240. A return on this form shall be filed even though all the tax was prepaid during the period. Also, unless exemption is granted by the Regional Regulatory Administrator, Bureau of Alcohol, Tobacco and Firearms a return shall be filed on this form for each semimonthly period even though no tax is incurred.

2. If no transactions are to be reported in schedule A, B, or C, the word "None" shall be entered in the applicable schedule. Schedule A will be used to report the taxable quantities of wine and the amount of tax thereon. Schedule B will be used to report unintentional errors which resulted in underpayments of tax on prior returns. Schedule C will be used to report unintentional errors which resulted in overpayments of tax on prior returns. An adequate explanation of each error shall be supplied in the appropriate schedule or in a statement attached to the return. Credit for tax prepaid (on ATF Form 2052) during the period shall be reported in schedule C, columns (a) and (b), as an adjustment decreasing the amount due. Likewise, credit for drawback of tax (allowed on ATF Form 2639) and credit for tax on unmerchantable wine returned to bond (allowed on ATF Form 2635) shall be reported in schedule C, columns (a) and (b), as an adjustment decreasing the amount due. The remainder of an allowed credit may be taken on the next subsequent return if the amount of tax due on one return is not sufficient to exhaust the total credit. Allowances on ATF Form 2635 or 2639 shall be identified in schedule C, column (a), by form number, serial number of claim, and date of approval.

3. Schedule D will be used to report the daily taxable quantities of wine removed. Each day of the period covered by the return shall be accounted for in schedule D. If no wine was removed subject to tax on any day covered by the return, enter in schedule D the date, followed by "No wine removed subject to tax". If the tax was prepaid for any day covered by the return, enter in the appropriate columns in schedule D the date, serial number of the ATF Form 2052, and the gallons removed prepaid.

4. The original and three copies of ATF Form 2050 shall be filed as follows: The original and two copies, with payment for the net amount due, shall be mailed or delivered to the District Director of Internal Revenue in whose district the premises are located; the third copy of the form shall be forwarded by the taxpayer to the Regional Regulatory Administrator, Bureau of Alcohol, Tobacco and Firearms at the same time.

The District Director will stamp all copies of the return to show receipt of the return and payment. If the taxpayer mails the return and payment, the District Director will send all copies to the Service Center and the Center will return to the taxpayer a stamped copy; if the taxpayer (or his messenger) delivers the return in person, the District Director will return one copy to the taxpayer (or his messenger) and send the original and one copy to the Service Center.

5. The law provides for payment of interest on underpayments or overpayments of tax and imposes penalties for late payment of tax or filing of tax returns. Interest is not allowable on credits for tax paid on unmerchantable wine returned to bond nor on credits for drawback of tax. Interest on overpayments or underpayments resulting from unintentional errors on prior returns will be computed and reported separately. Interest on underpayments will be entered in schedule B, column (c), and interest on overpayments will be entered in schedule C, column (c). Interest will be computed at the rate prescribed by 26 U.S.C. 6621. Interest on underpayments will be computed from the due date of the return in error to the date of filing the return (with payment) on which the error is adjusted. Interest on overpayments will be computed from the date of the overpayment (or due date if paid prior thereto) to the due date of the return on which the credit is taken. If a penalty has been imposed on a prior return, and such return has been found to be erroneous, the amount of penalty imposed may be adjusted on the return on which adjustment of the error is made. Underpayments of penalties should be adjusted by reporting such underpayments in schedule B, columns (a) and (b). Overpayments of penalties should be adjusted by reporting such overpayments in schedule C, columns (a) and (b).

ATF F 2050 (5120.27) (6–77)

FIG. 10.7.   ATF FORM 2050—WINE TAX RETURN (REVERSE SIDE)

	CONTINUING BOND		
· DEPARTMENT OF THE TREASURY BUREAU OF ALCOHOL, TOBACCO AND FIREARMS		REGISTRY NO.	F.A.A. ACT PERMIT
BOND COVERING DEFERRED PAYMENT OF WINE TAX *(File in duplicate. See instructions on reverse.)*	BONDED WINE CELLAR		
	BONDED WINERY		

PRINCIPAL *(See instructions 2, 3, and 4)*	BUSINESS ADDRESS *(Number, Street, City, State, ZIP Code)*
ADDRESS OF BONDED PREMISES	TRADE NAME UNDER WHICH OPERATED

SURETY(IES)	AMOUNT OF BOND $	EFFECTIVE DATE

KNOW ALL MEN BY THESE PRESENTS, That we, above-named principal and surety (or sureties), are held and firmly bound to the United States of America in the above-named amount, lawful money of the United States; for the payment of which we bind ourselves, our heirs, executors, administrators, successors, and assigns, jointly and severally, firmly by these presents.

This bond shall not in any case be effective before the above date, but if accepted by the United States it shall be effective according to its terms on and after that date without notice to the obligors: *Provided,* That if no date is inserted in the space provided therefor, the date of execution hereof shall be the effective date.

Whereas, the principal is operating, or intends to operate, a bonded wine cellar or bonded winery at the bonded premises specified above;

Now, therefore, the condition of this bond is such that if the principal shall pay, or cause to be paid, to the United States all taxes which have been determined on wine removed from the said bonded premises or transferred to a tax-paid wine room on the bonded premises, according to the laws of the United States and regulations made in conformity therewith, then this obligation, shall be null and void; otherwise, it shall remain in full force and effect: *Provided,* That the above obligation shall apply only with respect to taxes in excess of $100 which have been determined for deferred payment upon removal of the wine from said bonded premises or transfer to a tax-paid wine room on the bonded premises.

We, the obligors, for ourselves, our heirs, executors, administrators, successors, and assigns, do further covenant and agree that upon the breach of any of the covenants of this bond, the United States may pursue its remedies against the principal or surety independently, or against both jointly, and the said surety hereby waives any right or privilege it may have of requiring, upon notice or otherwise, that the United States shall first commence action, intervene in any action of any nature whatsoever already commenced, or otherwise exhaust its remedies, against the principal.

WITNESS our hands and seals this _____ day of _____ , 19___

Signed, sealed, and delivered in the presence of –

_____   _____ [SEAL]

_____   _____ [SEAL]

_____   _____ [SEAL]

_____   _____ [SEAL]

KIND OF BOND *(Check applicable box)*   ☐ ORIGINAL   ☐ STRENGTHENING   ☐ SUPERSEDING

ATF F 2053 (5120.26) (4-77)   EDITION OF 3/76 MAY BE USED

FIG. 10.8.   ATF FORM F 2053—BOND COVERING DEFFERED PAYMENT OF WINE TAX

DEPARTMENT OF THE TREASURY BUREAU OF ALCOHOL, TOBACCO AND FIREARMS **GAUGE REPORT** *(See footnotes and instructions on back)*		1. SERIAL NUMBER *(Begin with "1" each January "1")*	2. PLANT NUMBER DSP-
3. PURPOSE OF REPORT OR NOTICE AND TYPE OF GAUGE [1]		3a. RELATED FORM *(Number, serial number and date)* IF ANY.	
		3b. CONSIGNEE *(Name and Registry Number)*	
4. DATE	5. KIND OF SPIRITS [2]	6. PROOF OF DISTILLATION [3]	7. TYPE OF CONTAINER(S)
8. ORIGINAL ENTRY DATE [4]	9. AGE OF SPIRITS [5]	10. EARLIEST DATE OF MINGLING [6]	

11. THE SPIRITS WERE PRODUCED, BLENDED OR WAREHOUSED: [7]

11a. BY *(Real name)*	11b. AT DSP-	11c. UNDER THE TRADE NAME *(If any)*

12. GAUGE ELEMENTS

CONTAINERS			ENTRY PROOF [9]	PROOF GALLONS PER CONTAINER	TOTAL TAX GALLONS *(W.G. if denatured)*
SIZE *(If cases)* (a)	NUMBER (b)	IDENTIFICATION [8] (c)	(d)	(e)	(f)
TOTAL ➡					

Under penalties of perjury, I declare that I have examined this gauge report (and the accompanying ATF Form 2630 (5110.45), if any) and to the best of my knowledge and belief, it is correct and complete.

13. SERIAL NUMBER(S) OF SEALS USED TO SEAL BULK CONVEYANCES CONTAINING WINE SPIRITS

14. PROPRIETOR	14a. BY *(Signature and title)*
15. GAUGED BY	15a. ATF OFFICER *(Signature)* [10]

16. BONDED WINE CELLAR PROPRIETOR'S ACKNOWLEDGEMENT OF RECEIPT OF WINE SPIRITS [11]

16a. DATE REC.	16b. PROPRIETOR *(Name and registry number)*	16c. BY *(Signature and title)*

ATF FORM 2629 (5110.26) (1-78) PREVIOUS EDITIONS ARE OBSOLETE

FIG. 10.9.   ATF FORM 2629—GAUGE REPORT (FRONT SIDE)

FOOTNOTES

[1] Describe transaction *(see instruction 1)*.

[2] For denatured spirits show kind and formula number.

[3] Not required for denatured spirits, spirits for redistillation, or spirits of 190 degrees or more of proof.

[4] Not required for production gauge, denatured spirits, spirits for redistillation, or spirits of 190 degrees of more of proof.

[5] Not required when original entry date establishes the age of the spirits.

[6] Show only when spirits described in Item 12 have been previously stored in wooden packages and mingled under 27 CFR 201.297.

[7] For blended rums or brandies, enter the name and plant number of the blending warehouseman. For spirits of 190 degrees or more of proof, show the name and plant number of the producer or warehouseman, as appropriate; where such spirits are in cases or encased containers, or in packages or similar portable containers, the name and plant number of the proprietor marked thereon must be shown. For imported spirits, enter the name and plant number of the warehouseman who received the spirits from customs custody. For Virgin Islands or Puerto-Rican spirits, show the name of the producer in the Virgin Islands or in Puerto Rico.

[8] Designate wooden barrels as "C" for charred, "REC" for recharred, "P" for plain, "PAR" for paraffined, "G" for glued, and "R" for reused. In addition "PS" (pre-soaked) will follow the cooperage designation if a barrel has been steamed or water soaked before filling. Enter tank number, volumetric gauge factors, proof, and wine gallons in this column in blank space on the form.

[9] Show entry proof for "bourbon whisky", "rye whisky", "corn whisky", "malt whisky", or "rye malt whisky".

[10] Complete only when spirits are gauged by an ATF officer.

[11] Complete Item 16 only if wine spirits are received on bonded wine cellar premises to ATF Form 257.

INSTRUCTIONS

1. This form will be used as a *notice* for:
    (a) *Gauge and removal of wine spirits from distillery* (or bonded warehouse).
    (b) *Gauge and transfer of spirits* (or rinse water, or denatured spirits) *from warehouse to distillery for redistillation* (at same plant).
    (c) *Dumping and repackaging of packages of spirits of 190 degrees or more of proof for mingling.*

2. This form will be used as a *report* for:
    (a) *Production gauge and entry for deposit in bonded storage* (at the plant where produced).
    (b) *Packaging of spirits from storage tank* (for storage at same plant).
    (c) *Gauge and return of spirits* (or denatured spirits) *to bonded premises* (except tax determined spirits).
    (d) *Return to storage from bottling in bond* (for redeposit).
    (e) *Gauge of imported spirits received from Customs custody, for use of the United States.*

3. If operation involves a pipeline transfer, show in Item 7, "P/L". In such cases show the number of the tank or bulk conveyance into which the spirits are to be deposited in Item 12, column (c), except where wine spirits are being transferred by pipeline to a contiguous wine cellar, show the wine cellar registry number in Item 12, column (c).

4. No more than one transaction, or in the case of production gauge, one lot, may be reported on an individual ATF Form 2629 (5110.26). Show in Item 3A the related form that was prepared before the transaction being reported, i.e., ATF Form 257 for wine spirits being transferred, ATF Form 1515 (5110.36) for bottled-in-bond spirits (including alcohol) being redeposited, or ATF Form 1685 or 2323 (5110.29) for packages of spirits being redeposited.

5. Prior to the lading of spirits aboard any conveyance (or commencement of pumping operations in the case of pipeline transfers) for removal from bonded premises, or prior to dumping, gauging, packaging, or other operations listed in Instruction 1, the proprietor shall submit a copy of ATF Form 2629 (5110.26) to the ATF officer as notice of the appropriate operation. The form will be prepared by the proprietor, in duplicate, except that the form will be prepared in quintuplicate if wine spirits are being removed. When it is necessary to prepare documents describing spirits in packages, one copy will be prepared for each required copy of ATF Form 2629 (5110.26), except for removal of wine spirits in packages. If the form is used as a notice, the proprietor should follow the applicable provisions of 27 CFR Part 201 respecting any action which must be taken by the assigned officer prior to completing the transaction.

6. On completion of the transaction, the proprietor will sign the original and copies of ATF Form 2629 (5110.26) and any accompanying documents that require his signature. The assigned officer shall sign all forms when such action is required by regulations. The proprietor shall dispose of the forms as follows:

    (a) Where ATF Form 2629 (5110.26) covers wine spirits withdrawn pursuant to ATF Form 257, he will:
    (1) Forward two copies to the winemaker (who will acknowledge receipt of the spirits on both copies and forward one to his Regional Regulatory Administrator — see instruction on ATF Form 257;
    (2) Retain one copy with any related document for his files;
    (3) Forward the original with any related document to his Regional Regulatory Administrator; and
    (4) Forward a copy with any related document to the assigned officer.

    (b) Where ATF Form 2629 (5110.26) covers any other transaction, he will:
    (1) Retain one copy with any related document, and
    (2) Return the original with any related document to the assigned officer.

FIG. 10.10.   ATF FORM 2629—GAUGE REPORT (REVERSE SIDE)

Department of the Treasury - Bureau of Alcohol, Tobacco and Firearms

Application for Release of Wine Spirits and Report of Addition to Wine

(Prepare in Triplicate — See Instructions on back)

1. DATE

2. TYPE OF ESTABLISHMENT (Check one)
   ☐ BONDED WINE CELLAR   ☐ BONDED WINERY

3. REGISTRY NO.

4. OPERATED BY (Name)

5. AT (City or town and State)

PART I – APPLICATION

6. SCHEDULED OPERATIONS

6A. KIND OF WINE 1/   6B. PRODUCED AT 2/

6C. NO. OF WINE SPIRITS ADDITIONS

6D. MAXIMUM QUANTITY OF WINE SPIRITS TO BE USED   T.G.

Application is made for release of wine spirits covered by approved Form 257 for addition to wine, as shown under scheduled operations. I certify that the wine to be used has been produced in accordance with applicable laws and regulations and is eligible for the addition of wine spirits.

7. PROPRIETOR

7A. BY (Signature and title)

PART II – REPORT OF WINE AND WINE SPIRITS USED AND WINE PRODUCED

SERIAL NO. OF W.S.A. (a)	WINE USED					WINE SPIRITS USED			WINE PRODUCED	
	TANK NO. (b)	KIND (PORT, SHERRY, ORANGE, BLACK-BERRY, ETC.) (c)	GALLONS (d)	PERCENT ALCOHOL (e)	DESIRED PERCENT ALCO-HOL (f)	GAUGE PRIOR TO USE 3/		WITHDRAWAL GAUGE 4/ (TAX GALLONS) (i)	GALLONS (j)	PER-CENT ALCO-HOL (k)
						WINE GALLONS (g)	TAX GALLONS (h)			
TOTALS										

8. REMARKS

9. PROPRIETOR

9A. BY (Signature and title)

10. DATE AUDITED

11. ATF OFFICER (Signature)

PART III – REPORT OF WINE SPIRITS RELEASED

12. QUANTITY OF WINE SPIRITS RELEASED FOR USE IN WINE PRODUCTION.   T.G.

13. DATE

14. ATF OFFICER (Signature)

1/ Such as grape, apple, orange, etc.
2/ Enter registry number of bonded premises where wine was produced, if different from item 3.
3/ Details on Form 2629 (and Form 2630, if required) attached hereto, unless the wine spirits were withdrawn from adjacent premises for immediate use,

4/ when the withdrawal gauge will be reported in columns (g) and (h).
Where variation from the withdrawal gauge is shown, the proprietor will state under "Remarks" the amount of variation, the apparent reason for variation, and the surrounding facts and circumstances.

ATF Form **275** (5120.28) (8-77)   EDITION 12/72 MAY BE USED

FIG. 10.11. ATF FORM 275 (REVISED AS ATF 5120.28)—APPLICATION FOR RELEASE OF WINE SPIRITS AND REPORT OF ADDITION TO WINE

Figure 10.12 provides a reduced reprint of ATF Form 2058, special natural wine production record.

Record of Distilling Material or Vinegar Stock: Form 2059

Few small wineries become concerned with the production and inventory of distilling material or vinegar stock. While occasionally an estate winery may wish to sell a wine in bulk that is unworthy of a label, this wine will not generally have been aged with such an end in mind. These wines may be recorded on ATF Form 2056. However, the distinction must be made between wines that are made expressly for distilling material or vinegar stock, and those that are not. Part 240, CFR, Section 240.911 is explicit in the proper use of *Form 2059* with regard to this difference.

Monthly Report of Wine Cellar Operations: Form 702 (Revised as ATF F 5120.17)

All official ATF record-keeping revolves around *Form 702*. For this reason, the ATF *monthly report* shall be considered in detail.

Figures 10.13 and 10.14 are reduced reprints of the front and reverse sides, respectively, of Form 702.

Note how the tally lines across the bottom of ATF Forms 2056 and 2621, in particular, closely match the ATF monthly report items listed in Part I, "Summary of Wines in Bond": Section A, "Bulk Wines" and Section B, "Bottled Wines". This design is intended to help avoid errors and simplify posting efforts. Other ATF recording forms, such as Forms 2057 and 2058, are also designed to correlate as much as possible with the Form 702 monthly report.

The item entitled "On Hand Last Of Month" in the monthly report must be a reported gallonage that corresponds exactly to the following month's "On Hand First Of Month". In other words, if a vintner had 9241.4 gal. of bulk wine inventory reported in Item 31 of Part I for July, 1980, there must be the same 9241.4 gallons reported in Item 1 of Part I for August, 1980. The difference between "On Hand First of Month" and "On Hand Last of Month" will be justified by other entries which, in total, shall compensate for the remainder.

Once Forms 2056, 2621 and 2057 are entered in the monthly report, the next portion of Form 702 will be concerned with the status of wines in the winery taxpaid room(s). The information entered in this section should be the synopsis of the taxpaid room log entries for the appropriate month.

The reverse side of the ATF monthly report commences with Part III, "Summary Of Distilled Spirits (Tax Gallons)". Tax gallons are the same as proof gallons. In regulating distilled spirits, the ATF utilizes tax gallon measurements rather than wine gallons. A tax gallon is calculated by multiplying wine gallons by the proof factor. For example, 100 wine gal. at 191° proof would equal 191 tax gal. (100 × 1.91), or 234 wine gal. at 189.2° proof would result in 442.728 tax gal. In executing this portion of the ATF monthly report, it is required that tax gallons be calculated and recorded.

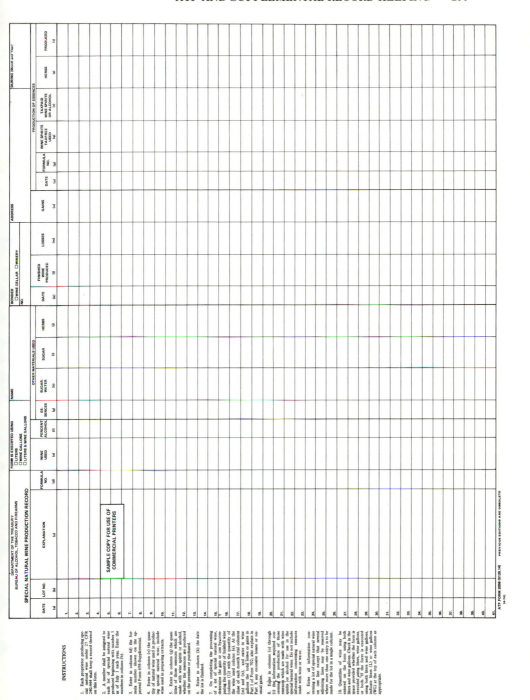

FIG. 10.12.   ATF FORM 2058—SPECIAL NATURAL WINE PRODUCTION RECORD

DEPARTMENT OF THE TREASURY — BUREAU OF ALCOHOL, TOBACCO AND FIREARMS
MONTHLY REPORT OF WINE CELLAR OPERATIONS

DATE *(Month and year)*

☐ BONDED WINE CELLAR   ☐ BONDED WINERY   NO.

OPERATED BY   *(Name and Address)*

INSTRUCTIONS

1. Prepare this form in duplicate. Forward the original to the Regional Regulatory Administrator, Bureau of Alcohol, Tobacco and Firearms, by the fifteenth day after the end of the month for which the report is made. Keep the copy on your bonded winery premises for inspection by Alcohol, Tobacco and Firearms Officers.

2. Explain in Part X any unusual operations.

3. Bottled wine shortages reported in Part 1, Section B, Line 19 must be fully explained in either Part X or in a separate signed statement submitted with this report. Failure to satisfactorily explain shortages of bottled wine may result in the assessment of taxes applicable to those shortages.

4. The quantities "on hand last of month" will ordinarily be "book inventory" figures, that is, the quantity required to balance each summary. On reports for any month when an actual physical inventory is taken the difference will be reported, as losses for bulk wine and shortages for bottled wine, or as gains, as the case may be.

5. If the quantity of wine previously reported on ATF F 5120.17 is affected by adjustments made on a tax return, ATF Form 2050 (5120.27), adjust the current Form 5120.17 in Section A (and Section B, if bottled wine is involved). Explain the entries in Part X.

PART I - SUMMARY OF WINES IN BOND (GALLONS)

ITEM	ALCOHOL CONTENT BY VOLUME			ARTIFICIALLY CARBONATED WINE (d)	SPARKLING WINE (e)
	NOT OVER 14 PERCENT (a)	OVER 14 TO 21 PERCENT *(Inclusive)* (b)	OVER 21 TO 24 PERCENT *(Inclusive)* (c)		
SECTION A - BULK WINES					
1. ON HAND FIRST OF MONTH					
2. PRODUCED BY FERMENTATION 1/					BF / BP
3. PRODUCED BY SWEETENING					
4. PRODUCED BY ADDITION OF WINE SPIRITS					
5. PRODUCED BY BLENDING					
6. PRODUCED BY AMELIORATION					
7. RECEIVED IN BOND					
8. BOTTLED WINE DUMPED TO BULK					
9. INVENTORY GAINS					
10.					
11.					
12. TOTAL					
13. BOTTLED 2/					BF / BP
14. REMOVED TAXPAID					
15. TRANSFERS IN BOND					
16. REMOVED FOR DISTILLING MATERIAL					
17. REMOVED TO VINEGAR PLANT					
18. USED FOR SWEETENING					
19. USED FOR ADDITION OF WINE SPIRITS					
20. USED FOR BLENDING					
21. USED FOR AMELIORATION					
22. USED FOR EFFERVESCENT WINE					
23. USED FOR TESTING					
24.					
25.					
26.					
27.					
28.					
29. LOSSES (OTHER THAN INVENTORY)					
30. INVENTORY LOSSES					
31. ON HAND LAST OF MONTH					
32. TOTAL					
SECTION B - BOTTLED WINES					
1. ON HAND FIRST OF MONTH					
2. BOTTLED 2/					BF / BP
3. RECEIVED IN BOND					
4. TAXPAID WINE RETURNED TO BOND					
5.					
6.					
7. TOTAL					
8. REMOVED TAXPAID					
9. TRANSFERRED IN BOND					
10. DUMPED TO BULK					
11. USED FOR TASTING					
12. REMOVED FOR EXPORT					
13. REMOVED FOR FAMILY USE					
14. USED FOR TESTING					
15.					
16.					
17.					
18. BREAKAGE					
19. INVENTORY SHORTAGE 3/					
20. ON HAND LAST OF MONTH					
21. TOTAL					

PART II - SUMMARY OF DOMESTIC WINES IN TAXPAID ROOMS (GALLONS)

ITEM	NOT OVER 14 PERCENT *(Inclusive)* (a)	OVER 14 TO 21 PERCENT *(Inclusive)* (b)	OVER 21 TO 24 PERCENT *(Inclusive)* (c)	ARTIFICIALLY CARBONATED WINE (d)	SPARKLING WINE (e)
1. RECEIVED					
2. REMOVED					
3. LOSSES					
4. ON HAND LAST OF MONTH					

AFT F 5120.17 (702) (6-79)   PREVIOUS EDITIONS ARE OBSOLETE.

1/ Enter in col. (e) on line marked "BF" the quantity of sparkling wine produced by fermentation in bottles, and on line marked "BP" the quantity of sparkling wine produced by bulk process.

2/ Section A line 13 and Section B line 2 should show the same quantities. Enter in col. (e) on line marked "BF" the quantity of finished bottle fermented sparkling wine bottled, and on line marked "BP" the quantity of finished bulk process wine bottled.

3/ Fully explain in either Part X, or on a separate signed statement submitted with this report.

FIG. 10.13. ATF FORM 702—MONTHLY REPORT OF WINE CELLAR OPERATIONS (FRONT SIDE)

PART III-SUMMARY OF DISTILLED SPIRITS (Tax Gallons)

ITEM	WINE SPIRITS				FOR PREPARATION OF DOSAGES OR ESSENCES (e)	DISTILLATES CONTAINING ALDEHYDES (Name of fruit from which spirits made, e.g., "blackberry," "apple ," etc.)		
	FOR ADDITION TO WINE (Name of fruit from which spirits made, e.g., "blackberry," "apple," etc.)							
	GRAPE (a)	(b)	(c)	(d)		(f)	(g)	(h)
1. ON HAND FIRST OF MONTH								
2. RECEIVED								
3. INVENTORY GAIN								
4. TOTAL								
5. USED								
6. TRANS. TO COL. (e)								
7.								
8. LOSSES								
9. ON HAND LAST OF MONTH								
10. TOTAL								

PART IV-SUMMARY OF MATERIALS RECEIVED AND USED

ITEM	GRAPE MATERIAL				KINDS OF MATERIALS OTHER THAN GRAPE (Pounds or Gallons)			SUGAR	
	GRAPES		JUICE (Gallons) (b)	CONCENTRATE (Gallons) (c)	(d)	(e)	(f)	DRY (Pounds) (g)	LIQUID (Gallons) (h)
	UNCRUSHED (Pounds) (a)	FIELD CRUSHED (Gallons) (b)							
1. ON HAND FIRST OF MONTH									
2. RECEIVED									
3. JUICE OR CONCENTRATE PRODUCED									
4. TOTAL									
5. USED IN WINE PRODUCTION									
6. USED IN JUICE OR CONCENTRATE PRODUCTION									
7. USED IN ALLIED PRODUCTS									
8. REMOVED									
9. ON HAND LAST OF MONTH									
10. TOTAL									

PART V-(RESERVED)

PART VI-SUMMARY OF DISTILLING MATERIAL AND VINEGAR STOCK (Gallons) 5/

ITEM	STATE WHETHER GRAPE, BLACKBERRY, ETC.			
	DISTILLING MATERIAL		VINEGAR STOCK	
	(a)	(b)	(c)	(d)
1. ON HAND FIRST OF MONTH (Storage Tanks)				
2. PRODUCED DURING MONTH				
3. RECEIVED FROM OTHER BONDED WINE CELLARS				
4.				
5. TOTAL				
6. REMOVED TO DISTILLED SPIRITS PLANTS				
7. REMOVED TO OTHER BONDED WINE CELLARS				
8. REMOVED TO VINEGAR PLANTS				
9.				
10. ON HAND LAST OF MONTH (Storage Tanks)				
11. TOTAL				

PART VII - IN FERMENTERS LAST OF MONTH (Gallons)

TOTAL	(State kind-grape, blackberry, etc.)					TOTAL
	(a)	(b)	(c)	(d)	(e)	
1. IN FERMENTERS (ESTIMATED QUANTITY OF LIQUID)						

PART VIII SUMMARY OF NONBEVERAGE WINES (Gallons)

ITEM	NOT OVER 14 PERCENT ALCOHOL (a)	OVER 14 TO 21 PERCENT ALCOHOL (Inclusive) (b)	TOTAL (c)
1. PRODUCED			
2. WITHDRAWN			

PART IX-SPECIAL NATURAL WINES (Gallons)

ITEMS	VERMOUTH (a)	OTHER SPECIAL NATURAL WINES				TOTAL (f)
		NOT OVER 14 PERCENT ALCOHOL (b)	OVER 14 PERCENT ALCOHOL (c)	ARTIFICIALLY CARBONATED (d)	SPARKLING (e)	
1. PRODUCED						
2. TAXABLE REMOVALS						
3. ON HAND LAST OF MONTH						

PART X-REMARKS

Under penalties of perjury I declare that I have examined this report, including the documents submitted in support thereof, and to the best of my knowledge and belief, it is true, correct, and complete.

PROPRIETOR	BY (Signature and Title)	DATE

5/ Distilling Material Such as Lees, Filter Wash and Other Residues, Used for Production of Wine Spirits. See Section 240.761 and 240.911.

ATF F 5120.17 (702) (6-79)

FIG. 10.14. ATF FORM 702—MONTHLY REPORT OF WINE CELLAR OPERATIONS (REVERSE SIDE)

The information for Part III is gathered from the reconciled wine spirits book inventory, ATF Forms 2629 (ATF 5110.26), 275 (ATF 5120.28), and other records which may be involved.

Part IV, "Summary Of Materials Received And Used" is completed with data taken entirely from supplemental record-keeping in the winery. The tally sheets, lot originations, fermentation control records and the sugar portion of the cellar materials book inventory are summarized and entered appropriately in this portion of the ATF Form 702, monthly report.

As was stated earlier with regard to ATF Form 2059, estate wineries do not normally become involved with distilling material and vinegar stock production. However, should such inventories be recorded, the tally figures from ATF Form 2059 will be entered in the monthly report in Part VI. There may also be some information from Form 2056 that will need to be posted in this section.

The next portion of Form 702, Part VII, "In Fermenters Last Of Month (Gallons)", will contain the total volume of juice and must (estimated juice) that is fermenting. This volume will agree with the total found in the fermentation control records at the end of the month. As was explained in Chapter 4, when new wines are racked from the fermenter and removed from the fermentation control record, such transactions will be totaled each month and subtracted from Item 1, "In Fermenters (Estimated Quantity Of Liquid)". In turn, these gallonages are picked up in column (c) of ATF Form 2056 as each fermenter is racked and transferred to the bulk wine book inventory. The tally of column (c) is recorded in Part I, Item 2, on the front side of the monthly report, as has already been discussed.

When all wines have been made and no juice or must remains fermenting in the winery, the vintage season will be fully completed and nothing will remain in Part VII of Form 702. If juice or must is being held in cold storage, unfermented, it will be accounted for in Part IV of the ATF monthly report.

Part VIII, "Summary Of Nonbeverage Wines (Gallons)" shall not be considered in this text as there is little application for the small winery.

The special natural wines that were considered with regard to ATF Form 2058 earlier in this chapter are summarized, tallied and entered in Part IX, "Special Natural Wines (Gallons)" of the ATF monthly report. This is straightforward and should require little effort for proper calculation and recording.

Any explanatory or supplemental data that may be felt appropriate may be added in Part X, "Remarks", of Form 702, monthly report. This should be reserved only for items directly associated with the report—it is not meant for casual correspondence. Questions and comments about reporting should be prepared in a serialized cover letter and attached to the ATF 702 when submitted.

Form 702 (ATF 5120.17), monthly report of wine cellar operations forms are provided upon request from regional ATF offices. These must be prepared, reconciled and submitted monthly prior to the tenth day of the following month. Part 240, CFR, Section 240.900 describes this monthly report and its preparation.

Inventory of Wine: Form 702-C

Throughout this text it has been stressed that the bulk wine physical inventory and the cased goods physical inventory should be taken at the close of business each month. This is highly recommended, but not required by the ATF. So much can transpire after a few months that reconciliation of book and physical inventories could become an unmanageable task.

Nevertheless, the ATF requires only that a physical inventory of bulk wine and cased goods be taken twice annually, at the close of business ending June and December each year. The form that this inventory is submitted on is called the *inventory of wine*, ATF *Form 702-C.*

This form follows a simple format—just a tank-by-tank, item-by-item and class-by-class inventory of all wines (not fermenters, however) on the bonded premises. The tally of this report will necessarily agree exactly with the total entered on the ATF Form 702 monthly report for the same month.

Figures 10.15 and 10.16 are reduced reprints of the front and reverse sides of ATF Form 702-C, inventory of wine.

Winery Data Analysis Worksheet: ATF 5120.22

The *winery data analysis worksheet,* ATF *5120.22,* is one of the newer forms introduced by the ATF. It is designed to help account for losses that occur in the normal operations of the winery. Any unusual loss should be reported to the appropriate ATF Regional Administrator's office immediately in order that a casualty investigation may commence as soon as practicable.

Normal manufacturing losses of wine are not generally reported in each ATF Form 702 Monthly Report, despite the fact that several spaces are provided for such reporting. It has been the custom of most regional offices to allow an accrual of losses in the winery over each six-month period from July 1 – December 31 and January 1 – June 30, the ATF fiscal year. Prior to ATF 5120.22, such losses were accounted for by supplemental record-keeping in the winery. The new format provides a simple documentation of cellar and warehouse losses. This form should be attached to each June and December ATF 702, monthly report and 702-C, inventory of wine.

The instructions for preparation and submission of ATF 5120.22, winery data analysis worksheet, are explained on the form (see Fig. 10.17).

Record-keeping Flow Chart

The record-keeping flow card (see Fig. 10.18) is an updated version of the supplemental record-keeping flow chart (Fig. 10.1). This complete version contains the official ATF recording and reporting forms just discussed.

DEPARTMENT OF THE TREASURY - BUREAU OF ALCOHOL, TOBACCO AND FIREARMS						DATE OF INVENTORY	SHEET NO.
INVENTORY OF WINE (See Instructions on back)							
TYPE OF ESTABLISHMENT (Check one)	REGISTRY NO.		OPERATED BY (Name)			AT (City or town and State)	
☐ BONDED WINE CELLAR ☐ BONDED WINERY							

PART I - DETAILS AS TO INVEN-TORY OF (Check one)   ☐ STILL WINE   ☐ BULK PROCESS SPARKLING WINE   ☐ BOTTLE FERMENTED SPARKLING WINE   ☐ ARTIFICIALLY CARBON-ATED WINE

SERIAL NO. OF CONTAINER[1]	KINDS OF WINE		GALLONS			SERIAL NO. OF CONTAINER[1]	KINDS OF WINE		GALLONS		
	GRAPE[2]	OTHER WINE[3]	NOT OVER 14 PERCENT ALCOHOL	OVER 14 TO 21 PERCENT ALCOHOL	OVER 21 TO 24 PERCENT ALCOHOL		GRAPE[2]	OTHER WINE[3]	NOT OVER 14 PERCENT ALCOHOL	OVER 14 TO 21 PERCENT ALCOHOL	OVER 21 TO 24 PERCENT ALCOHOL

[1]Where barrels and puncheons are involved, list as "barrel" or "puncheon." Total gallonage of one kind of wine in bottles or cases may be entered as one item, appropriately identified.

[2]State whether port, sherry, vermouth, claret, retsina, etc., or if wine cannot be identified by a specific class or type, list as white table, red table, white dessert or red dessert, as the case may be. Wine intended for marketing as vintage wine should be appropriately identified as for example "Port—'38 Vintage."
[3]State whether orange, blackberry, apple, honey, etc.

ATF FORM 702-C (8-73)   EDITION OF 5-69 MAY BE USED

FIG. 10.15.   ATF FORM 702-C—INVENTORY OF WINE (FRONT SIDE)

KINDS	GALLONS				
	NOT OVER 14 PERCENT ALCOHOL	OVER 14 TO 21 PERCENT ALCOHOL	OVER 21 TO 24 PERCENT ALCOHOL	ARTIFICIALLY CARBONATED	SPARKLING

PART II - SUMMARY

TOTAL

Under penalties of perjury, I declare that I have examined this report of inventory and, to the best of my knowledge and belief, it is a true, correct and complete report of all wines required to be inventoried.

DATE	PROPRIETOR	BY (Signature and title)

INSTRUCTIONS

Use this form to report semiannual inventories of wine on hand June 30 and December 31 of each year, and special inventories taken at any other time in connection with losses reported on Form 702. Prepare in duplicate, using separate sheets for still wine, bulk process sparkling wine, bottle fermented sparkling wine, and artificially carbonated wine. Wine in fermenters will NOT be included in these inventories. If more than one sheet is required for showing details of inventory (Part I), number the sheets consecutively and show the summary (Part II) on the last sheet only.

In preparing the summary (Part II), report still wine separately by taxable grade as well as by kind, as shown in Part I. Whenever inventory is taken at the end of a month, the totals of all kinds of wine by tax class, as shown in the summary, should be entered by the proprietor in Form 702 as wine on hand last of month.

The original of this inventory will be attached to the original of Form 702 for the appropriate month and sent to the Regional Director, Bureau of Alcohol, Tobacco and Firearms. The copy will be retained at the bonded premises with the copy of Form 702.

FIG. 10.16   ATF FORM 702-C—INVENTORY OF WINE (REVERSE SIDE)

DEPARTMENT OF THE TREASURY
BUREAU OF ALCOHOL, TOBACCO AND FIREARMS
WINERY DATA ANALYSIS WORKSHEET
(Bottled Wines in Bond)

YEAR COVERED	NAME AND ADDRESS OF PROPRIETOR

INSTRUCTIONS

1. With the recording of losses and overages for June and December forward a copy of this form to the appropriate area supervisor through the Chief, Field Operations. When the figures for a complete year have been recorded, place the original in the winery's monthly report file.

2. Entries to lines 1a. through 12a., will be made by taking figures from ATF Form 702, Monthly Report of Wine Cellar Operations, Part I, Section B.

3. Case Count — Convert wine gallons into 4/5 Qt. 12 bottle case equivalents by dividing wine gallons by 2.4 wine gallons per case.

4. Compute the tax liability for each class of tax where losses are indicated, and enter the total liability for all classes. For wine under 14% alcohol, multiply the wine gallons by $.17/gal; for over 14% multiply by $.67/gal; for artifically carbonated wine multiply by $2.25/gal; and for sparkling wine multiply by $3.40/gal.

5. Enter under REMARKS a summary of the winemaker's explanation of the losses. Also use this section to note when a copy of the form is referred to the area supervisor, or when an inspection assignment is initiated as a result of the reported losses.

LINE		TAX CLASS (-) Losses/(+) Overages	NOT OVER 14%	14 to 21%	ARTIFICALLY CARBONATED	SPARKLING	TOTAL	CUMULATIVE TAX LIABILITY (JAN - JUN)
1	JAN	a. WINE GALLONS						
		b. CASE COUNT						
		c. TAX LIABILITY	$	$	$	$	$	
2	FEB	a. WINE GALLONS						
		b. CASE COUNT						
		c. TAX LIABILITY	$	$	$	$	$	$
3	MAR	a. WINE GALLONS						
		b. CASE COUNT						
		c. TAX LIABILITY	$	$	$	$	$	$
4	APR	a. WINE GALLONS						
		b. CASE COUNT						
		c. TAX LIABILITY	$	$	$	$	$	$
5	MAY	a. WINE GALLONS						
		b. CASE COUNT						
		c. TAX LIABILITY	$	$	$	$	$	$
6	JUN	a. WINE GALLONS						
		b. CASE COUNT						
		c. TAX LIABILITY	$	$	$	$	$	$
7	JUL	a. WINE GALLONS						CUMULATIVE TAX LIABILITY (JUL - DEC)
		b. CASE COUNT						
		c. TAX LIABILITY	$	$	$	$	$	$
8	AUG	a. WINE GALLONS						
		b. CASE COUNT						
		c. TAX LIABILITY	$	$	$	$	$	$
9	SEP	a. WINE GALLONS						
		b. CASE COUNT						
		c. TAX LIABILITY	$	$	$	$	$	$
10	OCT	a. WINE GALLONS						
		b. CASE COUNT						
		c. TAX LIABILITY	$	$	$	$	$	$
11	NOV	a. WINE GALLONS						
		b. CASE COUNT						
		c. TAX LIABILITY	$	$	$	$	$	$
12	DEC	a. WINE GALLONS						
		b. CASE COUNT						
		c. TAX LIABILITY	$	$	$	$	$	$

REMARKS (Use reverse if more space is needed)

ATF F 5120.22 (10-76)

FIG. 10.17. ATF FORM 5120.22—WINERY DATA ANALYSIS WORKSHEET

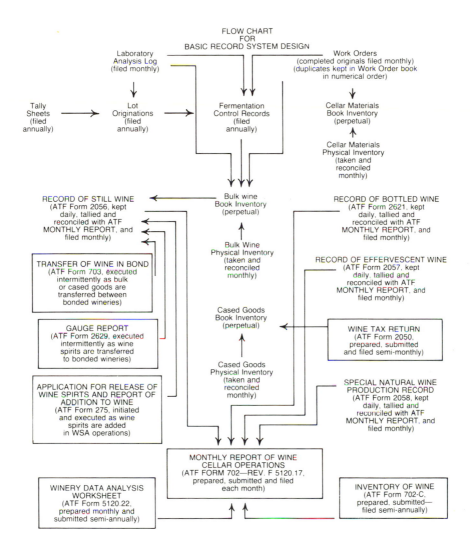

FIG. 10.18. FLOW CHART FOR BASIC RECORD SYSTEM DESIGN
(Forms in boxes are submitted to ATF as required.)

Certificate Of Label Approval Under The Federal Alcohol Administration Act: Form 1649

No cased goods may be sold or removed from a bonded winery or bonded wine cellar unless there is an approved ATF Form 1649 for that label on file in the winery. Title 27, CFR, Sections 4.30 through 4.39 contain details about what may and may not comprise the front, back and neck labels and other materials on wine bottles. It is advised that the vintner make himself fully aware of these regulations prior to designing and printing the labels. When specific design concepts are in doubt, the ATF in Washington, D.C. will usually provide an opinion on submitted proofwork prior to the actual print run. It must be understood, though, that such opinions are not binding and that sometimes a lack of color contrast or other printing flaws can necessitate redesigning and completely new printing.

SUMMARY

Despite the fact that there seems to be a multitude of forms and formats described in this chapter, there can be far more paperwork in larger wineries where specialized operations can create overwhelming volumes of record-keeping. An effort has been made in this chapter, and throughout this text, to minimize repetition in record-keeping without threatening the integrity of a system.

The progression of grapes from juice or must to shipment, can be seen in the flow chart in Fig. 10.18, fully expanded from the earlier version. The beginner may be overwhelmed by this diagram, but investigation and involvement with winemaking in the small commercial winery will show this system to be relatively simple and logical.

New regulations continually promulgated by the ATF result in intermittent changes in the form required for record-keeping. Some of the newer changes dispense with standard inventory formats such as ATF 2056, ATF 2057, etc. Winemasters are encouraged to continue with the same formats, adjusted if necessary, to fit individual needs.

Appendix A

Analytical Methods and Procedures in the Small Winery Laboratory

1. Acetaldehyde Determination
2. Agar Slant Preparation
3. Alcohol Determination by Distillation
4. Alcohol Determination by Salleron-DuJardin Ebulliometer
5. Balling Determination
6. Brix Determination by Hydrometer
7. Brix Determination by Refractometer
8. Carbon Dioxide Determination by Piercing Device
9. Copper and Iron Determination by Spectrophotometry
10. Differential Stain Procedure
11. Extract Determination by Hydrometer
12. Extract Determination by Nomograph—Dessert Wines
13. Extract Determination by Nomograph—Table Wines
14. Gram Stain Procedure
15. Light Transmission (Color Intensity) by Spectrophotometry
16. Malo-Lactic Fermentation Determination by Paper Chromatography
17. Microscopy
18. Organoleptic Analysis
19. Oxygen Determination
20. pH Determination
21. Plating Procedure
22. Sulfur Dioxide—Free
23. Sulfur Dioxide—Total
24. Total Acidity Determination by Titration
25. Total Acidity Determination by Titration—pH Meter
26. Viable Microorganisms in Bottled Wines—Millipore Method
27. Viable Yeasts in Bottled Wines—Rapid Method of Detection
28. Volatile Acidity Determination by Cash Volatile Acid Apparatus

1.   ACETALDEHYDE DETERMINATION

When analyzing wines for total acetaldehyde content, a small percentage (3–4% in wines containing 20% ethanol and less than 1% in table wine containing 12% ethanol) is bound as acetal. This is not recovered in the usual procedures. The procedure given below is that of Jaulmes and Hamelle as tested by Guymon and Wright and is an official method of the AOAC.

Modifications to consider the acetal concentration can be made. The air oxidative changes taking place during the alkaline titration step are prevented by addition of a chelating agent (EDTA) to bind copper present. Copper is a catalyst for the oxidation reaction. Addition of a small amount of isopropyl alcohol also inhibits the oxidation.

Procedure

Potassium Metabisulfite Solution.—Dissolve 15 g of $K_2S_2O_5$ in water, add 70 ml of concentrated hydrochloric acid, and dilute to 1 liter with water. The titer of 10 ml of this solution should not be less than 24 ml of 0.1 N iodine solution.

Phosphate—EDTA Solution.—Dissolve any of the following combinations and 4.5 g of the disodium salt of EDTA in water and dilute to 1 liter.

200 g of $Na_3PO_4 \cdot 12H_2O$, or
188 g of $Na_2HPO_4 \cdot 12H_2O$ + 21 g of NaOH, or
72.6 g of $NaH_2PO_4 \cdot H_2O$ + 42 g of NaOH, or
71.7 g of KH_2PO_4 + 42 g of NaOH.

Sodium Borate Solution.—Dissolve 100 g of boric acid plus 170 g of sodium hydroxide in water and dilute to 1 liter.

Pipet 50 ml of the wine sample containing less than 30 mg of acetaldehyde into the distilling flask and add 50 ml of a saturated borax solution to bring the pH to about 8. Distill 50 ml into a 750-ml Erlenmeyer flask containing 300 ml of water and 10 ml each of the potassium metabisulfite and the phosphate-EDTA solutions; the pH of the solution in the Erlenmeyer flask prior to distillation should be in the range of 7.0–7.2.

After the distillation add 10 ml of 3 N hydrochloric acid and 10 ml of a freshly prepared 0.2% starch solution to the Erlenmeyer flask. Mix and then immediately titrate with 0.1 N iodine solution just to a faint blue end point. Add 10 ml of the sodium borate solution and rapidly titrate the liberated bisulfite with 0.02 N iodine solution, using a 25-ml buret, to the same blue end point. Avoid direct sunlight. The pH of the solution should be 8.8–9.5.

$$\text{Acetaldehyde, mg/liter} = \frac{(V)\ (N)\ (22.0)\ (1000)}{v}$$

where V = volume of iodine solution used for titration after the addition of the sodium borate solution, in ml
N = normality of the iodine solution
v = volume of wine sample, in ml

If the first titration with iodine, to remove the excess bisulfite, takes only a few drops, then either the bisulfite solution was not up to proper strength, or the wine contains excessive amounts of acetaldehyde; if the second titration consumes excessive amounts of iodine and yellow iodoform formation is noted, the buffer was improperly prepared and the solution is too alkaline.

2. AGAR SLANT PREPARATION

Agar slants are glass test tubes filled about halfway with agar having been allowed to solidify while the test tube was held in a slanting position. This increases the surface area upon which microbial cultures can be grown. For the most part, agar slants are used for the storage of pure yeast and bacteria cultures in the winery laboratory.

Procedure

Clean test tubes thoroughly, rinsing several times with hot water.

Prepare and add enough yeast dextrose agar to fill test tubes about half full or slightly less. In the upright position, stuff test tubes closed with cotton, or apply screw cap if so equipped. Sterilize in autoclave.

Carefully remove sterilized test tubes from autoclave while they are still hot and lay tubes down so that tops are raised about 2 in. or so—enough that a good surface area is made, but not so as to allow the agar to touch the cotton or the screw top. Cover with a clean towel and allow to cool.

After agar has sufficiently cooled and solidified, the slants may be inoculated with cultures administered by a sterile loop. As the cotton or screw top is removed the tube should be "flamed", and flaming should be repeated after the closure is re-applied. Paraffin film may then be applied over the closure and flamed carefully to form a seal. Following the desired level of growth, the slants may be refrigerated. Figure A.1 pictures agar slants in storage.

3. ALCOHOL DETERMINATION BY DISTILLATION

Wine of almost any type may be accurately analyzed for alcohol content by the distillation method as long as any remaining carbon dioxide gas is removed by careful agitation or filtration.

The rationale for the distillation method is to separate the alcohol from the wine into an inert liquid medium that has no dissolved solids to affect

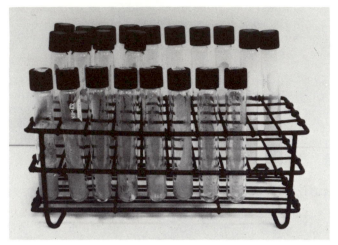

FIG. A.1.   AGAR SLANTS

the Tralle scale hydrometer. Figure 7.2 illustrates the apparatus that is
needed.

Apparatus and Reagents

200 ml Kohlrausch flask
1000 ml Erlenmeyer flask
Berl saddles
400 ml Allihn or Graham condenser
¼ in. ID Tygon tubing
¼ in. ID glass tubing
Rubber stopper to fit Erlenmeyer flask
Rubber stopper to fit condenser
Water trap
Tripod and wire gauze to support Erlenmeyer flask
Support rods with clamps for condenser
Thermometer
Cold running water for condenser
Distilled or deionized water
Tralle scale hydrometer (in the range of the wine to be tested)
Hydrometer jar or cylinder
Heat source
Refrigerated storage

Procedure

1. Adjust sample temperature to 68°F or to whatever temperature is
   indicated on the 200 ml Kohlrausch flask.

2. Fill clean 200 ml Kohlrausch flask so that the bottom meniscus rests exactly on the fill line. Shake out any bubbles that may be adhering to the sides of the flask.

3. Carefully empty contents of Kohlrausch flask into the 1000 ml Erlenmeyer flask and rinse out the Kohlrausch flask three times, each with about 25−30 ml distilled or deionized water. Empty all rinsings into the same 1000 ml Erlenmeyer flask. Add several Berl saddles into the 1000 ml Erlenmeyer flask.

4. Place the same clean Kohlrausch flask under the condenser outlet.

5. Insert connecting tube from the 1000 ml Erlenmeyer flask to condenser. Twist rubber stoppers slightly to be sure that they seal properly and will not leak.

6. Apply cold water at a proper rate through condenser and apply heat at a proper rate under Erlenmeyer flask.

7. Distill over approximately 150−175 ml of distillate. Be sure distillate runs out of condenser at a cool temperature. If not, start over cleaning all equipment three times with warm water, once with distilled water, and drain. Next time, increase condenser water flow rate appropriately but do not run too fast or the condenser may break from the pressure.

8. Add distilled water to distillate in Kohlrausch flask up to about 1 cm below the fill line. Mix by placing mouth of Kohlrausch flask on base of thumb in palm of hand so that no leakage takes place, inverting 6−8 times back and forth. Slide mouth of Kohlrausch flask off palm so no liquid is lost and place in refrigerated water bath. Cool to approximately 55°F.

9. Take Kohlrausch flask from refrigerated water bath and carefully dry the outside of the container with a clean towel. Adjust temperature to about 60°F and add distilled water in Kohlrausch flask so that the bottom meniscus rests exactly on the fill line. Remix contents in the same manner as Step 8.

10. Carefully empty Kohlrausch flask into a clean, dry hydrometer jar or cylinder.

11. Insert Tralle scale hydrometer carefully holding top of hydrometer stem in a pendulum effect.

12. Spin hydrometer carefully to free the instrument from the surface tension of the sides of the hydrometer jar or cylinder.

13. Read instrument at bottom meniscus. When using the hydrometer, take the temperature before and after reading. Preferably get two readings, as the sample is warming up and passing 60°F. Also note that it is very important to use a clean hydrometer in order to achieve accurate results.

   Calibration of hydrometers is desirable. Even factory calibrations should be checked by comparing hydrometers of the same range, since variations of 0.5% may be encountered. Post reading on laboratory analysis log.

14. Remove hydrometer from hydrometer jar carefully holding top of instrument in pendulum effect and rinse three times with warm water, once with distilled water and place in upright hydrometer rack.

15. Recheck temperature of distillate in hydrometer jar and make temperature corrections as indicated in Table A.1.

TABLE A.1.   CORRECTION OF ALCOHOL HYDROMETERS (TRALLE SCALE)
Calibrated at 60°F in Percent by Volume of Alcohol

Observed Alcohol Content Percent by Volume	Add To			Subtract From			
	57°F	58°F	59°F	61°F	62°F	63°F	64°F
1	0.14	0.10	0.05	0.05	0.10	0.16	0.22
2	0.14	0.10	0.05	0.05	0.11	0.17	0.23
3	0.14	0.10	0.05	0.06	0.12	0.18	0.24
4	0.14	0.10	0.05	0.06	0.12	0.19	0.25
5	0.15	0.10	0.05	0.07	0.13	0.20	0.26
6	0.17	0.11	0.06	0.07	0.14	0.20	0.27
7	0.18	0.12	0.06	0.07	0.14	0.21	0.29
8	0.19	0.13	0.06	0.08	0.16	0.23	0.31
9	0.21	0.14	0.07	0.08	0.16	0.24	0.32
10	0.23	0.16	0.08	0.08	0.17	0.25	0.34
11	0.25	0.16	0.08	0.09	0.18	0.27	0.37
12	0.27	0.18	0.09	0.10	0.20	0.29	0.39
13	0.29	0.19	0.10	0.10	0.21	0.31	0.42
14	0.32	0.21	0.11	0.11	0.22	0.32	0.44
15	0.35	0.23	0.12	0.12	0.24	0.35	0.48
16	0.37	0.24	0.12	0.13	0.26	0.38	0.52
17	0.40	0.26	0.13	0.14	0.27	0.41	0.54
18	0.44	0.29	0.14	0.14	0.29	0.44	0.58
19	0.47	0.32	0.16	0.15	0.30	0.46	0.62
20	0.51	0.34	0.17	0.16	0.32	0.49	0.66
21	0.53	0.35	0.18	0.17	0.34	0.51	0.68
22	0.56	0.38	0.19	0.17	0.36	0.53	0.71
23	0.58	0.41	0.20	0.18	0.37	0.55	0.74
24	0.60	0.40	0.20	0.18	0.38	0.56	0.77

16. Save contents remaining in 1000 ml Erlenmeyer flask for extract analysis (Procedure No. 11 in this Appendix), if desired.

17. Destroy sample from hydrometer jar and clean apparatus three times with warm tap water, once with distilled or deionized water, and drain dry.

4. ALCOHOL DETERMINATION BY SALLERON-DUJARDIN EBULLIOMETER

This analytical procedure should not be used for wines that are in excess of 14% alcohol and/or have a Balling of more than 0.0°.

The rationale for this procedure is to compare the boiling point of water with the boiling point of the wine being tested, the difference being due to the alcohol present in the wine. (See Fig. 4.9.)

Apparatus and Reagents

Salleron-DuJardin ebulliometer with thermometer and circular slide rule
100 ml graduated cylinder
Distilled or deionized water
Cold tap water

Procedure

1. Rinse entire inside surfaces of the ebulliometer with distilled or deionized water. Drain valve dry and close.

2. Fill upper reflux condenser jacket with cold tap water.

3. Measure 50 ml of distilled water into a clean 100 ml graduated cylinder and carefully pour into lower chamber inlet.

4. Very carefully insert thermometer into lower chamber inlet holding top of thermometer in one hand in pendulum effect and holding rubber stopper portion in the other hand. Twist rubber stopper position slightly to insure a tight fit.

5. Ignite ethanol burner and carefully position under lower chamber in proper position.

6. Watch thermometer mercury rise until it stops and holds for 15−20 sec at the same mark. Remove ethanol burner and close carefully to extinguish flame.

7. Remove thermometer very carefully in reverse manner to Step No. 4. Hold in vertical position until the mercury drops from the capillary. Clean with towel carefully and secure in cloth cover inside metal container provided.

8. Set circular slide rule at temperature reading noted for water boiling point (should be within the 99.5° to 100.5°C range). The water reading should be redone at least every 2 hours during the day, if testing is to continue.

9. Empty instrument carefully and rinse inner surfaces with 25–50 ml of wine sample to be tested. Drain instrument. Fill upper reflux condenser with cold tap water. Be careful that no water goes down inner tube.

10. Rinse the 100 ml graduate with about 25 ml of sample to be tested. Measure 50 ml of wine sample.

11. Repeat Steps 4, 5, 6, and 7.

12. Compare reading of thermometer to corresponding alcohol percentage on the circular slide rule. Post analysis on laboratory analysis log.

    Example: Water at 99.8°C, wine at 91.1°C = alcohol by volume at 12.1%.

13. Rinse instrument and graduated cylinder three times with warm water, once with distilled or deionized water, and drain dry.

5. BALLING DETERMINATION

The determination of Balling measures "mouth-feel", "body", specific gravity, or a number of other expressions. The Balling serves in the analysis of the combined effects contributed by alcohol and dissolved solids in wine. Figure 4.8 illustrates a Balling test being made.

Apparatus and Reagents

Brix-Balling hydrometer in the range of wine to be tested
Hydrometer jar or cylinder
Thermometer

Procedure

1. If the sample has been taken from a fermenter and contains carbon dioxide gas, the gas should be removed by careful agitation or filtration.

2. Adjust sample temperature to that required as indicated by the stem of the hydrometer.

3. Pour about 50 ml of the sample to be tested into the hydrometer jar, rinse and discard.

4. Pour sample into rinsed hydrometer jar up to about 2 in. from the top.

5. Insert clean and dry hydrometer carefully holding top of hydrometer stem in a pendulum effect.

6. Spin hydrometer carefully to free the instrument from the surface tension of the hydrometer jar.

7. Read instrument directly at the bottom meniscus; retake temperature immediately and make any adjustment necessary according to Table A.2, temperature corrections for Brix-Balling hydrometers. Post final results on laboratory analysis log.

8. Clean instruments and hydrometer jar three times with warm water, once with distilled or deionized water and drain dry, storing the hydrometer upright in the hydrometer rack.

6. BRIX DETERMINATION BY HYDROMETER

The Brix test is a measurement of dissolved solids in a wine being tested. Should there be any alcohol in the wine sample, the test would be properly called a Balling. The actual testing procedure is, however, identical. An illustration of this test being made is provided in Fig. 4.8.

Apparatus and Reagents

Identical to Procedure No. 5, *Balling Determination*

Procedure

Identical to Procedure No. 5, *Balling Determination*

7. BRIX DETERMINATION BY REFRACTOMETER

The rationale for this procedure is to apply a clean representative juice sample to a clean refractometer prism. Exposure to a light source refracts the incident light in a quantitative manner appropriate to the amount of dissolved sugar solids in the sample. For refractometers, temperature corrections are also available and there are hand refractometers that automat-

TABLE A.2. TEMPERATURE CORRECTIONS FOR BRIX-BALLING HYDROMETERS (CALIBRATED AT 20°C)

Temperature		Observed Percent of Sugar														
°C	°F	0	5	10	15	20	25	30	35	40	45	50	55	60	65	70
		Correction to be Subtracted from Observed Percent														
0	32.0	0.30	0.49	0.65	0.77	0.89	0.99	1.08	1.16	1.24	1.31	1.37	1.41	1.44	1.47	1.49
5	41.0	0.36	0.47	0.56	0.65	0.73	0.80	0.86	0.91	0.97	1.01	1.05	1.08	1.10	1.12	1.14
10	50.0	0.32	0.38	0.43	0.48	0.52	0.57	0.60	0.64	0.67	0.70	0.72	0.74	0.75	0.76	0.77
11	51.8	0.31	0.35	0.40	0.44	0.48	0.51	0.55	0.58	0.60	0.63	0.65	0.66	0.68	0.69	0.70
12	53.6	0.29	0.32	0.36	0.40	0.43	0.46	0.50	0.52	0.54	0.56	0.58	0.59	0.60	0.61	0.62
13	55.4	0.26	0.29	0.32	0.35	0.38	0.41	0.44	0.46	0.48	0.49	0.51	0.52	0.53	0.54	0.55
14	57.2	0.24	0.26	0.29	0.31	0.34	0.36	0.38	0.40	0.41	0.42	0.44	0.45	0.46	0.47	0.47
15	59.0	0.20	0.22	0.24	0.26	0.28	0.30	0.32	0.33	0.34	0.36	0.36	0.37	0.38	0.38	0.39
16	60.8	0.17	0.18	0.20	0.22	0.23	0.25	0.26	0.27	0.28	0.28	0.29	0.30	0.31	0.32	0.32
17	62.6	0.13	0.14	0.15	0.16	0.18	0.19	0.20	0.20	0.21	0.21	0.22	0.23	0.23	0.23	0.24
18	64.4	0.09	0.10	0.10	0.11	0.12	0.13	0.13	0.14	0.14	0.14	0.15	0.15	0.15	0.16	0.16
19	66.2	0.05	0.05	0.05	0.06	0.06	0.06	0.07	0.07	0.07	0.07	0.08	0.08	0.08	0.08	0.08
20	68.0	0	0	0	0	0	0	0	0	0	0	0	0	0	0	0
		Correction to be Added to Observed Percent														
21	69.8	0.04	0.05	0.06	0.06	0.06	0.07	0.07	0.07	0.07	0.08	0.08	0.08	0.08	0.08	0.09
22	71.6	0.10	0.10	0.11	0.12	0.12	0.13	0.14	0.14	0.15	0.15	0.16	0.16	0.16	0.16	0.16
23	73.4	0.16	0.16	0.17	0.17	0.19	0.20	0.21	0.21	0.22	0.23	0.24	0.24	0.24	0.24	0.24
24	75.2	0.21	0.22	0.23	0.24	0.26	0.27	0.28	0.29	0.30	0.31	0.32	0.32	0.32	0.32	0.32
25	77.0	0.27	0.28	0.30	0.31	0.32	0.34	0.35	0.36	0.38	0.38	0.39	0.39	0.40	0.39	0.39
26	78.8	0.33	0.34	0.36	0.37	0.40	0.40	0.42	0.44	0.46	0.47	0.47	0.48	0.48	0.48	0.48
27	80.6	0.40	0.41	0.42	0.44	0.46	0.48	0.50	0.52	0.54	0.54	0.55	0.56	0.56	0.56	0.56
28	82.4	0.46	0.47	0.49	0.51	0.54	0.56	0.58	0.60	0.61	0.62	0.63	0.64	0.64	0.64	0.64
29	84.2	0.54	0.55	0.56	0.59	0.61	0.63	0.66	0.68	0.70	0.70	0.71	0.72	0.72	0.72	0.72
30	86.0	0.61	0.62	0.63	0.66	0.68	0.71	0.73	0.76	0.78	0.78	0.79	0.80	0.80	0.80	0.81
35	95.0	0.99	1.01	1.02	1.06	1.10	1.13	1.16	1.18	1.20	1.21	1.22	1.22	1.23	1.22	1.22
40	104.0	1.42	1.45	1.47	1.51	1.54	1.57	1.60	1.62	1.64	1.65	1.65	1.65	1.66	1.65	1.65
45	113.0	1.91	1.94	1.96	2.00	2.03	2.05	2.07	2.09	2.10	2.10	2.10	2.10	2.10	2.09	2.08
50	122.0	2.46	2.48	2.50	2.53	2.56	2.57	2.58	2.59	2.59	2.58	2.58	2.57	2.56	2.54	2.52
55	131.0	3.05	3.07	3.09	3.12	3.12	3.12	3.12	3.11	3.10	3.08	3.07	3.05	3.03	3.00	2.97
60	140.0	3.69	3.72	3.73	3.73	3.72	3.70	3.67	3.65	3.62	3.60	3.57	3.54	3.50	3.47	3.43
70	158.0	5.1	5.1	5.1	5.0	5.0	5.0	4.9	4.8	4.8	4.7	4.7	4.6	4.6	4.5	4.4
80	176.0	7.1	7.0	7.0	6.9	6.8	6.7	6.6	6.4	6.3	6.2	6.1	6.0	5.9	5.7	5.6

EXAMPLE: If hydrometer reads 50° Brix at 24° C, the corrected reading would be 50 + .32 or 50.32° Brix.
Note: The table should be used with caution and only for approximate results when the temperature differs much from the standard temperature or from the temperature of the surrounding air.

ically adjust for temperature. Alcohol in the sample has a greater effect in refractometry than with hydrometry, so this method is best used only with juice. See Fig. 3.3.

Apparatus and Reagents

Refractometer
Light source

Procedure

1. Adjust sample temperature to that required by operating instruction of the instrument.

2. Open prism cover and rinse prism surface with several drops of sample, but do not flood. Gently wipe dry with absorbent lens paper.

3. Apply several drops of sample again and close prism cover. Point instrument towards light source and hold in the same manner as a telescope. Adjust focus and read Brix at light-dark dividing line. Post result on laboratory analysis log.

4. Rinse instrument prism surface and prism cover three times with warm water, once with distilled or deionized water and wipe dry with absorbent lens paper. Do not scratch prism surface.

8. CARBON DIOXIDE VOLUMES DETERMINATION BY PIERCING DEVICE

The rationale for this method of carbon dioxide determination is to measure the equalized head-space pressure at a given temperature. The two factors are plotted on a table in order to find the CO_2 volumes.

Apparatus and Reagents

Zahm and Nagel piercing device
Thermometer

Procedure

1. Raise crossbar and place sample bottle directly under piercing tip, being sure that valve is closed.

2. With a single firm motion push the crossbar down forcefully so as to be sure that the crown cap is fully pierced, releasing crossbar in that position.

3. Holding device and bottle in a horizontal position securely shake briskly for about 30 sec so as to render an equilibrium between dissolved CO_2 in the wine and CO_2 gas in the bottle head space.

4. Replace device and bottle to upright position and read gauge, recording result on laboratory analysis log. Release valve and allow foam to discharge in sink drain. Remove crown cap and immediately take temperature, recording result on laboratory analysis log.

5. Find carbon dioxide volumes by plotting pressure and temperature on Table A.3. Post carbon dioxide volumes determined upon laboratory analysis log.

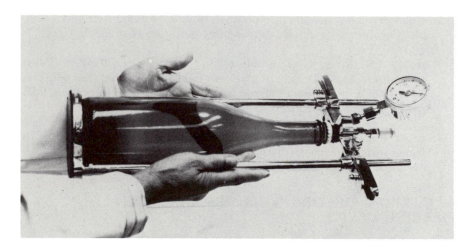

FIG. A.2. USE OF ZAHM AND NAGEL PIERCING DEVICE

9. COPPER AND IRON DETERMINATION BY SPECTROPHOTOMETRY

These analyses are considered together as they are very similar in procedure and wine samples are normally analyzed for copper and iron together.

The rationale of these procedures is to separate the copper and iron from the wine sample into a solution that can be measured for color intensity by the spectrophotometer.

Apparatus and Reagents

1/5 N ammonium hydroxide
Amyl acetate

Methyl alcohol, absolute
Sodium diethyldithiocarbamate
Hydrochloric-citric acid reagent
Sodium sulfate, anhydrous
Copper standard solutions: 0.5 ppm, 1.0 ppm and 2.5 ppm
Iron standard solutions: 1.0 ppm, 5.0 ppm and 10.0 ppm
Hydrochloric acid, dilute 1:3, $HCl:H_2O$
Hydrogen peroxide, 30%
Potassium thiocyanate, 50%
1 ml volumetric pipets
5 ml volumetric pipets
10 ml volumetric pipets
Separatory funnels
Filter paper (9 or 11 cm diameter), dense quality No. 2
Spectrophotometer, cuvettes and associated equipment
50 ml graduated cylinders
100 ml Griffin beakers
Cold tap water
Distilled or deionized water

Preparation of Copper Standards and Sample

1. Pipet 10 ml each of the copper standards into clean separatory funnels.

2. Add 1 ml of hydrochloric-citric acid reagent to each funnel; mix.

3. Add 2 ml of 1/5 N ammonium hydroxide to each funnel; mix.

4. Add 1 ml of 1% sodium diethyldithiocarbamate solution to each funnel, mix and allow 1 min for development of color.

5. In a graduated cylinder measure 10 ml amyl acetate, then add methyl alcohol up to 15 ml, mix, and add to each of the separatory funnels. (Optional): Mix a batch of 2:1 amyl acetate and methanol for a day's analysis; pipet 20 ml to each separatory funnel for extraction.

6. Shake funnels for 30 sec, and allow the 2 phases to separate.

7. Draw the lower portion from the funnel stopcocks and discard.

8. Transfer the upper (yellow) phases into clean small beakers.

9. While mixing, add small amounts of anhydrous sodium sulfate until the liquid is no longer cloudy.

10. Filter through paper into cuvettes.

TABLE A.3.   VOLUMES OF CO_2 GAS DISSOLVED IN WINE

Pressure Pounds Per Square Inch

Temperature Degrees Fahrenheit

	0	2	4	6	8	10	12	14	16	18	20	22	24	26	28	30	32	34	36	38	40	42	44	46	48	50
32°	1.71	1.9	2.2	2.4	2.6	2.9	3.1	3.3	3.5	3.8	4.0	4.2	4.4	4.7	4.9	5.2	5.4	5.6	5.8	6.1	6.3	6.5	6.7	7.0	7.2	7.4
33	1.68	1.9	2.1	2.4	2.6	2.8	3.0	3.2	3.5	3.7	3.9	4.1	4.3	4.6	4.8	5.1	5.3	5.5	5.7	5.9	6.2	6.4	6.6	6.8	7.1	7.3
34	1.64	1.9	2.1	2.3	2.5	2.7	2.9	3.2	3.4	3.6	3.8	4.1	4.3	4.5	4.7	4.9	5.2	5.4	5.6	5.8	6.0	6.2	6.5	6.7	7.0	7.2
35	1.61	1.8	2.0	2.3	2.5	2.7	2.9	3.1	3.3	3.5	3.8	4.0	4.2	4.4	4.8	4.8	5.1	5.2	5.5	5.7	5.9	6.1	6.3	6.6	6.8	7.0
36	1.57	1.8	2.0	2.2	2.4	2.6	2.8	3.0	3.3	3.5	3.7	3.9	4.1	4.3	4.5	4.7	5.0	5.2	5.4	5.6	5.8	6.0	6.2	6.4	6.6	6.9
37	1.54	1.7	2.0	2.2	2.4	2.6	2.8	3.0	3.2	3.4	3.6	3.8	4.0	4.2	4.4	4.6	4.9	5.1	5.3	5.5	5.7	5.9	6.1	6.3	6.5	6.7
38	1.51	1.7	1.9	2.1	2.3	2.5	2.7	2.9	3.1	3.3	3.5	3.7	3.9	4.1	4.3	4.5	4.8	5.0	5.2	5.4	5.6	5.8	6.0	6.2	6.4	6.6
39	1.47	1.7	1.9	2.1	2.3	2.5	2.7	2.9	3.1	3.3	3.5	3.7	3.9	4.0	4.3	4.5	4.7	4.9	5.1	5.3	5.4	5.7	5.9	6.1	6.2	6.4
40°	1.45	1.6	1.8	2.0	2.2	2.4	2.6	2.8	3.0	3.2	3.4	3.6	3.8	4.0	4.2	4.3	4.5	4.7	4.9	5.1	5.3	5.5	5.7	5.9	6.1	6.3
41	1.42	1.6	1.8	2.0	2.2	2.4	2.6	2.8	2.9	3.1	3.3	3.5	3.7	3.9	4.1	4.2	4.4	4.6	4.8	5.0	5.2	5.4	5.6	5.8	6.0	6.2
42	1.40	1.6	1.8	2.0	2.1	2.3	2.5	2.8	2.9	3.1	3.3	3.5	3.6	3.8	4.0	4.2	4.4	4.6	4.7	4.9	5.1	5.3	5.5	5.7	5.9	6.1
43	1.37	1.6	1.7	1.9	2.1	2.3	2.5	2.7	2.8	3.0	3.2	3.4	3.6	3.8	3.9	4.1	4.3	4.5	4.7	4.8	5.0	5.2	5.4	5.6	5.8	6.0
44	1.35	1.5	1.7	1.9	2.1	2.2	2.4	2.6	2.8	3.0	3.1	3.3	3.5	3.7	3.9	4.0	4.2	4.4	4.6	4.8	5.0	5.1	5.3	5.5	5.7	5.9
45	1.32	1.5	1.7	1.8	2.0	2.2	2.4	2.5	2.7	2.9	3.1	3.3	3.4	3.6	3.8	4.0	4.1	4.3	4.5	4.7	4.8	5.0	5.2	5.4	5.6	5.7
46	1.29	1.5	1.6	1.8	2.0	2.2	2.3	2.5	2.7	2.8	3.0	3.2	3.4	3.5	3.7	3.9	4.0	4.2	4.4	4.6	4.7	4.9	5.1	5.3	5.4	5.6
47	1.26	1.4	1.6	1.8	1.9	2.1	2.3	2.4	2.6	2.8	2.9	3.1	3.3	3.5	3.6	3.8	4.0	4.1	4.3	4.5	4.6	4.8	5.0	5.2	5.3	5.5
48	1.24	1.4	1.6	1.7	1.9	2.1	2.2	2.4	2.6	2.7	2.9	3.1	3.2	3.4	3.6	3.7	3.9	4.1	4.2	4.4	4.6	4.7	4.9	5.1	5.2	5.4
49	1.21	1.4	1.5	1.7	1.9	2.0	2.2	2.4	2.5	2.7	2.8	3.0	3.2	3.3	3.5	3.7	3.8	4.0	4.1	4.3	4.5	4.6	4.8	5.0	5.1	5.3
50°	1.19	1.4	1.5	1.7	1.8	2.0	2.2	2.3	2.5	2.6	2.8	2.9	3.1	3.3	3.4	3.6	3.7	3.9	4.0	4.2	4.4	4.5	4.7	4.9	5.0	5.2
51	1.17	1.3	1.5	1.6	1.8	2.0	2.1	2.3	2.4	2.6	2.7	2.9	3.1	3.2	3.4	3.5	3.7	3.8	4.0	4.2	4.3	4.5	4.8	4.8	5.0	5.1
52	1.15	1.3	1.5	1.6	1.8	1.9	2.1	2.2	2.4	2.5	2.7	2.8	3.0	3.2	3.3	3.5	3.6	3.8	3.9	4.1	4.2	4.4	4.5	4.7	4.9	5.0
53	1.13	1.3	1.4	1.6	1.7	1.9	2.0	2.2	2.3	2.5	2.6	2.8	2.9	3.1	3.3	3.4	3.6	3.7	3.8	4.0	4.2	4.3	4.4	4.6	4.8	4.9
54	1.11	1.3	1.4	1.6	1.7	1.9	2.0	2.2	2.3	2.4	2.6	2.7	2.9	3.0	3.2	3.3	3.5	3.6	3.8	3.9	4.1	4.2	4.4	4.5	4.7	4.8
55	1.10	1.2	1.4	1.5	1.7	1.8	2.0	2.1	2.3	2.4	2.6	2.7	2.8	3.0	3.1	3.3	3.4	3.6	3.7	3.9	4.0	4.1	4.3	4.4	4.6	4.7
56	1.08	1.2	1.4	1.5	1.6	1.8	1.9	2.1	2.2	2.4	2.5	2.6	2.8	2.9	3.1	3.2	3.3	3.5	3.7	3.8	3.9	4.1	4.2	4.4	4.5	4.7
57	1.06	1.2	1.3	1.5	1.6	1.8	1.9	2.0	2.2	2.3	2.5	2.6	2.7	2.9	3.0	3.2	3.3	3.5	3.6	3.7	3.9	4.0	4.4	4.3	4.4	4.8
58	1.04	1.2	1.3	1.5	1.6	1.7	1.9	2.0	2.1	2.3	2.4	2.6	2.7	2.8	3.0	3.1	3.3	3.4	3.5	3.7	3.8	3.9	4.1	4.2	4.4	4.5
59	1.02	1.2	1.3	1.4	1.6	1.7	1.8	2.0	2.1	2.2	2.4	2.5	2.7	2.8	2.9	3.1	3.2	3.3	3.5	3.6	3.7	3.9	4.0	4.2	4.3	4.4
60°	1.00	1.1	1.3	1.4	1.5	1.7	1.8	1.9	2.1	2.2	2.3	2.5	2.6	2.7	2.9	3.0	3.1	3.3	3.4	3.5	3.7	3.8	3.9	4.1	4.2	4.3
61	0.98	1.1	1.2	1.4	1.5	1.6	1.8	1.9	2.0	2.2	2.3	2.4	2.6	2.7	2.8	3.0	3.1	3.2	3.3	3.5	3.6	3.7	3.9	4.0	4.1	4.3
62	0.97	1.1	1.2	1.4	1.5	1.6	1.7	1.9	2.0	2.1	2.3	2.4	2.5	2.6	2.8	2.9	3.0	3.2	3.3	3.4	3.6	3.7	3.8	4.0	4.1	4.2
63	0.95	1.1	1.2	1.3	1.5	1.6	1.7	1.8	2.0	2.1	2.2	2.4	2.5	2.6	2.7	2.9	3.0	3.1	3.2	3.4	3.5	3.6	3.8	3.9	4.0	4.2
64	0.93	1.1	1.2	1.3	1.4	1.6	1.7	1.8	1.9	2.1	2.2	2.3	2.4	2.6	2.7	2.8	2.9	3.1	3.2	3.3	3.5	3.6	3.7	3.8	3.9	4.1
65	0.92	1.1	1.2	1.3	1.4	1.5	1.7	1.8	1.9	2.0	2.2	2.3	2.4	2.5	2.6	2.8	2.9	3.0	3.1	3.4	3.5	3.6	3.8	3.9	4.0	
66	0.90	1.0	1.2	1.3	1.4	1.5	1.6	1.8	1.9	2.0	2.1	2.2	2.4	2.5	2.6	2.7	2.8	3.0	3.1	3.2	3.3	3.5	3.6	3.7	3.8	3.9
67	0.89	1.0	1.1	1.2	1.4	1.5	1.6	1.7	1.9	2.0	2.1	2.2	2.3	2.4	2.6	2.7	2.8	2.9	3.0	3.2	3.3	3.4	3.5	3.6	3.7	3.8
68	0.88	1.0	1.1	1.2	1.3	1.5	1.6	1.7	1.8	1.9	2.0	2.2	2.3	2.4	2.5	2.6	2.7	2.9	3.0	3.1	3.2	3.3	3.5	3.6	3.7	3.8
69	0.86	1.0	1.1	1.3	1.3	1.4	1.5	1.6	1.8	1.9	2.0	2.1	2.2	2.4	2.5	2.6	2.7	2.8	2.9	3.0	3.2	3.3	3.4	3.5	3.6	3.8
70°	0.85	1.0	1.1	1.2	1.3	1.4	1.5	1.6	1.7	1.9	2.0	2.1	2.2	2.3	2.4	2.5	2.7	2.8	2.9	3.0	3.1	3.2	3.3	3.5	3.6	3.7
71	0.84	0.9	1.1	1.2	1.3	1.4	1.5	1.6	1.7	1.8	1.9	2.1	2.2	2.3	2.4	2.5	2.6	2.7	2.8	2.9	3.1	3.2	3.3	3.4	3.5	3.6
72	0.83	0.9	1.0	1.2	1.3	1.4	1.5	1.6	1.7	1.8	1.9	2.0	2.1	2.2	2.4	2.5	2.6	2.7	2.8	2.9	3.0	3.1	3.2	3.4	3.5	3.6
73	0.81	0.9	1.0	1.1	1.2	1.4	1.5	1.6	1.7	1.8	1.9	2.0	2.1	2.2	2.3	2.4	2.5	2.6	2.8	2.9	3.0	3.1	3.2	3.3	3.4	3.5
74	0.79	0.9	1.0	1.1	1.2	1.3	1.4	1.5	1.6	1.8	1.9	2.0	2.1	2.2	2.3	2.4	2.5	2.6	2.7	2.8	2.9	3.0	3.1	3.2	3.3	3.5
75	0.78	0.9	1.0	1.1	1.2	1.3	1.4	1.5	1.6	1.7	1.8	2.0	2.2	2.3	2.4	2.5	2.6	2.7	2.8	2.9	3.0	3.1	3.2	3.3	3.4	
76	0.77	0.9	1.0	1.1	1.2	1.3	1.4	1.5	1.6	1.7	1.8	1.9	2.0	2.1	2.2	2.4	2.5	2.6	2.7	2.8	2.9	3.0	3.1	3.3	3.4	
77	0.76	0.9	1.0	1.1	1.2	1.3	1.4	1.5	1.6	1.7	1.8	1.9	2.0	2.1	2.2	2.3	2.4	2.5	2.6	2.7	2.8	2.9	3.0	3.1	3.2	3.3
78	0.75	0.9	0.9	1.0	1.1	1.2	1.3	1.4	1.5	1.6	1.7	1.8	1.9	2.0	2.1	2.2	2.3	2.4	2.5	2.6	2.7	2.8	2.9	3.0	3.1	3.3
79	0.74	0.8	0.9	1.0	1.1	1.2	1.3	1.4	1.5	1.6	1.7	1.8	1.9	2.0	2.1	2.2	2.4	2.5	2.6	2.7	2.8	2.9	3.0	3.1	3.2	
80°	0.73	0.8	0.9	1.0	1.1	1.2	1.3	1.4	1.5	1.6	1.7	1.8	1.9	2.0	2.1	2.2	2.3	2.4	2.5	2.6	2.7	2.8	2.9	3.0	3.1	3.2
81	0.72	0.8	0.9	1.0	1.1	1.2	1.3	1.4	1.5	1.6	1.7	1.8	1.9	2.0	2.1	2.2	2.3	2.4	2.5	2.6	2.7	2.8	2.9	3.0	3.1	
82	0.71	0.8	0.9	1.0	1.1	1.2	1.3	1.4	1.5	1.6	1.6	1.7	1.8	1.9	2.0	2.1	2.2	2.3	2.4	2.5	2.6	2.7	2.8	2.9	3.0	3.1
83	0.70	0.8	0.9	1.0	1.1	1.2	1.3	1.4	1.4	1.5	1.6	1.7	1.8	1.9	2.0	2.1	2.2	2.3	2.4	2.5	2.6	2.6	2.7	2.8	2.9	3.0
84	0.69	0.8	0.9	1.0	1.1	1.1	1.4	1.4	1.4	1.5	1.6	1.7	1.8	1.9	2.0	2.1	2.2	2.3	2.4	2.5	2.6	2.7	2.8	2.9	3.0	
85	0.68	0.8	0.9	0.9	1.0	1.1	1.2	1.3	1.4	1.5	1.6	1.7	1.8	1.9	1.9	2.0	2.1	2.2	2.3	2.4	2.5	2.6	2.7	2.7	2.8	2.9
86	0.67	0.8	0.8	0.9	1.0	1.1	1.2	1.3	1.4	1.5	1.5	1.6	1.7	1.8	1.9	2.0	2.1	2.2	2.3	2.4	2.4	2.5	2.6	2.7	2.8	2.9
87	0.66	0.7	0.8	0.9	1.0	1.1	1.2	1.3	1.4	1.4	1.5	1.6	1.7	1.8	1.9	2.0	2.1	2.1	2.2	2.3	2.4	2.5	2.6	2.7	2.8	
88	0.65	0.7	0.8	0.9	1.0	1.1	1.2	1.2	1.4	1.4	1.5	1.6	1.7	1.8	1.9	1.9	2.0	2.1	2.2	2.3	2.4	2.5	2.6	2.7	2.8	
89	0.64	0.7	0.8	0.9	1.0	1.1	1.1	1.2	1.3	1.4	1.5	1.6	1.7	1.7	1.8	1.9	2.0	2.1	2.2	2.3	2.3	2.4	2.5	2.6	2.7	2.8
90°	0.63	0.7	0.8	0.9	1.0	1.0	1.1	1.2	1.3	1.4	1.5	1.6	1.6	1.7	1.8	1.9	2.0	2.1	2.1	2.2	2.3	2.4	2.5	2.6	2.7	2.7
91	0.62	0.7	0.8	0.9	0.9	1.0	1.1	1.2	1.3	1.4	1.5	1.6	1.6	1.7	1.8	1.9	2.0	2.0	2.1	2.2	2.3	2.4	2.4	2.5	2.6	2.7
92	0.61	0.7	0.8	0.8	0.9	1.0	1.1	1.2	1.3	1.4	1.4	1.5	1.6	1.7	1.8	1.9	1.9	2.0	2.1	2.2	2.3	2.4	2.5	2.6	2.7	
93	0.60	0.7	0.8	0.8	0.9	1.0	1.1	1.2	1.3	1.4	1.4	1.5	1.6	1.7	1.7	1.8	1.9	2.0	2.1	2.1	2.2	2.3	2.4	2.5	2.5	2.6
94	0.60	0.7	0.8	0.8	0.9	1.0	1.1	1.2	1.2	1.3	1.4	1.5	1.6	1.6	1.7	1.8	1.9	2.0	2.0	2.1	2.2	2.3	2.3	2.4	2.5	2.6
95	0.59	0.7	0.7	0.8	0.9	1.0	1.1	1.1	1.2	1.3	1.4	1.5	1.5	1.6	1.7	1.8	1.9	1.9	2.0	2.1	2.2	2.3	2.3	2.4	2.5	2.6
96	0.58	0.7	0.7	0.8	0.9	0.9	1.0	1.1	1.2	1.3	1.4	1.4	1.5	1.6	1.7	1.8	1.8	1.9	2.0	2.1	2.2	2.2	2.3	2.4	2.5	2.5
97	0.57	0.7	0.7	0.8	0.9	1.0	1.0	1.1	1.2	1.3	1.3	1.4	1.5	1.6	1.7	1.7	1.8	1.9	2.0	2.0	2.1	2.2	2.3	2.3	2.4	2.5
98	0.57	0.6	0.7	0.8	0.9	0.9	1.0	1.1	1.2	1.2	1.3	1.4	1.5	1.5	1.6	1.7	1.8	1.9	2.0	2.0	2.1	2.2	2.3	2.3	2.4	2.5
99	0.56	0.6	0.7	0.8	0.8	0.9	1.0	1.1	1.1	1.2	1.2	1.3	1.4	1.5	1.6	1.6	1.7	1.8	1.9	1.9	2.0	2.1	2.2	2.3	2.3	2.4
100°	0.56	0.6	0.7	0.8	0.8	0.9	1.0	1.1	1.1	1.2	1.3	1.4	1.4	1.5	1.6	1.7	1.7	1.8	1.9	2.0	2.0	2.1	2.2	2.3	2.3	2.4

52	54	56	58	60	62	64	66	68	70	72	74	76	78	80	82	84	86	88	90	92	94	96	98	100	
7.7	7.9	8.2	8.4	8.6	8.8	9.0	9.3	9.5	9.7	10.0	10.2	10.4	10.7	10.9	11.2	11.5	11.7	12.0	12.2	12.4	12.7	12.9	13.2	13.4	32°
7.5	7.8	8.0	8.2	8.4	8.6	8.9	9.1	9.3	9.5	9.8	10.0	10.2	10.4	10.7	11.0	11.3	11.5	11.7	11.9	12.2	12.4	12.6	12.9	13.1	33
7.4	7.6	7.8	8.0	8.2	8.4	8.7	8.9	9.1	9.3	9.6	9.8	10.0	10.2	10.5	10.8	11.0	11.2	11.5	11.7	11.9	12.2	12.4	12.6	12.8	34
7.2	7.4	7.6	7.8	8.0	8.3	8.5	8.7	8.9	9.2	9.4	9.6	9.8	10.0	10.3	10.6	10.8	11.0	11.3	11.5	11.7	11.9	12.1	12.3	12.5	35
7.1	7.3	7.5	7.7	7.9	8.1	8.3	8.6	8.8	9.0	9.2	9.4	9.6	9.8	10.0	10.4	10.6	10.8	11.0	11.2	11.4	11.7	11.9	12.1	12.3	36
6.9	7.1	7.4	7.6	7.8	8.0	8.2	8.4	8.6	8.8	9.0	9.2	9.4	9.6	9.8	10.1	10.3	10.6	10.8	11.0	11.2	11.4	11.6	11.8	12.0	37
6.8	7.0	7.2	7.4	7.6	7.8	8.0	8.2	8.4	8.6	8.8	9.0	9.2	9.4	9.6	9.9	10.1	10.3	10.5	10.7	10.9	11.1	11.4	11.6	11.8	38
6.6	6.8	7.0	7.2	7.4	7.6	7.8	8.0	8.2	8.4	8.6	8.8	9.0	9.2	9.4	9.7	9.9	10.1	10.3	10.5	10.7	10.9	11.1	11.3	11.5	39
6.5	6.7	6.9	7.1	7.3	7.5	7.7	7.9	8.1	8.3	8.5	8.7	8.8	9.0	9.2	9.5	9.7	9.9	10.1	10.3	10.5	10.7	10.9	11.1	11.3	40°
6.4	6.6	6.8	7.0	7.1	7.3	7.5	7.7	7.9	8.1	8.3	8.5	8.7	8.9	9.1	9.4	9.6	9.8	10.0	10.2	10.3	10.5	10.7	10.9	11.1	41
6.3	6.4	6.6	6.8	7.0	7.2	7.4	7.6	7.8	8.0	8.2	8.3	8.5	8.7	8.9	9.2	9.4	9.6	9.8	10.0	10.1	10.3	10.5	10.7	10.9	42
6.1	6.3	6.5	6.7	6.9	7.0	7.2	7.4	7.6	7.8	8.0	8.2	8.3	8.5	8.7	9.0	9.2	9.4	9.6	9.8	10.0	10.2	10.4	10.6	10.7	43
6.0	6.2	6.4	6.6	6.7	6.9	7.1	7.3	7.5	7.6	7.8	8.0	8.2	8.4	8.6	8.8	9.1	9.3	9.5	9.6	9.8	10.0	10.2	10.3	10.5	44
5.9	6.1	6.2	6.4	6.6	6.8	6.9	7.1	7.3	7.5	7.7	7.8	8.0	8.2	8.4	8.7	8.9	9.0	9.3	9.4	9.6	9.8	10.0	10.1	10.3	45
5.8	6.0	6.1	6.3	6.4	6.6	6.8	7.0	7.2	7.4	7.5	7.7	7.9	8.0	8.2	8.4	8.6	8.8	9.0	9.2	9.4	9.6	9.7	9.9	10.1	46
5.7	5.9	6.0	6.2	6.3	6.5	6.7	6.9	7.0	7.2	7.4	7.6	7.7	7.9	8.0	8.3	8.5	8.7	8.9	9.0	9.2	9.4	9.5	9.7	9.9	47
5.6	5.7	5.9	6.1	6.2	6.4	6.6	6.8	6.9	7.1	7.2	7.4	7.6	7.7	7.9	8.1	8.3	8.5	8.7	8.8	9.0	9.2	9.3	9.5	9.7	48
5.5	5.6	5.8	6.0	6.1	6.3	6.4	6.6	6.8	6.9	7.1	7.2	7.4	7.6	7.8	8.0	8.2	8.3	8.5	8.7	8.9	9.0	9.2	9.3	9.5	49
5.4	5.5	5.7	5.9	6.0	6.2	6.3	6.5	6.6	6.8	7.0	7.1	7.3	7.4	7.6	7.9	8.0	8.2	8.4	8.5	8.7	8.9	9.0	9.2	9.3	50°
5.3	5.4	5.6	5.7	5.9	6.1	6.2	6.4	6.5	6.7	6.8	7.0	7.2	7.3	7.5	7.7	7.9	8.0	8.2	8.4	8.5	8.7	8.8	9.0	9.2	51
5.2	5.3	5.5	5.6	5.8	5.9	6.1	6.3	6.4	6.6	6.7	6.9	7.0	7.2	7.3	7.6	7.8	7.9	8.1	8.2	8.4	8.5	8.7	8.9	9.0	52
5.1	5.2	5.4	5.5	5.7	5.9	6.0	6.1	6.3	6.4	6.6	6.7	6.9	7.0	7.2	7.4	7.6	7.8	7.9	8.1	8.2	8.4	8.5	8.7	8.9	53
5.0	5.2	5.3	5.4	5.6	5.7	5.9	6.0	6.2	6.3	6.5	6.6	6.8	6.9	7.1	7.3	7.5	7.6	7.8	8.0	8.1	8.3	8.4	8.6	8.7	54
4.9	5.1	5.2	5.3	5.5	5.6	5.8	5.9	6.1	6.2	6.3	6.5	6.6	6.8	6.9	7.2	7.4	7.5	7.7	7.8	8.0	8.1	8.3	8.4	8.6	55
4.8	5.0	5.1	5.2	5.4	5.5	5.7	5.8	6.0	6.1	6.2	6.4	6.5	6.7	6.8	7.0	7.2	7.4	7.5	7.7	7.8	8.0	8.1	8.3	8.4	56
4.7	4.9	5.0	5.2	5.3	5.4	5.6	5.7	5.9	6.0	6.1	6.3	6.4	6.6	6.7	6.9	7.1	7.2	7.4	7.5	7.7	7.8	8.0	8.1	8.3	57
4.6	4.7	4.9	5.1	5.2	5.3	5.5	5.6	5.8	5.9	6.0	6.2	6.3	6.4	6.6	6.8	7.0	7.1	7.3	7.4	7.5	7.7	7.8	8.0	8.1	58
4.6	4.7	4.8	5.0	5.1	5.3	5.4	5.5	5.7	5.8	5.9	6.1	6.2	6.3	6.5	6.7	6.8	7.0	7.1	7.3	7.4	7.5	7.7	7.8	8.0	59
4.5	4.6	4.7	4.9	5.0	5.2	5.3	5.4	5.5	5.7	5.8	6.0	6.1	6.2	6.3	6.6	6.7	6.8	7.0	7.1	7.2	7.4	7.5	7.7	7.8	60°
4.4	4.5	4.7	4.8	4.9	5.1	5.2	5.3	5.5	5.6	5.7	5.9	6.0	6.1	6.2	6.4	6.6	6.7	6.9	7.0	7.1	7.3	7.4	7.6	7.7	61
4.3	4.4	4.6	4.7	4.8	5.0	5.1	5.3	5.4	5.5	5.6	5.8	5.9	6.1	6.1	6.3	6.5	6.6	6.8	6.9	7.0	7.2	7.3	7.4	7.6	62
4.3	4.4	4.5	4.6	4.8	4.9	5.0	5.2	5.3	5.4	5.5	5.7	5.8	5.9	6.1	6.2	6.4	6.5	6.7	6.8	6.9	7.0	7.2	7.3	7.4	63
4.2	4.3	4.4	4.6	4.7	4.9	5.1	5.2	5.3	5.4	5.5	5.7	5.8	6.0	6.1	6.3	6.4	6.5	6.7	6.8	6.9	7.2	7.3	7.4		64
4.1	4.2	4.5	4.4	4.5	4.7	4.8	4.9	5.0	5.2	5.4	5.5	5.6	5.8	5.9	6.0	6.2	6.3	6.4	6.5	6.7	6.8	6.9	7.0	7.2	65
4.0	4.2	4.3	4.4	4.5	4.7	4.8	4.9	5.0	5.2	5.3	5.4	5.5	5.7	5.8	5.9	6.1	6.2	6.3	6.4	6.5	6.7	6.8	6.9	7.0	66
4.0	4.1	4.2	4.3	4.4	4.6	4.7	4.8	4.9	5.1	5.2	5.3	5.4	5.5	5.7	5.8	6.0	6.1	6.2	6.3	6.5	6.6	6.7	6.8	6.9	67
3.9	4.0	4.2	4.3	4.4	4.5	4.6	4.7	4.8	5.0	5.1	5.2	5.3	5.4	5.6	5.7	5.9	6.0	6.1	6.2	6.4	6.5	6.6	6.7	6.8	68
3.9	4.0	4.1	4.2	4.3	4.4	4.5	4.7	4.8	4.9	5.0	5.1	5.3	5.4	5.5	5.7	5.8	5.9	6.0	6.1	6.3	6.4	6.5	6.6	6.7	69
3.8	3.9	4.0	4.1	4.2	4.3	4.5	4.6	4.7	4.8	4.9	5.1	5.2	5.3	5.4	5.6	5.7	5.8	6.0	6.1	6.2	6.3	6.4	6.5	6.6	70°
3.7	3.9	4.0	4.1	4.2	4.3	4.4	4.5	4.6	4.7	4.9	5.0	5.1	5.2	5.3	5.5	5.6	5.7	5.9	6.0	6.1	6.2	6.3	6.4	6.5	71
3.7	3.8	3.9	4.0	4.1	4.2	4.3	4.4	4.6	4.7	4.8	4.9	5.0	5.1	5.2	5.4	5.5	5.6	5.8	5.9	6.0	6.1	6.2	6.3	6.4	72
3.6	3.7	3.8	3.9	4.0	4.1	4.2	4.4	4.5	4.6	4.7	4.8	4.9	5.0	5.1	5.3	5.4	5.5	5.7	5.8	5.9	6.0	6.1	6.2	6.3	73
3.6	3.7	3.8	3.9	4.0	4.1	4.2	4.3	4.4	4.5	4.6	4.7	4.8	4.9	5.0	5.2	5.3	5.4	5.5	5.6	5.7	5.8	5.9	6.0	6.1	74
3.5	3.6	3.7	3.8	3.9	4.0	4.1	4.2	4.3	4.4	4.5	4.7	4.8	4.9	5.0	5.1	5.2	5.3	5.4	5.5	5.6	5.7	5.8	5.9	6.0	75
3.5	3.6	3.7	3.8	3.9	4.0	4.1	4.2	4.3	4.4	4.5	4.6	4.7	4.8	4.9	5.0	5.1	5.2	5.3	5.4	5.5	5.6	5.7	5.8	6.0	76
3.4	3.5	3.6	3.7	3.8	3.9	4.0	4.1	4.2	4.3	4.4	4.5	4.6	4.7	4.8	5.0	5.1	5.2	5.3	5.4	5.5	5.6	5.7	5.8	5.9	77
3.4	3.5	3.6	3.7	3.8	3.9	4.0	4.1	4.2	4.3	4.4	4.5	4.6	4.7	4.8	4.9	5.0	5.1	5.2	5.3	5.4	5.5	5.6	5.7	5.8	78
3.3	3.4	3.5	3.6	3.6	3.7	3.9	4.0	4.1	4.2	4.3	4.4	4.4	4.6	4.7	4.9	5.0	5.1	5.2	5.3	5.4	5.5	5.6	5.7	5.8	79
3.3	3.4	3.5	3.6	3.6	3.7	3.8	3.9	4.0	4.1	4.2	4.3	4.4	4.4	4.6	4.8	4.9	5.0	5.1	5.2	5.3	5.4	5.5	5.6	5.7	80°
3.2	3.3	3.4	3.5	3.5	3.6	3.7	3.8	3.9	4.0	4.1	4.2	4.3	4.4	4.5	4.7	4.8	4.9	5.1	5.2	5.3	5.3	5.4	5.5	5.6	81
3.2	3.3	3.4	3.5	3.5	3.6	3.7	3.8	3.9	4.0	4.1	4.2	4.3	4.4	4.5	4.6	4.7	4.8	4.9	5.1	5.2	5.3	5.4	5.5	5.6	82
3.1	3.2	3.3	3.4	3.5	3.6	3.7	3.8	3.9	4.0	4.1	4.1	4.2	4.3	4.4	4.5	4.6	4.7	4.8	4.9	5.0	5.1	5.2	5.3	5.4	83
3.1	3.1	3.2	3.3	3.4	3.5	3.6	3.7	3.8	3.9	4.0	4.1	4.2	4.3	4.4	4.5	4.6	4.7	4.8	4.9	5.0	5.1	5.2	5.3		84
3.0	3.1	3.2	3.3	3.4	3.5	3.6	3.7	3.7	3.8	3.9	4.0	4.1	4.2	4.3	4.4	4.5	4.6	4.7	4.8	4.9	5.0	5.1	5.2	5.3	85
3.0	3.1	3.2	3.2	3.3	3.4	3.5	3.6	3.7	3.8	3.9	4.0	4.0	4.1	4.2	4.3	4.4	4.5	4.6	4.7	4.8	4.9	5.0	5.1	5.2	86
2.9	3.0	3.1	3.2	3.3	3.4	3.5	3.6	3.7	3.8	3.8	3.9	4.0	4.1	4.2	4.3	4.4	4.5	4.6	4.7	4.9	5.0	5.1			87
2.9	3.0	3.1	3.1	3.2	3.3	3.4	3.5	3.6	3.7	3.8	3.9	3.9	4.0	4.1	4.2	4.3	4.4	4.5	4.5	4.6	4.7	4.8	4.9		88
2.9	2.9	3.0	3.1	3.2	3.3	3.4	3.5	3.5	3.6	3.7	3.8	3.9	4.0	4.0	4.1	4.2	4.3	4.4	4.5	4.5	4.6	4.7	4.8	4.9	89
2.8	2.9	3.0	3.1	3.2	3.3	3.4	3.5	3.5	3.6	3.7	3.7	3.8	3.9	4.0	4.0	4.1	4.2	4.3	4.4	4.5	4.6	4.7	4.8	4.9	90°
2.8	2.9	2.9	3.0	3.1	3.2	3.3	3.3	3.4	3.5	3.6	3.7	3.8	3.9	3.9	4.0	4.1	4.2	4.3	4.4	4.4	4.5	4.6	4.7	4.8	91
2.7	2.8	2.9	3.0	3.1	3.2	3.2	3.3	3.4	3.5	3.6	3.6	3.7	3.8	3.9	4.0	4.0	4.1	4.2	4.3	4.4	4.5	4.5	4.6	4.8	92
2.7	2.8	2.9	2.9	3.0	3.1	3.2	3.3	3.3	3.5	3.5	3.6	3.7	3.8	3.8	3.9	4.0	4.1	4.2	4.2	4.3	4.4	4.5	4.6	4.7	93
2.7	2.8	2.8	2.9	3.0	3.1	3.1	3.2	3.3	3.4	3.5	3.6	3.6	3.7	3.8	3.9	4.0	4.1	4.1	4.2	4.3	4.4	4.5	4.6	4.7	94
2.6	2.7	2.8	2.9	3.0	3.1	3.2	3.3	3.4	3.4	3.5	3.6	3.7	3.8	3.8	3.9	4.0	4.1	4.2	4.3	4.4	4.4	4.5	4.6		95
2.6	2.7	2.8	2.8	2.9	3.0	3.1	3.2	3.3	3.4	3.5	3.6	3.6	3.7	3.8	3.9	4.0	4.1	4.2	4.3	4.4	4.4	4.5	4.6		96
2.6	2.6	2.7	2.8	2.9	3.0	3.0	3.1	3.2	3.3	3.3	3.4	3.5	3.6	3.7	3.8	3.9	4.0	4.0	4.1	4.2	4.3	4.4	4.4	4.5	97
2.6	2.6	2.7	2.8	2.8	2.9	3.0	3.1	3.2	3.3	3.4	3.5	3.5	3.6	3.7	3.7	3.8	3.9	4.0	4.0	4.1	4.2	4.3	4.3	4.4	98
2.5	2.6	2.6	2.7	2.8	2.9	3.0	3.0	3.1	3.2	3.3	3.4	3.4	3.5	3.6	3.6	3.7	3.7	3.8	3.9	4.0	4.1	4.1	4.2	4.3	99
2.5	2.6	2.6	2.7	2.8	2.9	3.0	3.1	3.2	3.2	3.3	3.4	3.5	3.6	3.6	3.7	3.8	3.9	3.9	4.0	4.1	4.2	4.2	4.3		100°

Courtesy: Zahm and Nagel
Buffalo, New York

11. Immediately measure percent transmittance of each cuvette on the spectrophotometer. Use distilled water as a "blank" to adjust at 100% transmittance when wavelength control is set at 440μ.

12. Line up and insert sample cuvettes; read and record percent transmittance.

13. For wines, white or red, use 10 ml sample and follow exactly the same procedure as for the standards. Most wines should contain less than 1 ppm copper. If a percent transmittance lower than 15% is obtained, that wine should be diluted exactly in half, analyzed again, and the end result multiplied by two.

Preparation of Standard Curve

1. The standards should be analyzed each time new sodium diethyldithiocarbamate solution is made. The standard curve is drawn using percent transmittance readings recorded for each of the standard copper solution.

2. Figure A.3 provides a sample standard curve drawn with the following points determined from the standard copper solutions:

0.5 ppm standard = 1st sample at 85% LT
                   2nd sample at 83% LT

1.0 ppm standard = 1st sample at 74% LT
                   2nd sample at 69% LT

2.5 ppm standard = 1st sample at 32% LT
                   2nd sample at 34% LT

Determination

Using the standard curve developed, take the recorded reading of % transmittance from the wine samples and find the corresponding parts per million results on the x axis of the curve. Record the result on the laboratory analysis log.

Preparation of Iron Standards and Sample

1. Pipet 5 ml each of the iron standards into clean separatory funnels.

2. Add 1 ml of dilute hydrochloric acid, 1 ml potassium thiocyanate solution, and 3 drops of hydrogen peroxide to each funnel, mix. (A

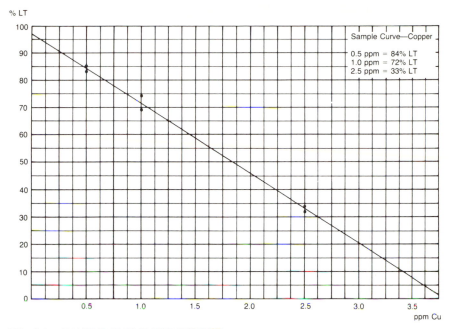

FIG. A.3.   SAMPLE CURVE FOR COPPER

duplicate sample without the addition of peroxide will give the ferric (Fe^{3+}) level—with peroxide, total inorganic ($Fe^{3+} + Fe^{2+}$) is determined. The ferric ion is the active form that precipitates, while the total indicates the potential for iron precipitates. Ferric is generated upon oxidation and (without precipitation) slowly reverts to the ferrous (Fe^{2+}) form.)

3. Add 10 ml amyl acetate and 5 ml methyl alcohol to each funnel, mix for 15 sec. An increasing amount of red color should be noted.

4. Draw the lower portion from the funnel stopcocks and discard.

5. Transfer the upper phases into clean small beakers.

6. Filter through paper into cuvettes.

7. Immediately measure percent transmittance of each on the spectrophotometer. Use distilled water as a "blank" to adjust at 100% transmittance when the wavelength control is set at 580μ.

8. Line up and insert sample cuvettes; read and record percent transmittance.

9. For wines, white or red, use 5 ml sample and follow exactly the same procedure as for the standards. Most wines should contain less than 10 ppm iron. If a percent transmittance lower than 15% is obtained, that wine should be diluted exactly in half, analyzed again, and the end result multiplied by two.

Preparation of Standard Curve

1. The standards should be analyzed each time new solutions are made. The standard curve is drawn using the percent transmittance readings recorded for each of the standard iron solutions.

2. Figure A.4 provides a sample standard curve drawn with the following points determined from the standard iron solutions:

    1.0 ppm standard = 1st sample at 69% LT
                       2nd sample at 71% LT

    5.0 ppm standard = 1st sample at 47% LT
                       2nd sample at 46% LT

    10.0 ppm standard = 1st sample at 21% LT
                        2nd sample at 23% LT

Determination

Using the standard curve developed, take the recorded reading of percent transmittance from the wine sample and find the corresponding parts per million results on the x axis of the curve. Record the result on the laboratory analysis log.

10. DIFFERENTIAL STAIN PROCEDURE

Most observations of yeast are made by the use of a light microscope with stained smears. Chemical stains not only help to reveal cellular arrangement, shape and style, but also aid in the investigation of internal details.

Differential staining of yeast cells involves applying a stain and then observing the effects of that stain, in order to distinguish between cells and to determine whether they are viable, sporated or dead.

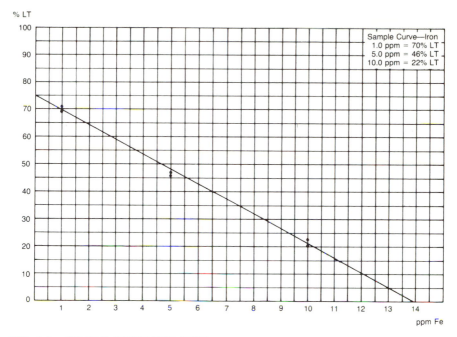

FIG. A.4.   SAMPLE CURVE FOR IRON

A. DIFFERENTIATING YEAST ASCOSPORES AND
VEGETATIVE CELLS From Evans, King, and Bartholomew (1949)

Apparatus and Reagents

Light microscope and light source
Slides
Tap water
Open-flame burner
Aniline crystal violet preparation:

crystal violet (C.C.)	5 g
95% alcohol	10 ml
aniline	2 ml
distilled water	20 ml

95% alcohol containing 3% hydrochloric acid
Safranin preparation:

2.5% safranin 0 (C.C.) in 95% alcohol	10 ml
distilled water	100 ml

Procedure

1. Smear slide with medium—1 drop is sufficient.
2. Air dry, and lightly heat-fix with open flame.
3. Flood slide with aniline crystal violet solution and heat gently for 3 min, replenishing the stain as it evaporates. The slide should be heated to the point where steam is given off but the stain should not be allowed to boil.
4. Rinse in tap water.
5. Decolorize 15 sec with 95% alcohol which contains 3% hydrochloric acid.
6. Rinse in tap water.
7. Stain 10–15 sec in Safranin preparation.
8. Rinse in tap water.
9. Carefully blot dry and examine in microscope.
10. Vegetative (viable) cells are light pink color; sporated cells are deep violet color.

B. DIFFERENTIATING BETWEEN VIABLE AND DEAD YEAST CELLS

Apparatus and Reagents

Light microscope and light source
Slides
Tap water
Gentian violet stain

Procedure

1. Smear slide with medium—1 drop is sufficient.
2. Air dry, and lightly heat-fix with open flame.
3. Flood slide with gentian violet stain and hold for 30 to 60 sec.
4. Rinse in tap water.
5. Carefully blot dry and examine in microscope.
6. Viable yeast cells are light pink color; dead yeast cells are deep violet color (Fig. A.5).

11. EXTRACT DETERMINATION BY HYDROMETER

The rationale for this method of extract determination is to separate the dissolved solids from the wine into an inert liquid medium that has no alcohol to affect the Brix-Balling hydrometer.

Apparatus and Reagents

Identical to Procedure No. 5, *Balling Determination*

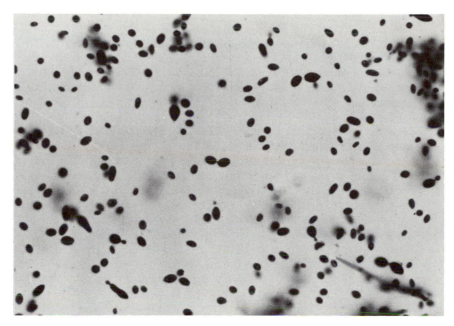

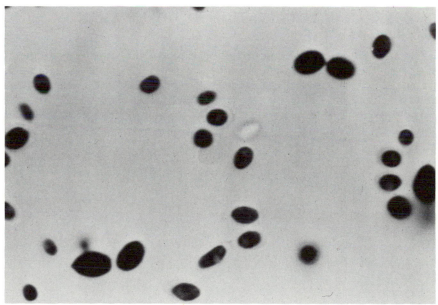

Courtesy of Dr. Bruce Glick,
Mississippi State University

FIG. A.5.   LIGHTER SHADED CELLS APPEARING TRANSLUCENT ARE VIABLE, WHILE DARKER SHADED CELLS APPEARING SOLID IN COLOR ARE DEAD
(Smaller cells: 430× magnification; larger cells: 1000× magnification)

Procedure

1. Take contents from 1000 ml Erlenmeyer flask remaining in Step No. 16 of Procedure No. 3, *Alcohol Determination by Distillation,* and cool by running cold tap water on the outside of the flask. Holding flask upright at a moderate angle, carefully swirl extract contents until temperature is reduced to about 68°F.

2. Carefully pour extract from the 1000 ml Erlenmeyer flask into the same 200 ml Kohlrausch flask (clean and dry) used in the alcohol test. Rinse Erlenmeyer flask carefully with about 25 ml distilled water; pour rinsing into Kohlrausch flask. Repeat rinsing and pouring twice more. Be careful that the Berl saddles do not fall into the Kohlrausch flask. Adjust fill height in Kohlrausch flask with distilled or deionized water so that the bottom meniscus rests exactly on fill line. Shake out any bubbles that may be adhering to sides of the flask.

3. Mix Kohlrausch flask in the same method as described in Step No. 8 of Procedure No. 3, *Alcohol Determination by Distillation.*

4. Proceed with Steps 2–8 of Procedure No. 5, *Balling Determination.*

12. EXTRACT DETERMINATION BY NOMOGRAPH— DESSERT WINES

The determination of extract by nomograph requires the knowledge of alcohol content and Balling of the wine to be tested. The procedure is simply to find the extract with a straight-edge (a ruler is ideal) so as to intersect the known alcohol and Balling levels.

The nomograph for dessert and aperitif wine usage is provided in Fig. A.6.

Examples

Alcohol = 19.0% by volume
Balling = 5.0°
Extract = 10.8°

Balling = 2.5°
Extract = 8.0°
Alcohol = 17.5% by volume

Extract = 12.5°
Alcohol = 15.5% by volume
Balling = 7.6°

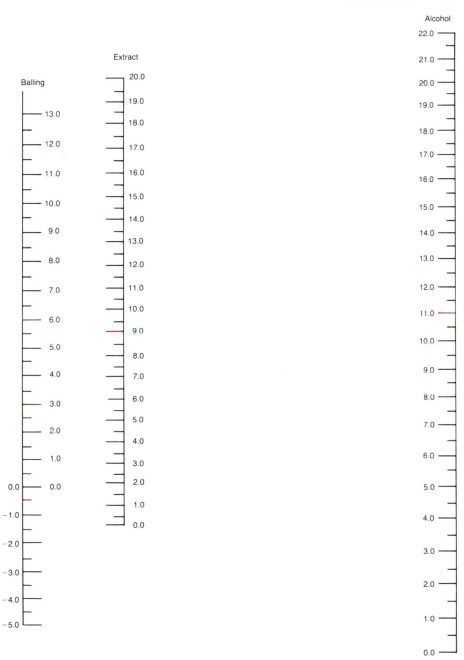

FIG. A.6.   BALLING ALCOHOL EXTRACT NOMOGRAPH FOR DESSERT WINES
(Adapted from "Balling-Alcohol-Extract Nomograph", Wine Institute, San Francisco)

13. EXTRACT DETERMINATION BY NOMOGRAPH— TABLE WINES

The determination of extract by nomograph for table wines is found in the same way as in Procedure No. 12, *Extract Determination by Nomograph: Dessert Wines,* except that table wine extract analyses require a different scale, as provided in Fig. A.7.

Examples

Alcohol = 12.0% by volume
Balling = −2.0°
Extract = 2.0°

Balling = 0.0°
Extract = 3.9°
Alcohol = 11.4% by volume

Extract = 4.8°
Alcohol = 11.0% by volume
Balling = 1.0°

14. GRAM STAIN PROCEDURE

Apparatus and Reagents

Gentian violet solution
    (10 ml Gentian violet, saturated alcoholic solution and 40 ml ammonium oxalate 1.0% aqueous solution)
Gram's iodine solution
    (1 g iodine crystals, 2 g potassium iodide and 300 ml distilled or deionized water)
Safranin solution, saturated aquaeous solution
Burner
Slide and cover slip
Distilled or deionized water
95% alcohol
Bibulous paper

Procedure

1. Place a drop of the medium or culture on the center of a clean slide, spreading the material over an area of about ½ in.^{2}

2. Allow the sample material to dry and then fix by quickly passing the slide over the burner flame several times. Do not allow the slide to become too hot or the sample material may burn.

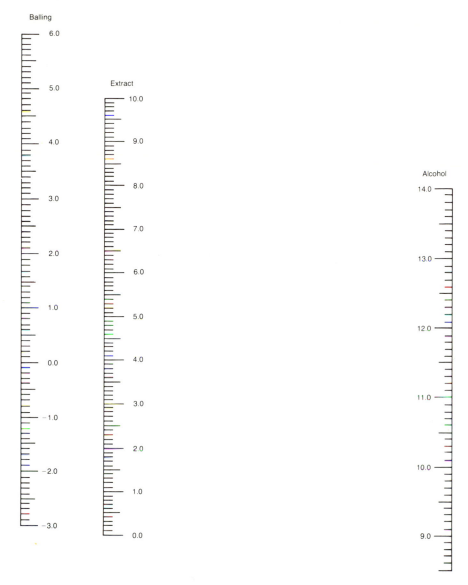

FIG. A.7. BALLING ALCOHOL EXTRACT NOMOGRAPH FOR TABLE WINES
(Adapted from Vahl, J.M. Am. J. Enol. Vitic., *30* (3) 1979.

3. Flood slide with Gentian violet and allow to remain 30 sec, then rinse
    with distilled or deionized water.

4. Cover the sample with Gram's iodine and allow it to react for 30 sec,
    then rinse with distilled or deionized water.

5. Decolorize in 95% alcohol for 20–30 sec, then rinse with distilled or deionized water.

6. Counterstain with Safranin, allowing a reaction of 10 sec, then rinse with distilled or deionized water.

7. Dry with bibulous paper.

8. Observe under microscope (see Procedure No. 17, *Microscopy*). Gram-positive organisms will stain a purple-black; Gram-negative organisms will stain pink or red.

15. LIGHT TRANSMISSION (COLOR INTENSITY) BY SPECTROPHOTOMETRY

This test method involves sending a beam of light, at a desired standard wavelength, through a sample of distilled water. This is immediately followed by comparing the amount of light from the same beam which will pass through a brilliantly clear wine sample. The final result is expressed as "percent light transmission".

The difference between the amount of light which is transmitted through the distilled water (100%) and that through the wine sample is attributed to color intensity. Figure 5.5 portrays a spectrophotometer in use.

Apparatus and Reagents

Spectrophotometer (colorimeter) with a minimum range of 400 to 700 nm
Voltage regulator for constant power source
Cuvette(s) appropriate to the spectrophotometer
Distilled water

Procedure

1. Turn on machine and allow to warm-up for 15–30 min, or as directed by manufacturer's instructions.

2. Rinse a cuvette with distilled water and then fill to the prescribed fill height. Dry outside with absorbent lens cloth or paper. (Remember that cuvettes are optical devices and must not be scratched or else results will be in error.)

3. In the same manner as Step 2, rinse a second cuvette with a brilliantly clear, room-temperature wine sample and then fill to the prescribed fill height.

4. Insert distilled water cuvette into machine and standardize to 100% transmission at the desired wavelength. Percent transmittance values

may be limited in application and one should also consider absorbance. Absorbance values are proportional to actual depth of color (or concentration of pigment, colored ion complexes, etc.) Use of 420 and 520 nm or 425 and 525 is more common.) Recommendations follow:

White wines at 425 nm
Rosé wines at 475 nm
Red wines at 525 nm

5. After spectrophotometer will repeat the 100% distilled water reading several times (inserting and removing sample cuvette), insert wine cuvette taking, likewise, several readings.

6. Average reading results, if necessary. The machine should, however, remain stable with the proper use of the voltage regulator.

7. Record results upon laboratory analysis log.

8. Clean cuvettes with distilled water only, being sure that only optical cleaning and drying devices are used.

16. MALO-LACTIC FERMENTATION DETERMINATION BY PAPER CHROMATOGRAPHY

The use of paper chromatograms (Fig. A.8) in the detection of malo-lactic fermentation distinguishes qualitatively between malic acid and lactic acid in the wine sample being tested. The disappearance of malic acid is an indication of this bacterial fermentation. The formation of lactic acid, by itself, is not valid evidence, as this acid could also result from other microbial activity.

This method determines the ratio of distance from a "base-line" to an acid "spot", called the R_f, upon each paper chromatogram by the use of standard acid solutions. Once these standards, or controls, are made, then a simple comparison is made with the chromatograms run with the wine sample.

Apparatus and Reagents

Chromatographic grade filter paper cut into 20 × 30 cm rectangles
1.2 × 75 mm micropipets
Separatory funnel
1-gal. wide-mouth glass jars with covers
Solvent constituents:

distilled water	100 ml
n-butyl alcohol	100 ml
concentrated formic acid	10.7 ml
1% water soluble bromcresol green	15 ml

2% standard solutions:
  tartaric acid
  citric acid
  malic acid
  lactic acid
  succinic acid
  fumaric acid

Procedure

1. Wine sample (or standard) is spotted on a pencil line (base line) approximately 2.5 cm parallel to the long edge, about 2.5 cm apart. Each spot is made 4 times (allowed to dry in between) at a volume of 10 microliters from the micropipet.

2. A cylinder is made from the paper by stapling the short ends, without overlapping.

FIG. A.8. PAPER
CHROMATOGRAPHY

3. Place solvent constituents in separatory funnel and mix. After about 20 min the lower aqueous phase is drawn off and discarded.

4. Transfer 70 ml of the upper layer into the wide-mouth jar and place spotted edge (base line) of the chromatogram in the solvent; cover jar.

5. Chromatogram should develop in about 6 hours, but may be extended to overnight.

6. Remove yellow chromatogram and store in a well ventilated area until dry and the formic acid has vaporized, leaving a blue-green background with yellow spots of acid having the following approximate R_f values:

tartaric acid	0.28
citric acid	0.45
malic acid	0.51
lactic acid	0.78
succinic acid	0.78
fumaric acid	0.91

7. Standards and wine samples should be run simultaneously, if possible, or one immediately following another.

8. Solvent may be used repeatedly if care is taken to remove any aqueous layer which may have separated after each run.

Adapted from: Kunkee, R.E. 1974. *In* Malo-Lactic Fermentation and Winemaking, Part 7 of Chemistry of Winemaking, A.D. Webb (Editor). American Chemical Society, Washington, D.C.

17. MICROSCOPY

The microscope is a very delicate, precision instrument, and should be placed in the hands of an operator who will accept responsibility for properly using and maintaining it. Given that, the operation of a microscope can be rather simple and rewarding.

Figure A.9 depicts a modern microscope with its optical and mechanical features outlined.

The following procedure is provided for the analyst. However, if manufacturer's operating instructions are available, they should take precedence in the use and care of the microscope.

Procedure

1. Open the aperture diaphragm of the condenser.

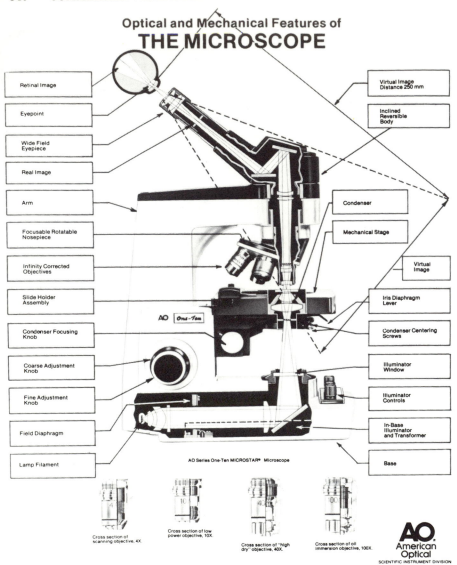

Optical and Mechanical Features of
THE MICROSCOPE

Retinal Image

Eyepoint

Wide Field Eyepiece

Real Image

Arm

Focusable Rotatable Nosepiece

Infinity Corrected Objectives

Slide Holder Assembly

Condenser Focusing Knob

Coarse Adjustment Knob

Fine Adjustment Knob

Field Diaphragm

Lamp Filament

Virtual Image Distance 250 mm

Inclined Reversible Body

Condenser

Mechanical Stage

Virtual Image

Iris Diaphragm Lever

Condenser Centering Screws

Illuminator Window

Illuminator Controls

In-Base Illuminator and Transformer

Base

AO Series One-Ten MICROSTAR® Microscope

Cross section of scanning objective, 4X.

Cross section of low power objective, 10X.

Cross section of "high dry" objective, 40X.

Cross section of oil immersion objective, 100X.

AO
American Optical
SCIENTIFIC INSTRUMENT DIVISION

FIG. A.9. A MODERN MICROSCOPE

2. Turn on illuminator or adjust mirror to external light source.

3. Turn coarse adjustment knob to raise nosepiece and objectives sufficiently for insertion of slide.

4. Place specimen slide on the stage.

5. Position the low-power objective over slide and lower the body with the coarse adjustment knob until the objective is about ⅛ in. from the slide. It is important that the objective does not actually touch the slide, to avoid scratching the objective and causing other damage to the instrument. This operation is best accomplished by keeping the eye level with the stage.

6. Slowly elevate the body with the coarse adjustment knob while looking through the eyepiece. Once the image appears in approximate focus, stop adjustment. Adjust only upward with the adjustment knobs when viewing through the eyepiece. Adjust downward only when the eye is level with the stage.

7. Fine tune focus by adjustment with the fine adjustment knob. Further adjustment may be necessary with the light source in order to achieve the optimum image.

8. Adjust diaphragm by moving the iris diaphragm level to the desired opening while viewing the image through the eyepiece.

9. Place the desired object or specimen in the exact center of the image by manipulation of the stage.

10. Turn coarse adjustment so as to move body upward allowing positioning of 40× objective over slide. Be careful not to move slide on stage or the centered image will be lost.

11. With eye at stage level, lower body with coarse adjustment knob about ⅛ in. from the cover slip.

12. Repeat Step 6.

13. Repeat Step 7.

14. Repeat Step 8.

15. Repeat Step 9.

16. Turn coarse adjustment so as to move body upward allowing positioning of 100× oil-immersion objective over slide. Be careful not to move slide on stage or the centered image will be lost.

17. With eye at stage level lower body until objective is approximately ¼ in. from the slide.

18. With extreme patience and care, place one drop of cedar oil on slide-cover slip just below objective, again being sure not to move cover slip.

19. With eye at stage level, lower body with coarse adjustment knob until the lens of the 100× objective comes in contact with the oil. Continue to lower body very carefully until objective nearly touches cover slip. Do not, however, allow the objective to actually come in contact with the slide.

20. Viewing through the eyepiece, slowly elevate body by means of fine adjustment knob until a focus is achieved.

21. Make adjustments of light source and diaphragm that are necessary for the optimum image.

22. When observation is finished, raise body so that objective is about 1 in. from cover slip and return 10× objective to the focusing position.

23. The oil-immersion objective should be polished dry with dry lens paper after every use. The other lower power objectives should never be used with the oil-immersion technique. If, however, some cedar oil should come in contact with the 10× or 40× objectives, they will require immediate cleaning with lens paper moistened with xylol, and polished with dry lens paper.

24. Remove slide from stage. If any oil spills or other liquids are on the stage, they should be cleaned with a cheesecloth moistened with xylol and then dried with an untreated cheesecloth.

General Operating Instructions

1. The microscope should never be forced. All moving parts should do so freely. If something binds or becomes inoperable, the microscope should be serviced by a qualified person only.

2. The lenses in the objectives should never be touched with anything but appropriate cleaning materials. Other items may deposit film, oils, or even scratch the precision surfaces.

3. The objectives should never touch slides, cover slips or the stage.

4. Specimens should be examined first with low power objective, increasing magnification as necessary.

5. The body of the microscope should never be adjusted downward with the coarse knob while the operator is looking through the eyepiece. At

eye level, the objective can be brought close to the cover slip and then adjusted for focus upward while viewing through the eyepiece. This will aid in preventing the objective from touching the slide and perhaps ruining the lens.

6. The microscope should be stored with the low power, 10× objective in the focusing position.

7. The microscope should be carried only by the arm, insuring that it is maintained in the upright position.

8. Eye strain can be kept minimal by keeping both eyes open when using the microscope. Squinting causes the eyes to tire very quickly and, apart from the discomfort, much can be lost or overlooked in specimen observation.

9. One should become totally familiar with a new, or different, microscope. It may help to "dummy" with the instrument without a slide in position on the stage, in order to get the feel and position of adjustment locations and operation.

10. Securing the microscope from operation should be preceeded with a careful dusting and lubrication as may be instructed by the owner's manual, if available; or as may be advised by a qualified serviceman. Cover with dust cover and store in wooden storage box or cupboard.

18. ORGANOLEPTIC ANALYSIS

Organoleptic analysis and criticism require serious concentration and the ability to relate sensory memory to the evaluation of an individual wine sample. There is no substitute for experience in wine judging, the key being, of course, much practice.

The beginning winemaker may wish to sample a large number of both positive and negative wine constituents and chemical components in water and/or water-alcohol solutions. This will help in learning to identify such compounds without any masking effects that may be provided in a wine base. The large amounts of acids and ethanol in wine can render identification of volatile compounds much more difficult than in prepared laboratory samples. Recall of these sensations should be practiced to help develop a good eye, nose and palate. The very best training is often available at an institution of higher learning which offers curricula for training in organoleptic evaluations.

Apparatus and Conditions

Wine glasses, preferably several dozen, all identical, of the all-purpose tulip shape; cleaned with hot water, but without the use of detergents; and stored in a neutral environment free of varnish or paint

White background, preferably a solid, stark white Formica-type counter-top

Palate-neutralizers such as salt-free soda crackers, mild cheese such as Muenster, and room-temperature pure water which has not been chlorinated or fluoridated

Wine samples stabilized at cool room temperature, without regard to type

Noiseless and odorless environment, preferably a separate room where interference of any type is minimal

Temperature maintained at 68–72°F with moderate humidity

Daylight or incandescent light of ample supply; fluorescent illumination and other types of light will interfere with the best color judgement

Procedure

Color.—The first portion of the visual stage. Wine color is often referred to as the "robe" by the French which is loosely translated into English as "gown", although these terms do not have precisely the same meaning. Effervescent wines may prove difficult to judge for color and clarity at first, and if they are to be judged frequently, a special technique may be required to reduce the amount of bubbles.

Generally, all white wines fall into a category of either *water-white, pale straw, straw-gold, dark straw* or *very dark gold,* in increasing order of color intensity. Each of these categories may be appropriate for one or another wine. For example, dry French-type vermouth may be most marketable and proper as a near water-white product; anything darker would be grounds for deduction of quality points in judgement. Old French sauternes vinified from "noble-mold" grapes may be a dark straw or even very dark straw-gold in color; anything lighter would bring criticism.

Red wines range from a light crimson hue found typically in Beaujolais wines to a heavy ruby value found in the Medoc wines of Bordeaux and the port wines of the Duoro of Portugal. Tawny, *madeirized* wines range from a very pale straw found in some dry sherry-types to the very dark amber of Oloroso, or cream sherry types. The amber may be a caramel-ruby hue found in tawny port types. The red may also be a deep chocolate brown such as found in sweet Italian type vermouth.

The neophyte may be best advised to obtain several good examples of the wines mentioned in the previous paragraph. This, of course, will be a very expensive education. Any effort to learn about wine colors with water and food coloring is simply unacceptable as a proper foundation for color appraisal.

Typical criticisms of color in white wines are "browning" due to oxidation, "yellowing" due to excessive leucoanthocyanin pigmentation, and "tinting" which may occur in white wines made from red grapes. Rosé table wines often lose their attractive pink color, or light-redness, because of sulfur-dioxide additions and/or aging, resulting in rather "orange" wines. This may or may not be a serious fault, depending upon the particular rosé wine type being judged. Young red wines may have a pronounced purple hue which is almost always criticized, even with the "nouveau" wines famous to Beaujolais. Good red table wine color balance usually shows some browning about the edges of the wine in the tasting glass. Except for purposely over-oxidized wines such as the sherry types and tawny port types, excessive browning in red wines may be a criticism worthy of penalty.

Clarity.—The second portion of the visual stage. The surface of a wine is often called a "disc", especially in France, and should appear absolutely brilliant in a finished wine. Wines in unfinished production stages may, of course, not exhibit "bottle-brightness" and should be accounted for in a manner appropriate to age and progress. There are four general classes of clarity usually referred to in wine judging: *cloudy, hazy, clear* and *brilliant*, in ascending order of brilliance value.

An 8-oz wine glass filled with 4 oz of pure water and 2 drops of whole milk will represent a cloudy condition. A mixture of 4 oz of pure water and 1 drop of milk creates a hazy condition. Diluting the hazy mixture with 4 more oz of water should result in a clear liquid. The pure water with no addition of milk constitutes a comparison to a "brilliant", or "bottle-bright" wine.

The disc can appear rather "dusty", or dull, which may indicate bacterial action, especially if the wine develops an unpleasant, "vinegary" nose and taste. The disc may be "ropy" or iridescent, which can result from bacterial infection, a condition called "Vins Filant" by the French. A "shiny" or "oily" disc may result from carelessness in the cellars with use of lubricants or other contaminants.

In the sight phase of organoleptic analysis there are two positions from which the wine should be observed. The first is at eye level, holding the wine glass by the stem or base. Behind the glass should be a candle, or open-filament, incandescent light bulb. The light source can be as close as just several inches from the wine glass, or perhaps more than a foot, depending upon the visual comfort and ability to focus of each individual judge. The important item in lining up the "candling" is that it serves the purpose for finite clarity evaluations. The light source should be used primarily as an aid in searching out particles of suspended solids, colloids and metal casse. Cloudiness and haziness are usually easily detectable without the assistance of a separate light source. The bulb, or candle, should not be used directly in the judgement of color.

The second position is with the wine glass at rest on a solid white surface at waist height, such as on a stark-white countertop, or a desk surface with a heavy white paper placed beneath the glass. The hue and intensity of color composition is most accurately observed with wine glasses in this position, with the judge looking directly downward through the wine to the white surface. The overhead light source should be common daylight incandescent illumination. Fluorescent and other such light sources will foul and interfere with color judgement.

Figure A.10 illustrates a good position and technique for the judgement of clarity, while Figure A.11 shows the proper angle for observing and criticizing color.

Nose.—Nose is the olfactory stage, which has to do with the odor of wines. The smell of wine is primarily a function of the human sensory epithelium, which covers about 1 sq. in. of the roof and walls in the nasal cleft just behind the nose. Figure A.15 illustrates olfactory nerves penetrating the cribriform plate to the olfactory bulb. Figure A.12 is a diagram of olfactory receptors which are common to most mammals. A stimulus, such as a wine odor, reacts with these sensors, or receptors, and its value is passed on from the olfactory system to the brain by means of a fiber tract.

The human brain, properly trained, will classify the stimuli according to the accuracy and breadth of stored and recalled information. In other words, the nose of the judge is applied to experience and ability to identify specific odorous compounds. Some people are more sensitive than others to certain volatile substances which may result in a better ability to judge various wine odors, as long as this condition is kept in perspective.

The first nose observation should be made while the wine is at rest, without the benefit of swirling. This will provide the judge with only the amount of wine vapors that are being given off naturally. Figure A.13 shows a good position for judging a wine's nose.

FIG. A.10. POSITION AND TECHNIQUE FOR JUDGMENT OF CLARITY OF WINE

FIG. A.11. THE PROPER ANGLE FOR OBSERVATION AND CRITICISM OF WINE COLOR

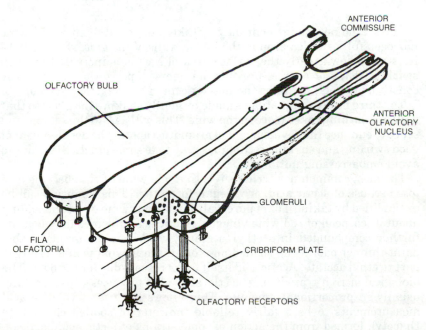

FIG. A.12. OLFACTORY RECEPTORS
(Adapted from Schultz, H.W., Day, E.A. and Libbey, L.M. 1967. The Chemistry and Physiology of Flavors. AVI Publishing Company, Westport, Conn.)

FIG. A.13.   JUDGING THE NOSE
OF WINE

The second nose judgement may be taken immediately after the wine has been fully swirled around the inside walls of the glass. Figure A.14 illustrates how greatly the surface area of a wine is increased by such swirling. This, of course, provides for more vaporization of aromatic constituents and magnifies the nose effects in the olfactory system.

The third nose examination is made by gently shaking the glass so that a little splashing takes place in the wine. This will aerate the sample and serve to enhance the development of aromatic vapors. The development of good swirling and shaking techniques may take some practice in order to avoid embarrassing untidiness.

The most common nose criticism is that of a wine having become *acetose* because of acetic acid, or vinegar, formation. This condition can be exemplified by taking 4 oz of pure water in an 8-oz wine glass and adding about a teaspoon of red wine vinegar. Acetic acid in wines may become further compounded into ethyl acetate, which has an unmistakable paint-thinner odor. (The spoiled nose of an acetic wine is due in larger part to ethyl acetate. Acetic acid itself is *not* responsible for the odor. The biological activity producing acetic acid also produces ethyl acetate, usually in proportional amounts. This allows the use of volatile acid measurements to be a fairly reliable indicator of spoiled character.) Diacetyl, formed from the action of some strains of lactic acid bacteria, has the aroma of butter or margarine.

Acetaldehyde formation in most table wines is a fault. It is detected as a rather "nutty" aroma, resulting primarily from the oxidation of ethanol. However, in sherry type wines, the formation of acetaldehyde is desired in very pronounced concentrations. Similarly, the "caramel-like" value of hydroxymethylfurfural is criticized in most wines, other than madeirized types such as sherry and Marsala.

The aromatic values of wines are made up of volatile acetals, acids, alcohols, amides, carbonyls and esters. The list is rather lengthy and is reserved for a more advanced text.

In addition to the items just discussed, the following is a list of major important terms that should be learned in order to properly judge the nose of wines. (However, definitions vary with different authors and different winemakers.)

FIG. A.14.   SWIRLING INCREASES
THE SURFACE AREA OF WINE

Aroma   The odor that is contributed by the grapes used for a wine.

Bouquet   The odor that is contributed by the grapes in combination with the odor assimilated in the cellaring procedures.

Character   The odor that is descriptive of a particular grape cultivar, geographic location, or cellar technique; or a combination of all three.

Delicate   A nose that is faint and rather difficult to gather in the nasal passages.

Flowery   A nose that has the effect of flowers, such as that of a well made wine from the cultivar, Johannisberg Riesling.

Fruity   The odor that is contributed by heavily flavored grapes, such as those from the species, *V. labrusca, V. rotundifolia,* and the Muscat cultivars of *V. vinifera.*

Full   A nose that is obvious and fills the nasal passages.

Grassy   A nose that is "green" from wines that are immature, or have been aged in redwood cooperage.

Heady   A nose of high alcohol content, often called "strong".

Moldy   A nose from wines made from grapes having been infected with mold, or wines that have been aged in mold-infected cooperage, or both.

Musty   A nose from wines which were aged in cooperage that has decayed or become waterlogged.

Off   A nose that does not exhibit a proper character, or exhibits the wrong character; a term very often misused.

Organic   A nose that is reminiscent of "rotten eggs", the result of hydrogen sulfide contamination.

Soapy   A self-descriptive term resulting from wines having been exposed to equipment and/or aging vessels that have not been properly cleaned or rinsed.

Spicy   A nose that has the effect of spice aromatics, such as a well made wine from the cultivar, Gewürztraminer.

Typical   A nose that is expected from the wine because of its type, production technique, varietal usage, origin, or some combination of these.

Varietal   A nose that is typical to a particular variety, or cultivar, of grape used in making the wine.

Woody   A nose that exhibits an aroma of wood, generally a wine that has been wood-aged too long.

Yeasty   A typical yeast-like aroma; a condition expected in bottle-fermented sparkling wines, but often criticized in other wines as a fault of leaving wines on the lees too long.

Taste.—The gustatory stage that many unfamiliarized people regard as the only real part of wine evaluation. It is actually the least informative of our sensory organs.

The primary human sensory organ involved in the tasting process is the tongue, although the throat is sometimes involved, particularly in aftertaste reactions. Over the upper surface of the tongue are about 3000 expanded skin protuberances, called *papillae*, which contain the taste buds. Each taste bud is connected directly to the brain by a nerve. Figure A.15 diagrams a taste bud.

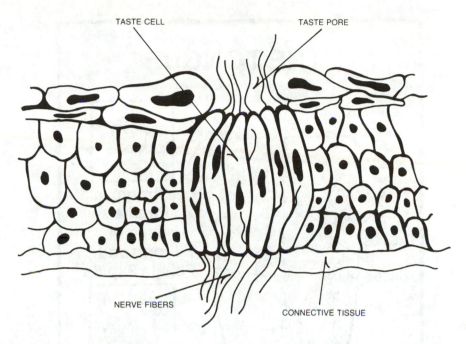

FIG. A.15.   A TASTE BUD

The four tastes, or gustatory sensations that can be determined by the human tongue are acidity, bitterness, saltiness, and sweetness. The distinction between flavors is actually an olfactory judgement, not a function of taste. Theory has it that, as wine flows between the papillae of the tongue, the reactions are passed on to the brain for judgement. The stimuli of acid (tart), bitter and sweet are evaluated by different types of papillae, as illustrated in Fig. A.16.

Some forms of stimulus, such as astringency, are preceived on trigeminal nerve endings located in the mouth, nasal and pharyngeal passages. Other influences, such as from exposure to ethanol and carbon dioxide gas, are burning or prickling sensations that do not involve either the gustatory or the olfactory receptors.

Excessively cold wines have a paralyzing effect upon the taste receptors, so that all wines should be critically tasted at cool, or normal, room temperature. (Wines being casually enjoyed may, of course, be chilled as desired.)

The body or "thickness" of wines is determined by "mouth-feel", the distinction between wines that are thin or light-bodied and those that are full, or heavy-bodied.

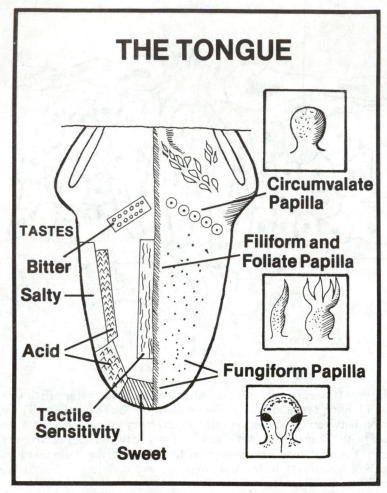

FIG. A.16. DIFFERENT TYPES OF PAPILLAE FOUND ON THE HUMAN
TONGUE
(Adapted from Puisais, J. and Chabanon, R.L. 1974. Initiation into the Art of
Wine Tasting, Interpublish, Madison, Wisconsin)

Despite all of the above distinctions between the olfactory and gustatory stages, the definition of flavor is made from a combination of nasal
and tongue reactions to the influence of wine stimuli. It is this flavor that
enables a wine evaluator to properly judge a wine's taste.

The tasting glass is filled to about ¼ capacity with the wine sample to
be evaluated, and raised to the lips with only a sip of the wine taken into
the mouth. This is washed around the mouth and spat. Another sip is

taken and held in the mouth, again washed around so that all sensitive surfaces in the olfactory and gustatory organs are fully exposed. After 10–20 sec, a small portion of the wine may be swallowed, with any remainder desired being spat. The swallowed portion is exposed to the pharyngeal passage for aftertaste impressions. This may be repeated if the judgement is not clear or some aspect of the wine was confusing at first exposure. After the tasting is completed, the taster should take a sip of pure, room-temperature water, rinse the mouth fully and spit. A bite of neutral, salt-free cracker or some other solid such as bland cheese may be taken and swallowed, followed by another water rinsing. The mouth should then be prepared for another evaluation. No more than 8 evaluations should be attempted consecutively at one sitting or else the palate will become overworked and less effective as an instrument of analysis.

A common fault in wines has to do with acid, either too little in insipid wines or too much in tart wines. Normally wines below .400 g/100 ml of total acidity will be criticized for being too bland. At the other end of the scale, wines in excess of .700 g/100 ml are routinely penalized for being too acid. Of course, there are exceptions. Sweetness can mask acidity and this interaction of stimuli must be carefully studied and experienced before formally judging such phenomena.

Bitterness is rather uncommon in most commercial wines and, unlike acidity or tartness, bitterness is not masked by sweetness. The most unfortunate form of bitterness in wines results from the formation of acrolein from glycerol which is produced by special strains of lactic acid bacteria. This infection is known as *Amertume*.

Much like acidity, wines are often criticized for being too dry (lacking in sweetness) or too sweet. Sweetness arises from natural or added dissolved sugar solids, primarily from glucose, sucrose and fructose, in ascending order of sweetening effect. Glucuronic acid is another sweetening compound found in wines that have been made from grapes that were allowed to mature with the "noble mold", *Botrytis cinerea*.

Recording the results of wine organoleptic analysis is usually done in one of two ways in the wine industry. The first is a numerical point system whereby high scores, or point totals, indicate high quality. Such point systems are usually based on a perfect 20, as first devised at the University of California at Davis. The other recording method is a simple check-reject system indicating approval or disapproval.

Figure A.17 is a reproduction of *Suggestions For Assignment Of Evaluation Points* as published by the American Wine Society. This format is used in conjunction with Fig. 6.2.

In the day-to-day evaluation of wines in production progress, the use of a simple check-mark may be made upon supplemental record forms, such as the laboratory analysis log illustrated in Fig. 2.3. The "Remarks" column may be used for wines that do not meet the check-mark standard.

SUGGESTIONS FOR ASSIGNMENT OF EVALUATION POINTS

Clarity (2)	Brilliant, 2; Clear, very slight haze, 1; Dull, slightly cloudy, 0; Distinctly cloudy, −1
Color (2)	Characteristic of grape and age, 2; Slightly off, 1; Distinctly off, 0
Aroma (4)	Characteristic of grape variety, 4; Distinct but not varietal, 2; Clean, 1; Lacking, 0
Bouquet (2)	Characteristic of grape variety, and age, 2; Faint, 1; Vinegary, 0; Other-odors, −1 or −2
Total Acidity (2)	Balanced, 2; Slightly high or low, 1; Distinctly high or low, 0
Tannin (2)	Smooth, no harshness or bitterness, 2; Slightly harsh & bitter, 1; Distinctly harsh & bitter, 0
Body (1)	Normal, 1; Too heavy or light, 0
Sugar (1)	Balanced, 1; Too high or low, 0
General Flavor (2)	Well balanced, smooth, 2; Some aftertaste, 1; Strong aftertaste, 0
Overall (2)	Very enjoyable, 2; Moderately enjoyable, 1; Distinctly unpleasant, 0

Courtesy: American Wine Society

FIG. A.17.   SUGGESTIONS FOR ASSIGNMENT OF EVALUATION POINTS

19. OXYGEN DETERMINATION

This method of oxygen measurement involves the use of a sensor with a membrane which allows oxygen from the sample to permeate. A polarizing voltage is applied, and dissolved oxygen reacting at the cathode is metered.

The Yellow Springs instrument, Model 54, is commercially available and is fully applicable to the analysis of dissolved oxygen in wine. Free SO_2 does not directly affect the analytical results, but can quickly deteriorate the silver anode in the probe, a condition that requires frequent probe cleaning. Results are affected by the presence of hydrogen sulfide. Figure 5.4 portrays an oxygen meter in use in a wine laboratory.

Apparatus and Reagents

Yellow Springs Oxygen Meter Model 54 with probe
Eyedropper
Magnetic stirring device
250 ml Griffin low-form beaker
Potassium chloride solution (50% saturated KCl and 50% distilled water; add 5 drops of Kodak Photo-Flo solution per 100 ml as a wetting agent)
　　Note: Be sure that distilled water is used or performance of the instrument will be affected by contaminants

Procedure

1. Carefully inspect probe tip for cleanliness and rinse with KCl solution to remove dirt, salts and/or other foreign materials.

2. Fill central well of the probe with KCl solution using a clean eyedropper avoiding trapped air. Add more KCl solution until a large drop accumulates above the probe surface.

3. Apply membrane to probe tip as follows:
   a. Be sure that probe is filled with KCl solution. Also wet "O" ring grove.
   b. With left hand grasp threaded section between thumb and forefinger, securing one end of the membrane under thumb.
   c. With right hand grasp free end of the membrane, and with a continuous motion *stretch* the membrane up, over, and down the other side. Stretching forms the membrane down the sides of the probe.
   d. Secure the membrane end under the left forefinger. Inspect to be sure that the membrane is wrinkle-free and tight like a drum head. Then slip on the "O" ring carefully.
   e. Trim off excess membrane near the "O" ring, leaving the temperature sensor exposed.

4. Rinse probe tip several times with distilled water. Then secure with probe holder so that the probe is properly immersed in distilled water "stand-by" storage.

5. Membranes may last indefinitely, depending upon usage. Average replacement, however, is 2–4 weeks. Should the electrolyte be allowed to evaporate and an excessive amount of bubbles form under the membrane, or the membrane become damaged, thoroughly flush the reservoir with KCl and install a new membrane.

6. It is important that the instrument be placed in the intended operating position—vertical, tilted, or on its back, before it is prepared for use and calibrated. Readjustment may be necessary when the instrument operating position is changed.

7. With switch in the OFF position, adjust the meter pointer to Zero with the screw in the center of the meter panel.

8. Switch to RED LINE and Adjust the RED LINE knob until the meter needle aligns with the red mark at the 31°C position.

9. Switch to ZERO and adjust to zero with zero control knob.

10. Attach the prepared probe to the PROBE connector of the instrument and adjust the retaining ring finger tight.

11. Before calibrating, allow 15 minutes for optimum probe stabilization. Repolarize whenever the instrument has been off or the probe has been disconnected.

12. Place the probe in moist air. Wait about 10 minutes for temperature stabilization. This may be done simultaneously while the probe is stabilizing.

13. Switch to TEMPERATURE and read °C. Refer to Table A.4.

14. Use probe temperature and true local atmospheric pressure (or feet above sea level) to determine calibration values from Table A.4 and Table A.5.

> Example: Probe temperature = 21°C; altitude = 1000 feet. From Table A.4 the calibration value for 21°C is 9.0 ppm. From Table A.5 the altitude factor for 1000 feet is approximately .96. The correction calibration value is then:

$$9.0 \text{ ppm} \times .96 \text{ factor} = 8.64 \text{ ppm}$$

15. Switch to 0-10 or 0-21 ppm range and adjust meter with CAL control to calibration value determined in Step No. 14.

16. Place probe in sample and stir.

TABLE A.4.   SOLUBILITY OF OXYGEN IN FRESH WATER

Temperature °C	PPM Dissolved Oxygen	Temperature °C	PPM Dissolved Oxygen
0	14.6	23	8.7
1	14.2	24	8.5
2	13.9	25	8.4
3	13.5	26	8.2
4	13.2	27	8.1
5	12.8	28	7.9
6	12.5	29	7.8
7	12.2	30	7.7
8	11.9	31	7.5
9	11.6	32	7.4
10	11.3	33	7.3
11	11.1	34	7.2
12	10.8	35	7.1
13	10.6	36	7.0
14	10.4	37	6.8
15	10.2	38	6.7
16	9.9	39	6.6
17	9.7	40	6.5
18	9.5	41	6.4
19	9.3	42	6.3
20	9.2	43	6.2
21	9.0	44	6.1
22	8.8	45	6.0

TABLE A.5. CORRECTION TABLE FOR EFFECTS OF ATMOSPHERIC PRESSURE OR ALTITUDE

Atmospheric Pressure mm Hg	or	Equivalent Altitude Ft.	=	Correction Factor
775		−540		1.02
760		0		1.00
745		542		.98
730		1094		.96
714		1688		.94
699		2274		.92
684		2864		.90
669		3466		.88
654		4082		.86
638		4756		.84
623		5403		.82
608		6065		.80
593		6744		.78
578		7440		.76
562		8204		.74
547		8939		.72
532		9694		.70
517		10472		.68
502		11273		.66

17. Allow sufficient time for probe to stabilize to sample temperature and dissolved oxygen.

18. Read dissolved oxygen on appropriate range, (0−10 or 0−20 ppm).

19. Leave instrument on between measurements to avoid the necessity for repolarization of the probe.

Adapted from: *Instruction Manual, YSI Models 54ARC and 54ABP Dissolved Oxygen Meters,* Yellow Springs Instrument Company, Inc:, 1979.

20. pH DETERMINATION

A pH probe consists of a hydrogen ion sensitive electrode and a reference electrode. Both produce a voltage when in contact with hydrogen ions. The value of voltage is a linear function of the pH. Figure 3.1 illustrates a pH meter in operation.

The following procedure is provided for the analyst. However, if manufacturer's operating instructions are available, they should take precedence in the use and care of the pH meter.

Apparatus and Reagents

pH meter with electrodes
50 ml polyethylene or polypropylene beakers
Thermometer
Wash bottle with distilled water
pH buffer solution

Procedure

1. Turn pH meter on and allow to become stable.

2. Pour about 20 ml of buffer solution into a clean 50 ml poly beaker.

3. Immerse electrodes into buffer solution.

4. Adjust temperature control knob of pH meter to buffer solution temperature.

5. Remove buffer solution and rinse electrode with some of the wine to be analyzed.

6. Pour about 20 ml of sample into a clean 50 ml poly beaker and immerse electrodes into sample while being gently agitated.

7. Read pH of sample directly from pH meter and post results on laboratory analysis log.

21. PLATING PROCEDURE

This method is designed to spread individual cells across a plate of agar inside a sterile Petri dish so that individual colonies may be identified, counted and "picked" with a sterile loop for propagation, if desired.

Apparatus and Materials:

2 Petri dishes, glass or disposable plastic, sterilized
Yeast dextrose agar medium, sterilized
1/.1 ml serological pipet(s), glass or disposable plastic, sterilized
Sterile cotton
75% alcohol solution
Burner with open flame
Incubator

Procedure

1. The sample to be plated should be handled so there is no exposure to contamination. Use sterile cotton moistened with alcohol solution to wipe opened bottles or other sample containers before inserting pipet in Step No. 4.

2. Melt the prepared agar solution with low heat so that the medium does not get excessively hot, yet so that it will not solidify immediately upon application to the Petri dish. Temperatures cannot be taken, as contamination will be introduced. It takes practice seeing what the agar looks like and how it flows in order to properly make temperature judgements. If it is too hot the microorganisms will be killed. If it is too

cold the medium will not distribute the cells properly in Step No. 4.

3. With the sample close at hand and ready for the pipet, flame the opening of the agar storage vessel and pour just enough of the medium to cover the bottom halves of the two covered sterile Petri dishes. Replace the Petri dish covers quickly and flame the opening of the agar storage vessel again. Reseal, flame once more and return to storage unless more Petri dishes are to be made.

4. Immediately pipet 0.1 ml of sample into one Petri dish and 1 ml into the other Petri dish in the same careful manner. Replace Petri dish covers quickly and gently swirl the 2 Petri dishes so that the samples are each evenly distributed across the plate. Properly prepared, the agar should commence to solidify shortly after the swirling has been completed.

5. Incubate for approximately 3 to 4 days at 30°C. Figure A.23 may be used to identify colonies. The 0.1 ml plate should, of course, show about 1/10 the number of colonies that the 1 ml sample plate exhibits. These two different dilutions are made in order to ease colony differentiation and the choosing of colonies for propagation, if desired. Agar slant planting, incubation and storage of microorganisms is provided in Procedure No. 2 in this appendix.

22. SULFUR DIOXIDE—FREE

Free SO_2 analysis is done by the titration of unbound SO_2 with an iodine reagent, using a starch solution as an indicator.

Figure A.19 provides an illustration of the simple apparatus required.

Apparatus and Reagents

20 ml volumetric pipet
250 ml wide-mouth Erlenmeyer flask
1/40 N iodine
1% starch indicator (preserved in a water solution with 15% ethanol)
Bicarbonate of soda
25% sulfuric acid

Procedure

1. Adjust sample temperature to 68°F, or to whatever temperature indicated on 20 ml volumetric pipet.

2. Pipet sample into clean 250 ml wide-mouth Erlenmeyer flask.

3. Add 5 ml 25% sulfuric acid, then a pinch of bicarbonate of soda and 5 ml of 1% starch indicator.

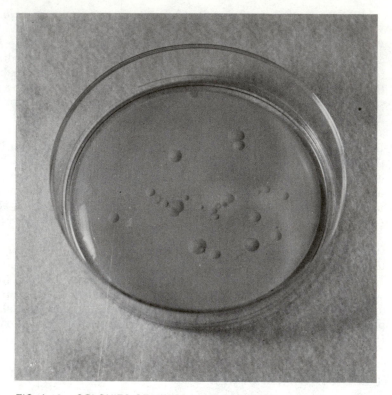

FIG. A.18.   COLONIES OF YEAST IN A PETRI DISH

4. Also, use of reflected light may help with some very dark or turbid samples. A yellow light source is useful for reds. Titration should be done rapidly since there are slow side reactions that interfere.

5. Titrate carefully with 1/40 N iodine from buret. When color changes (white wines to light blue, red wines to blue-green) and holds for 15 sec or more, take buret reading of ml 1/40 N iodine used and find free sulfur dioxide on Table A.6.

    Examples:

        1.4 ml =   56 ppm free sulfur dioxide
        2.5 ml = 100 ppm free sulfur dioxide
        3.6 ml = 144 ppm free sulfur dioxide
        4.7 ml = 188 ppm free sulfur dioxide

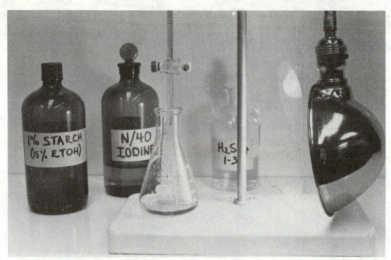

FIG. A.19. APPARATUS REQUIRED FOR ANALYSIS OF FREE SULFUR DIOXIDE

6. Post analysis on laboratory analysis log.

Note:
   Some red wines will need to be diluted in that they may be too dark to accurately read the titration end point. Add distilled water in amounts up to 80 ml as may be required.

23. SULFUR DIOXIDE—TOTAL

Apparatus and Reagents

   20 ml volumetric pipet
   250 ml wide-mouth Erlenmeyer flask with rubber stopper
   1/40 N iodine
   1% starch indicator (preserved in a water solution with 15% ethanol)
   Bicarbonate of soda
   25% sulfuric acid
   25% sodium hydroxide

Procedure

   1. Adjust sample temperature to 68°F, or to whatever temperature is indicated on 20 ml volumetric pipet.

TABLE A.6.   SULFUR DIOXIDE TABLE (20 ML SAMPLE × 1/40 N IODINE × 1% STARCH INDICATOR)

Ml Iodine	Ppm SO$_2$	Ml Iodine	Ppm SO$_2$
.1	4	4.1	164
.2	8	4.2	168
.3	12	4.3	172
.4	16	4.4	176
.5	20	4.5	180
.6	24	4.6	184
.7	28	4.7	188
.8	32	4.8	192
.9	36	4.9	196
1.0	40	5.0	200
1.1	44	5.1	204
1.2	48	5.2	208
1.3	52	5.3	212
1.4	56	5.4	216
1.5	60	5.5	220
1.6	64	5.6	224
1.7	68	5.7	228
1.8	72	5.8	232
1.9	76	5.9	236
2.0	80	6.0	240
2.1	84	6.1	244
2.2	88	6.2	248
2.3	92	6.3	252
2.4	96	6.4	256
2.5	100	6.5	260
2.6	104	6.6	264
2.7	108	6.7	268
2.8	112	6.8	272
2.9	116	6.9	276
3.0	120	7.0	280
3.1	124	7.1	284
3.2	128	7.2	288
3.3	132	7.3	292
3.4	136	7.4	296
3.5	140	7.5	300
3.6	144	7.6	304
3.7	148	7.7	308
3.8	152	7.8	312
3.9	156	7.9	316
4.0	160	8.0	320

2. Pipet sample into clean 250 ml wide-mouth Erlenmeyer flask and add 5 ml 25% sodium hydroxide. Seal firmly with rubber stopper. Mix and set aside for about 15 min.

3. Remove stopper, add 5 ml 25% sulfuric acid, a pinch of bicarbonate of soda and 5 drops of 1% starch solution to sample in flask.

4. Also, use of reflected light may help with some very dark or turbid samples. A yellow light source is useful for reds. Titration should be done rapidly since there are slow side reactions that interfere.

5. Titrate carefully with 1/40 N iodine from buret. When color changes (white wines to light blue, red wines to blue-green) and holds for 15 sec or more, take buret reading of ml 1/40 N iodine used and find total sulfur dioxide on the sulfur dioxide table.

   Examples:

   1.4 ml =   56 ppm total sulfur dioxide
   2.5 ml = 100 ppm total sulfur dioxide
   3.6 ml = 144 ppm total sulfur dioxide
   4.7 ml = 188 ppm total sulfur dioxide

6. Post analysis on laboratory analysis log.

Note:
Some red wines will need to be diluted, as they may be too dark to accurately read the titration end point. Add distilled water in amounts up to 80 ml as may be required.

24. TOTAL ACIDITY DETERMINATION BY TITRATION

This procedure describes how to neutralize the acids in the sample with the alkaline solution of sodium hydroxide, using phenolphthalein as an indicator. Figure 3.4 provides an illustration of the simple equipment required to perform a total acidity analysis by titration.

Apparatus and Reagents

10 ml volumetric pipet
250 ml wide-mouth Erlenmeyer flask
25 or 50 ml buret with stopcock
1/10 N sodium hydroxide
1% phenolphthalein indicator (preserved in water solution with 15% ethanol)
Distilled or deionized water

Procedure

1. If the sample has been taken from a freshly pressed lot of juice or must there may be too many suspended solids for the pipet to allow flow through the narrow capillary at the tip. In this case the juice sample should be carefully filtered (in a neutral medium so as not to influence acidity). Adjust sample temperature to 68°F or to whatever temperature is indicated on the 10 ml volumetric pipet.

2. Pipet sample into clean 250 wide-mouth Erlenmeyer flask.

3. Add 5 drops 1% phenolphthalein indicator to sample in 250 ml Erlenmeyer flask. (1% cresol red may be used instead of phenolphthalein. Color change is from yellow to purple, at a pH of 7.7 to 7.8. Red wines are not as deeply colored at this pH so the indicator is more visible.)

4. Titrate carefully with 1/10 N sodium hydroxide from buret. When slight pink color holds (green for red wines) for 15 sec or more, take buret reading of ml 1/10 N sodium hydroxide used.

5. Grams per 100 ml total acidity is found by using the total acidity table (Table A.7).

   Examples:

   6.7 ml = .503 g/100 ml total acidity
   9.0 ml = .675 g/100 ml total acidity

6. Post results on laboratory analysis log.

25. TOTAL ACIDITY DETERMINATION BY TITRATION—pH METER

Total acidity is determined by neutralizing the acids in the sample with the alkaline solution of sodium hydroxide, using a pH meter to determine the end point.

Figure A.20 provides an illustration of the equipment required to perform a total acidity analysis using a pH meter as part of the titration apparatus.

Apparatus and Reagents

10 ml volumetric pipet
25 or 50 ml buret with stopcock and special tip, so as to reach 50 ml poly beaker
1/10 N sodium hydroxide
pH meter with electrodes
50 ml polyethylene or polypropylene beakers
Thermometer
Wash bottle with distilled water
pH buffer solution

Procedure

1. If the sample has been taken from a freshly pressed lot of juice or must there may be too many suspended solids for the pipet to allow flow through the narrow capillary at the tip. In this case the juice sample should be carefully filtered (in a neutral medium so as not to influence acidity). Adjust sample temperature to 68°F or to whatever temperature is indicated on the 10 ml volumetric pipet.

TABLE A.7.   TOTAL ACIDITY TABLE (G/100 ML EXPRESSED AS TARTARIC ACID)

1/10 N NaOH	g/100 ml T.A.	1/10 N NaOH	g/100 ml T.A.	1/10 N NaOH	g/100 ml T.A.
5.1	.383	9.1	.683	13.1	.983
5.2	.390	9.2	.690	13.2	.990
5.3	.398	9.3	.698	13.3	.998
5.4	.405	9.4	.705	13.4	1.005
5.5	.413	9.5	.713	13.5	1.013
5.6	.420	9.6	.720	13.6	1.020
5.7	.428	9.7	.728	13.7	1.028
5.8	.435	9.8	.735	13.8	1.035
5.9	.443	9.9	.743	13.9	1.043
6.0	.450	10.0	.750	14.0	1.050
6.1	.458	10.1	.758	14.1	1.058
6.2	.465	10.2	.765	14.2	1.065
6.3	.473	10.3	.773	14.3	1.073
6.4	.480	10.4	.780	14.4	1.080
6.5	.488	10.5	.788	14.5	1.088
6.6	.495	10.6	.795	14.6	1.096
6.7	.503	10.7	.803	14.7	1.103
6.8	.510	10.8	.810	14.8	1.110
6.9	.518	10.9	.818	14.9	1.118
7.0	.525	11.0	.825	15.0	1.125
7.1	.533	11.1	.833	15.1	1.133
7.2	.540	11.2	.840	15.2	1.140
7.3	.548	11.3	.848	15.3	1.148
7.4	.555	11.4	.855	15.4	1.155
7.5	.563	11.5	.863	15.5	1.163
7.6	.570	11.6	.870	15.6	1.170
7.7	.578	11.7	.878	15.7	1.178
7.8	.585	11.8	.885	15.8	1.185
7.9	.593	11.9	.893	15.9	1.193
8.0	.600	12.0	.900	16.0	1.200
8.1	.608	12.1	.908	16.1	1.208
8.2	.615	12.2	.915	16.2	1.215
8.3	.623	12.3	.923	16.3	1.223
8.4	.630	12.4	.930	16.4	1.230
8.5	.638	12.5	.938	16.5	1.238
8.6	.645	12.6	.945	16.6	1.245
8.7	.653	12.7	.953	16.7	1.253
8.8	.660	12.8	.960	16.8	1.260
8.9	.668	12.9	.968	16.9	1.268
9.0	.675	13.0	.975	17.0	1.275

FIG. A.20.   EQUIPMENT USED FOR TOTAL ACIDITY ANALYSIS

2. Pipet sample into clean 50 ml poly beaker.

3. Follow Steps 1 through 5 of Procedure No. 20, *pH Determination.*

4. Place electrode into poly beaker with sample and commence titration very carefully, applying gentle agitation to the beaker as the sodium hydroxide drips into the sample. As the pH reaches about 6.5 the titration should be slowed down considerably, as the endpoint at pH 8.2–8.4 will be achieved very quickly once pH 6.5 is observed. (Use of cresol red or phenolphthalein helps one to judge how quickly an end point is approaching. This saves time since the indicator responds faster to pH than the meter. The stirring plate and magnetic round "star" stirrers are preferred to hand mixing.)

5. Hold for about 10 sec, continuing the gentle agitation to be sure that the end point of pH 8.2–8.4 remains constant. If pH drops back toward 8.0, add one more drop of sodium hydroxide reagent. If pH goes beyond 8.4, start over and repeat the test to this point. After the proper end point has been reached, take buret reading of ml 1/10 N sodium hydroxide used.

6. Grams per 100 ml total acidity can be found by using the Total Acidity Table.
    Examples:
       6.7 ml = .503 g/100 ml total acidity
       9.0 ml = .675 g/100 ml total acidity.

7. Post results on laboratory analysis log.

8. Clean glassware, beakers and electrode with distilled water only (pipet may be cleaned first with warm water).

26. VIABLE MICROORGANISMS IN BOTTLED WINES— MILLIPORE METHOD

The rationale for this method is to filter a measured amount of wine sample and culture medium through a prepared plastic receptacle (monitor case—See Fig. A.21) under aseptic conditions. The incubation of the medium discloses the growth of microbial colonies.

The operation is fairly simple and straightforward, as long as the analyst develops a good handling technique so that nothing is touched that comes in direct contact with the filter, or the material being filtered.

Apparatus, materials and detailed instructions are available in kits from Millipore Corporation.

Apparatus and Materials

Vacuum pump
Vacuum flask with appropriate tubing
Incubator
Millipore Flash-O-Lens
Millipore Whirl-Pak bag(s)
Millipore plastic monitor case(s)
Millipore green yeast and mold medium ampoule(s)
Millipore forceps
75% ethanol solution (approx 250 ml in a 500 ml beaker)
Sterile cotton

Procedure

1. Remove a monitor case from the Millipore kit. Using the flat end of the forceps pry off the blue-plugged top section carefully by using a

Courtesy of the Millipore
Corporation, Bedford, Mass.

FIG. A.21. THE MILLIPORE MONITOR CASE

little leverage on alternate sides. Place the top to one side, blue plug down. Place monitor case onto the adaptor on the vacuum flask, being certain not to touch the upper rim of the monitor case.

2. Dip wine bottle neck (capsule entirely removed) in the alcohol solution. If corked, the corkscrew should also be dipped in the alcohol. Do not, however, insert corkscrew through the cork and into the headspace of the bottle. Insert only as far as necessary to remove the cork. If capped, simply remove cap, being certain not to touch the lip of the bottle. Carefully wipe opening of the bottle with sterile cotton which has been moistened with the alcohol solution. Measure a 50–100 ml wine sample from the bottle into a graduated Whirl-Pak bag and filter through the monitor with vacuum.

3. Dip ampoule of green yeast and mold medium in alcohol while the plastic sleeve is grasped between thumb and two fingers. Carefully tap the sleeve of the ampoule with a fingernail so as to jog all of the medium down toward the pointed tip.

4. Dip forceps in alcohol solution and place the top part of the plastic sleeve, as close to the top as possible, between the blue sections of the forceps.

5. Crush the plastic sleeve firmly but carefully. Place forefinger on the top of the crushed sleeve, as in using a pipet. Keeping the forefinger in place, grasp the ampoule between thumb and middle finger. At the lower tip of an ampoule is a tiny groove. Dip this end of the ampoule into 70% alcohol, then insert the tip, up to the etched mark, into the open hold of an old, used, cleaned monitor. The glass tip will snap off at the etched mark and the medium will stay in the ampoule, as long as the other end is held by the finger.

6. Place the tip of the ampoule over the top of the monitor case at an angle of about 30° in order that the broken tip and glass will not fall into the top of the monitor. Remove forefinger and allow media to trickle into the monitor case. Discard empty ampoule.

7. Rotate the flask and monitor so that the medium spreads as evenly as possible across the membrane surface. Hold the monitor loosely above the adapter. Start the vacuum pump and by lowering the monitor onto the adapter, allow the free medium on top of the membrane to be drawn through. Just at the point where the liquid has passed through the surface, lift the monitor from the plastic adapter so that no more vacuum continues to draw the medium out of the absorbent pad, which could inhibit growth of microorganisms.

8. Take up the top of the monitor from where it was placed in Step 1. Replace it firmly on the monitor case. Remove the monitor from the vacuum flask. Take a red plug from the plastic bag and place it in the hole in the bottom of the monitor. Mark the monitor for identification as is appropriate.

9. Turn on incubator and maintain at 32°C. The monitor case should be placed in the incubator upside down with the red button on top. Maintain the 32°C incubation temperature for 3 to 4 days.

10. After the 3 to 4 day incubation period, any yeast, mold and bacteria present on the filter membrane will have grown into visible colonies which can be easily counted with the aid of the Flash-O-Lens. To aid in counting and identifying colonies, remove the top section of the monitor case. After observations replace the top of the case and dispose of the monitor.

11. Colony identification can be made by comparison to the illustrations provided in Fig. A.23.

27. VIABLE YEASTS IN BOTTLED WINES—RAPID METHOD OF DETECTION

This method involves the differential staining of a membrane which has been used to filter a bottling-line sample. Dead yeast cells appear microscopically as blue-colored while total cells are red.

Apparatus and Materials

Light microscope capable of 400× magnification with Schott KL 150 light source (goose-neck light lead) or equivalent high intensity lamp

FIG. A.22.   COLONY IDEN-
TIFICATION

Courtesy of the Millipore
Corporation, Bedford, Mass.

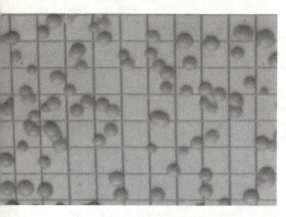
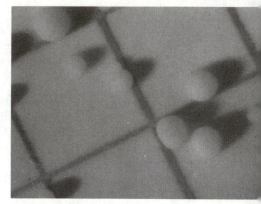

Fig. A.23a.   YEAST COLONIES

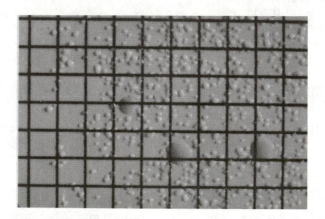

FIG. A.23b.   BACTERIA COLONIES

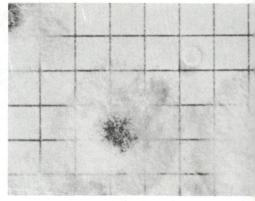

FIG. A.23c.   MOLD COLONIES

Courtesy of the Millipore Corporation, Bedford, Mass.

Vacuum pump
Vacuum filter with 13 mm funnel and appropriate tubing
Glass or Plexiglas slides—one with 11 mm hole drilled in the center
Burner with open flame
Sterile cotton
75% methanol solution
Boiling pure water (distilled or de-ionized)
Membrane filter (1.2 nm porosity or less)
0.01% methylene blue solution (in citrate buffer of pH 4.6)
Ash-free pad(s), 60 mm in diameter
Gummed tape
100 g Ponceau S solution (0.9 Ponceau S, 13.4 g trichloracetic acid, 13.4 g
    sulfosalicylic acid and 72.3 g pure water in a Petri dish) Note: The
    Ponceau S solution must be prepared fresh each time.
Acetic acid (5%)

Procedure

1. Thoroughly clean filter assembly by the usual methods and then
   sterilize by rinsing with the alcohol solution. The membrane, sterilized by 15 min in boiling water, is then placed into the filter holder
   under aseptic conditions. Care should be taken during opening and
   pouring the sample that no cork or debris reach the filter.

2. Depending upon the wine, about 8 to 15 min or perhaps longer, will be
   required for filtration. When filtration is completed, turn off vacuum
   pump and allow vacuum to equalize. Add 2 ml of the methylene blue
   solution to the filter, wait about 30 sec and then vacuum filter same to
   the flask.

3. The membrane is then carefully placed in the middle of a microscope
   slide (the lines of the grid being placed parallel, or vertical, to the edge
   of the slide). To prevent curling of the membrane during subsequent
   examination, the membrane is covered with a second glass slide, or a
   clear Plexiglas slide, which has had an 11 mm hole bored in the middle.
   In order to avoid a displacement of the slides, they are fastened together on both sides of the hole with transparent gummed tape. Thus,
   the moist membrane which is laid out smooth is held stretched by the
   edges (the working surface of the filter has a diameter of 10 mm). To
   prevent fogging of the objective and to obtain better contrast with a dry
   membrane, one can warm the slide very cautiously with a flame.

4. A 400× magnification with incident light and a microscope with a
   large field of view is recommended. In order that dead (blue-colored)
   yeast cells stand out and become easy to recognize, incident light
   diagonally from the side at an angle between 10 and 30 degrees is
   necessary. The Schott KL 150 light source is very well suited for this

purpose. This lamp has the advantage that the light can be delivered near the objective and at the correct angle.

5. A gridded filter surface of 78.5 mm^2 (10 mm diameter) has about 8.17 squares (the side of each grid is approximately 3.1 mm). At least 4 or 5 squares should be examined, since uneven distribution of the yeast cells on the membranes would give erroneous results. The examined squares are marked in an appropriate manner because they must be counted a second time after staining with Ponceau S.

6. Place the membrane on an absorbent, ash-free pad of 60 mm diameter in a Petri dish. The pad will have been previously saturated with Ponceau S solution. Tilt the Petri dish slightly so that the surplus Ponceau S solution can drain away. A second pad is soaked in the same manner with 5% acetic acid solution.

7. After the first microscopic examination, the membrane is carefully taken from between the two microscope slides and laid out flat and smooth on the pad soaked with the Ponceau S solution. After 4 min it is transferred onto the pad soaked with acetic acid. In order that the surplus red stain is removed as completely as possible by the acetic acid, the membrane is transferred in intervals of about 30 to 60 sec from one fresh uncolored place on the pad to another (5 or 6 times). When the membrane has assumed a white to pale pink color it is laid out smooth on a microscope slide, aligned with the slide. The slide is then passed 3 or 4 times briefly over the flame. The acetic acid evaporates and the membrane is lightly baked onto the slide and will not curl up.

8. Ponceau S is a stain for protein and is not removed from the yeast by the treatment with the diluted acetic acid, but is removed from the debris (cork cells, etc.) and the membrane itself. This considerably facilitates the second microscopic examination. After the final counting has given the difference between red and blue yeast cells, the total number of living cells present in the sample can be calculated from the numbers of squares examined and the volume of wine taken.

9. Note Fig. A.24, after blue-staining, and A.25, after red staining. Only one yeast cell has taken up the methylene blue stain in the first step. The other yeast cell, which became visible by the red staining, has therefore been determined a living yeast.

Precautionary Measures

Dust, dirt or other foreign particles in the sample aggravate microscopic examination, especially after staining with methylene blue. This may reduce the precision, and under some conditions, even the validity of

FIG. A.24.   SECTION AFTER METHYLENE BLUE STAINING (400×)

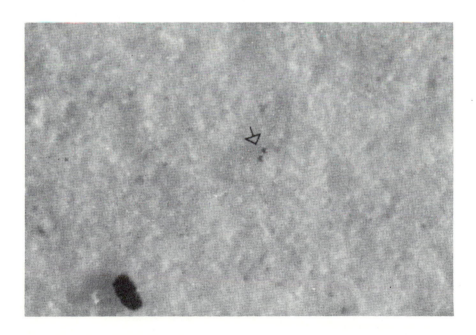

FIG. A.25.   IDENTICAL SECTION AFTER PONCEAU S STAINING

the results. It is therefore essential that the cleanest conditions be employed. In extreme cases, the use of bottles not well cleaned or of new bottles not well rinsed or otherwise cleaned, or the use of inferior corks could render this method useless. The sample bottles to be tested should not be taken until 30 min after the start of the filling operation because it is likely that during this time considerably more debris will be found in the wine. In order to avoid reinfections by splashing water, it is generally recommended not to cool the filter used in the bottling process. During steaming, particles are dislodged and are more numerous at the beginning of the filling operation.

If excessive cork dust makes the microscopic examination difficult, one may insert into the vacuum filter a sterile 100μm screen. If reinfection by the corking machine is virtually not possible, the test samples may be taken before the corker and closed with an artificial closure having been sterilized in the alcohol solution.

After longer shut-downs (weekends), the number of dead yeast cells in the bottled product can be rather high, because a longer time has been allowed for multiplication of the organisms in the equipment (filter and filler). The cells will be killed by steam sterilization and, with time, flushed out of the bottling system.

SOURCE: This method [Weinberg and Keller, *20*, 469−478 (1973) and Wines & Vines *55*(12), 36−39 (1974)] was developed by R.E. Kunkee, Department of Viticulture and Enology, University of California, Davis, and F. Neradt, Seitz-Werke GmbH, Bad Kreuznach, Federal Republic of Germany. It is reprinted with their permission.

28. VOLATILE ACIDITY DETERMINATION BY CASH VOLATILE ACID APPARATUS

The rationale for this method of determining volatile acids is to separate the volatile, or distillable, acids from the fixed acids in the cash still. The acids are then titrated with the alkaline solution of sodium hydroxide, with phenolphthalein as an indicator.

Figure A.26 illustrates a cash volatile acid still in the process of distilling the volatile acids from a wine sample. The titration apparatus is the same as provided in Fig. 3.4

Apparatus and Reagents

Cash volatile acid apparatus complete with condenser
Support rods and clamps for cash still
¼" Tygon tubing
10 ml volumetric pipet
250 ml Erlenmeyer flask
25 or 50 ml buret with stopcock (0.1 ml subdivisions)

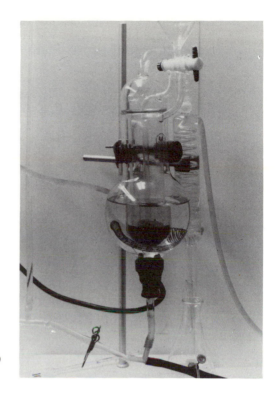

FIG. A.26. CASH VOLATILE ACID
STILL

1/10 N sodium hydroxide
1% phenolphthalein indicator (preserved in a water solution with 15% ethanol)
Distilled or deionized water
Cold tap water for condenser

Procedure

1. Turn stopcock on cash still so that passage runs from funnel to inner tube.

2. Adjust sample temperature to 68°F or to whatever temperature indicated on 10 ml volumetric pipet.

3. Start distilled water running into lower (outer) pot very slowly. Close outlet tube.

4. Pipet sample into funnel and drain into inner tube of cash still. Wash funnel into inner tube with 10–15 ml of distilled water.

5. Turn stopcock so that passage goes to outer tube. Allow distilled water running into lower pot to reach a level about ¾ in. over wire element. Close inlet tube. Place a clean 250 ml Erlenmeyer flask under condenser outlet.

6. Apply cold tap water at a moderate rate through condenser and turn on electricity to power heating coil in lower pot.

7. Small bubbles should begin to form first, then larger ones from the heating coil. Allow these to boil off, keeping the stopcock turned to outer tube until "spitting" starts. Then turn stopcock to the horizontal position (closing off both tubes).

8. Distill over approximately 100 ml of distillate. Be sure that distillate runs out of condenser cool. If not, start over, cleaning still pots three times with distilled water rinsings. Increase condenser water appropriately but do not run too fast or else condenser may break from the pressure.

9. After distillate is collected, turn off electricity to heating coil and turn stopcock to inner tube passage. Remove 250 ml Erlenmeyer flask carefully and close with stopper.

10. Open outlet tube and add approximately 50 ml distilled water into funnel. Apparatus should "cough", cleaning itself. Repeat twice more, spacing "coughings" about equally at beginning, middle and end of outer tube drainage. Allow apparatus to drain.

11. Remove stopper from distillate flask, add 5 drops phenolphthalein indicator solution. Titrate very carefully with 1/10 N sodium hydroxide solution from buret. When a slightly pink color holds for 15 sec or more, take buret reading of ml 1/10 N sodium hydroxide used and find volatile acidity, g/100 ml, from Table A.8.

   Examples:

       .4 ml = .024 g/100 ml volatile acidity
       .7 ml = .042 g/100 ml volatile acidity
      1.1 ml = .066 g/100 ml volatile acidity
      1.5 ml = .090 g/100 ml volatile acidity

12. Post results on laboratory analysis log.

13. Rinse flask and pipet three times with warm water, once with distilled or deionized water and drain dry.

Note: The cash apparatus should be adequately cooled before distilling another sample as in Step No. 4, the sample will be exhausted from the inner tube to the outer pot by contraction.

TABLE A.8.   VOLATILE ACIDITY TABLE (G/100 ML EXPRESSED AS ACETIC ACID)
(For use when distilling 10 ml sample in Cash VA apparatus and titrating using N/10 NaOH)

(Ml) NaOH	G/100 ml VA
.1	.006
.2	.012
.3	.018
.4	.024
.5	.030
.6	.036
.7	.042
.8	.048
.9	.054
1.0	.060
1.1	.066
1.2	.072
1.3	.078
1.4	.084
1.5	.090
1.6	.096
1.7	.102
1.8	.108
1.9	.114
2.0	.120
2.1	.126
2.2	.132
2.3	.138
2.4	.144
2.5	.150
2.6	.156
2.7	.162
2.8	.168
2.9	.174
3.0	.180

Appendix B

Charts and Tables

Charts

1. Brix (Soluble Solids) Table at 20°C
2. Traditional Bulk Wine Vessels
3. Alcohol Measurements
4. Approximate Dilution of Extract from the Addition of High-proof Brandy
5. Tons of Grapes Per Acre
6. Winery Hose Data
7. Flow Estimation
8. Sulfur Dioxide Addition
9. Legal U.S. Wine Bottle Sizes
10. U.S. Traditional Wine Bottle Sizes
11. Correction of Wine Spirits Volume to 60°F
12. Calculation of Partially Filled Horizontal Tanks
13. Wine Tank Capacities
14. Vat Capacity
15. Tank Chart for Straight-sided Horizontal Tank
16. Tank Chart for Straight-sided Vertical Tank

Figures

1. Filtration Porosity Comparisons
2. Flow Estimation from a Horizontal 2-in. hose
3. Periodic Table of the Elements
4. Periodic Properties of the Elements

TABLE B.1.   BRIX (SOLUBLE SOLIDS) AT 20°C; SUCROSE (CANE SUGAR) SCALE

Degrees Brix	Specific Gravity	Lb Total Weight per U.S. Gallon	Lb Solids per U.S. Gallon	G Solids per Liter	Lb Water per U.S. Gallon
0.1	1.00040	8.34878	.00834	1.000	8.34044
0.2	1.00079	8.35196	.01675	2.007	8.33521
0.3	1.00118	8.35514	.02513	3.011	8.33001
0.4	1.00156	8.35828	.03351	4.015	8.32477
0.5	1.00195	8.36145	.04189	5.020	8.31956
0.6	1.00233	8.36462	.05026	6.024	8.31436
0.7	1.00272	8.36779	.05864	7.028	8.30915
0.8	1.00310	8.37096	.06702	8.031	8.30394
0.9	1.00349	8.37413	.07539	9.035	8.29874
1.0	1.00387	8.37730	.08377	10.039	8.29353
1.1	1.00425	8.38047	.09219	11.047	8.28828
1.2	1.00464	8.38372	.10060	12.055	8.28312
1.3	1.00503	8.38698	.10903	13.065	8.27795
1.4	1.00542	8.39023	.11746	14.075	8.27277
1.5	1.00581	8.39348	.12590	15.086	8.26758
1.6	1.00620	8.39674	.13435	16.099	8.26239
1.7	1.00659	8.39999	.14280	17.111	8.25719
1.8	1.00698	8.40325	.15126	18.125	8.25199
1.9	1.00737	8.40650	.15972	19.139	8.24678
2.0	1.00776	8.40976	.16820	20.155	8.24156
2.1	1.00815	8.41301	.17667	21.170	8.23634
2.2	1.00854	8.41627	.18516	22.187	8.23111
2.3	1.00893	8.41952	.19365	23.205	8.22587
2.4	1.00933	8.42286	.20215	24.223	8.22071
2.5	1.00972	8.42611	.21065	25.242	8.21546
2.6	1.01011	8.42937	.21916	26.262	8.21021
2.7	1.01051	8.43271	.22768	27.282	8.20503
2.8	1.01090	8.43596	.23621	28.305	8.19975
2.9	1.01129	8.43922	.24474	29.327	8.19448
3.0	1.01169	8.44255	.25328	30.350	8.18927
3.1	1.01208	8.44581	.26182	31.373	8.18399
3.2	1.01248	8.44915	.27037	32.398	8.17878
3.3	1.01287	8.45240	.27893	33.423	8.17347
3.4	1.01327	8.45574	.28750	34.450	8.16824
3.5	1.01366	8.45899	.29606	35.476	8.16293
3.6	1.01406	8.46233	.30464	36.504	8.15769
3.7	1.01445	8.46559	.31323	37.534	8.15236
3.8	1.01485	8.46892	.32182	38.563	8.14710
3.9	1.01524	8.47218	.33042	39.593	8.14176
4.0	1.01564	8.47552	.33902	40.624	8.13650
4.1	1.01603	8.47877	.34763	41.656	8.13114
4.2	1.01643	8.48211	.35625	42.689	8.12586
4.3	1.01683	8.48545	.36487	43.722	8.12058
4.4	1.01723	8.48878	.37351	44.757	8.11527
4.5	1.01763	8.49212	.38215	45.792	8.10997
4.6	1.01802	8.49538	.39079	46.827	8.10459
4.7	1.01842	8.49871	.39944	47.864	8.09927
4.8	1.01882	8.50205	.40810	48.902	8.09395
4.9	1.01922	8.50539	.41676	49.939	8.08863
5.0	1.01962	8.50873	.42544	50.980	8.08329
5.1	1.02002	8.51207	.43412	52.020	8.07795
5.2	1.02042	8.51540	.44280	53.060	8.07260
5.3	1.02082	8.51874	.45149	54.101	8.06725
5.4	1.02122	8.52208	.46019	55.144	8.06189
5.5	1.02163	8.52550	.46890	56.187	8.05660
5.6	1.02203	8.52884	.47762	57.232	8.05122
5.7	1.02243	8.53218	.48633	58.276	8.04585
5.8	1.02283	8.53552	.49506	59.322	8.04046
5.9	1.02323	8.53885	.50379	60.368	8.03506

TABLE B.1.   *(Cont'd.)*

Degrees Brix	Specific Gravity	Lb Total Weight per U.S. Gallon	Lb Solids per U.S. Gallon	G Solids per Liter	Lb Water per U.S. Gallon
6.0	1.02364	8.54228	.51254	61.417	8.02974
6.1	1.02404	8.54561	.52128	62.464	8.02433
6.2	1.02444	8.54895	.53003	63.512	8.01892
6.3	1.02485	8.55237	.53880	64.563	8.01357
6.4	1.02525	8.55571	.54757	65.614	8.00814
6.5	1.02566	8.55913	.55634	66.665	8.00279
6.6	1.02606	8.56247	.56512	67.717	7.99735
6.7	1.02646	8.56581	.57391	68.770	7.99190
6.8	1.02687	8.56923	.58271	69.825	7.98652
6.9	1.02727	8.57257	.59151	70.879	7.98106
7.0	1.02768	8.57599	.60032	71.935	7.97567
7.1	1.02808	8.57933	.60913	72.991	7.97020
7.2	1.02849	8.58275	.61796	74.049	7.96479
7.3	1.02890	8.58617	.62679	75.107	7.95938
7.4	1.02931	8.58989	.63563	76.166	7.95396
7.5	1.02972	8.59301	.64448	77.227	7.94853
7.6	1.03012	8.59635˙	.65332	78.286	7.94303
7.7	1.03053	8.59977	.66218	79.348	7.93759
7.8	1.03094	8.60319	.67105	80.410	7.93214
7.9	1.03135	8.60662	.67992	81.473	7.92670
8.0	1.03176	8.61004	.68880	82.538	7.92124
8.1	1.03216	8.61338	.69768	83.602	7.91570
8.2	1.03257	8.61680	.70658	84.668	7.91022
8.3	1.03298	8.62022	.71548	85.735	7.90474
8.4	1.03339	8.62364	.72439	86.802	7.89925
8.5	1.03380	8.62706	.73330	87.870	7.89376
8.6	1.03421	8.63048	.74222	88.939	7.88826
8.7	1.03462	8.63390	.75115	90.009	7.88275
8.8	1.03503	8.63733	.76009	91.080	7.87724
8.9	1.03544	8.64075	.76903	92.151	7.87172
9.0	1.03585	8.64417	.77798	93.224	7.86619
9.1	1.03626	8.64759	.78693	94.296	7.86066
9.2	1.03667	8.65101	.79589	95.370	7.85512
9.3	1.03708	8.65443	.80486	96.445	7.84957
9.4	1.03750	8.65794	.81385	97.522	7.84409
9.5	1.03791	8.66136	.82283	98.598	7.83853
9.6	1.03832	8.66478	.83182	99.675	7.83296
9.7	1.03874	8.66829	.84082	100.754	7.82747
9.8	1.03915	8.67171	.84983	101.833	7.82188
9.9	1.03956	8.67513	.85884	102.913	7.81629
10.0	1.03998	8.67863	.86786	103.994	7.81077
10.1	1.04039	8.68205	.87689	105.076	7.80516
10.2	1.04081	8.68556	.88593	106.159	7.79963
10.3	1.04122	8.68898	.89496	107.241	7.79402
10.4	1.04164	8.69249	.90402	108.327	7.78847
10.5	1.04205	8.69591	.91307	109.411	7.78284
10.6	1.04247	8.69941	.92214	110.498	7.77727
10.7	1.04288	8.70283	.93120	111.584	7.77163
10.8	1.04330	8.70634	.94028	112.672	7.76606
10.9	1.04371	8.70976	.94936	113.760	7.76040
11.0	1.04413	8.71326	.95846	114.850	7.75480
11.1	1.04454	8.71669	.96755	115.940	7.74914
11.2	1.04496	8.72019	.97666	117.031	7.74353
11.3	1.04538	8.72370	.98578	118.124	7.73792
11.4	1.04580	8.72720	.99490	119.217	7.73230
11.5	1.04622	8.73071	1.00403	120.311	7.72668
11.6	1.04663	8.73413	1.01316	121.405	7.72097
11.7	1.04705	8.73763	1.02230	122.500	7.71533
11.8	1.04747	8.74114	1.03145	123.597	7.70969

TABLE B.1.   *(Cont'd.)*

Degrees Brix	Specific Gravity	Lb Total Weight per U.S. Gallon	Lb Solids per U.S. Gallon	G Solids per Liter	Lb Water per U.S. Gallon
11.9	1.04789	8.74464	1.04061	124.694	7.70403
12.0	1.04831	8.74815	1.04978	125.795	7.69837
12.1	1.04873	8.75165	1.05895	126.892	7.69270
12.2	1.04915	8.75516	1.06813	127.992	7.68703
12.3	1.04957	8.75866	1.07732	129.093	7.68134
12.4	1.04999	8.76217	1.08651	130.194	7.67566
12.5	1.05041	8.76567	1.09571	131.297	7.66996
12.6	1.05083	8.76918	1.10492	132.400	7.66426
12.7	1.05125	8.77268	1.11413	133.504	7.65855
12.8	1.05167	8.77619	1.12335	134.609	7.65284
12.9	1.05209	8.77969	1.13258	135.715	7.64711
13.0	1.05251	8.78320	1.14182	136.822	7.64138
13.1	1.05293	8.78670	1.15106	137.929	7.63564
13.2	1.05335	8.79021	1.16031	139.038	7.62990
13.3	1.05378	8.79379	1.16957	140.147	7.62422
13.4	1.05420	8.79730	1.17884	141.258	7.61846
13.5	1.05463	8.80089	1.18812	142.370	7.61277
13.6	1.05505	8.80439	1.19740	143.482	7.60699
13.7	1.05547	8.80790	1.20668	144.594	7.60122
13.8	1.05590	8.81149	1.21599	145.710	7.59550
13.9	1.05632	8.81499	1.22528	146.823	7.58971
14.0	1.05675	8.81858	1.23460	147.940	7.58398
14.1	1.05717	8.82208	1.24391	149.055	7.57817
14.2	1.05760	8.82567	1.25325	150.174	7.57242
14.3	1.05803	8.82926	1.26258	151.292	7.56668
14.4	1.05846	8.83285	1.27193	152.412	7.56092
14.5	1.05889	8.83644	1.28128	153.533	7.55516
14.6	1.05931	8.83994	1.29063	154.654	7.54931
14.7	1.05974	8.84353	1.30000	155.776	7.54353
14.8	1.06017	8.84712	1.30937	156.899	7.53775
14.9	1.06060	8.85071	1.31876	158.044	7.53195
15.0	1.06103	8.85430	1.32815	159.150	7.52615
15.1	1.06146	8.85788	1.33754	160.275	7.52034
15.2	1.06189	8.86147	1.34794	161.401	7.51453
15.3	1.06232	8.86506	1.35635	162.529	7.50871
15.4	1.06275	8.86865	1.36577	163.657	7.50288
15.5	1.06318	8.87224	1.37520	164.787	7.49704
15.6	1.06361	8.87583	1.38463	165.917	7.49120
15.7	1.06404	8.87941	1.39407	167.049	7.48534
15.8	1.06447	8.88300	1.40351	168.180	7.47949
15.9	1.06490	8.88659	1.41297	169.313	7.47362
16.0	1.06534	8.89026	1.42244	170.448	7.46782
16.1	1.06577	8.89385	1.43191	171.583	7.46194
16.2	1.06620	8.89744	1.44139	172.719	7.45605
16.3	1.06663	8.90103	1.45087	173.855	7.45016
16.4	1.06707	8.90470	1.46037	174.993	7.44433
16.5	1.06750	8.90829	1.46987	176.132	7.43842
16.6	1.06793	8.91188	1.47937	177.270	7.43251
16.7	1.06837	8.91555	1.48890	178.412	7.42665
16.8	1.06880	8.91914	1.49842	179.553	7.42072
16.9	1.06923	8.92272	1.50794	180.693	7.41478
17.0	1.06967	8.92640	1.51749	181.838	7.40891
17.1	1.07010	8.92998	1.52703	182.981	7.40295
17.2	1.07054	8.93366	1.53659	184.127	7.39707
17.3	1.07098	8.93733	1.54616	185.273	7.39117
17.4	1.07141	8.94092	1.55572	186.419	7.38520
17.5	1.07185	8.94459	1.56530	187.567	7.37929
17.6	1.07229	8.94826	1.57489	188.716	7.37337
17.7	1.07272	8.95185	1.58448	189.865	7.36737

TABLE B.1.   *(Cont'd.)*

Degrees Brix	Specific Gravity	Lb Total Weight per U.S. Gallon	Lb Solids per U.S. Gallon	G Solids per Liter	Lb Water per U.S. Gallon
17.8	1.07316	8.95552	1.59408	191.015	7.36144
17.9	1.07360	8.95919	1.60370	192.168	7.35549
18.0	1.07404	8.96286	1.61331	193.320	7.34955
18.1	1.07448	8.96654	1.62294	194.474	7.34360
18.2	1.07492	8.97021	1.63258	195.629	7.33763
18.3	1.07536	8.97388	1.64222	196.784	7.33166
18.4	1.07580	8.97755	1.65187	197.940	7.32568
18.5	1.07624	8.98122	1.66153	199.098	7.31969
18.6	1.07668	8.98489	1.67119	200.255	7.31370
18.7	1.07712	8.98857	1.68086	201.414	7.30771
18.8	1.07756	8.99224	1.69054	202.574	7.30170
18.9	1.07800	8.99591	1.70023	203.735	7.29568
19.0	1.07844	8.99958	1.70992	204.896	7.28966
19.1	1.07888	9.00325	1.71962	206.059	7.28363
19.2	1.07932	9.00693	1.72933	207.222	7.27760
19.3	1.07977	9.01068	1.73906	208.388	7.27162
19.4	1.08021	9.01435	1.74878	209.559	7.26557
19.5	1.08066	9.01811	1.75853	210.721	7.25958
19.6	1.08110	9.02178	1.76827	211.888	7.25351
19.7	1.08154	9.02545	1.77801	213.055	7.24744
19.8	1.08199	9.02921	1.78778	214.226	7.24143
19.9	1.08243	9.03288	1.79754	215.396	7.23534
20.0	1.08288	9.03663	1.80733	216.569	7.22930
20.1	1.08332	9.04031	1.81710	217.739	7.22321
20.2	1.08377	9.04406	1.82690	218.914	7.21716
20.3	1.08421	9.04773	1.83669	220.087	7.21104
20.4	1.08466	9.05149	1.84650	221.262	7.20499
20.5	1.08510	9.05516	1.85631	222.438	7.19885
20.6	1.08555	9.05891	1.86614	223.616	7.19277
20.7	1.08599	9.06259	1.87596	224.793	7.18663
20.8	1.08644	9.06634	1.88580	225.972	7.18054
20.9	1.08688	9.07001	1.89563	227.150	7.17438
21.0	1.08733	9.07377	1.90549	228.331	7.16828
21.1	1.08777	9.07744	1.91534	229.511	7.16210
21.2	1.08822	9.08120	1.92521	230.694	7.15599
21.3	1.08867	9.08495	1.93509	231.878	7.14986
21.4	1.08912	9.08871	1.94498	233.063	7.14373
21.5	1.08957	9.09246	1.95488	234.249	7.13758
21.6	1.09002	9.09622	1.96478	235.436	7.13144
21.7	1.09047	9.09997	1.97469	236.623	7.12528
21.8	1.09092	9.10373	1.98461	237.812	7.11912
21.9	1.09137	9.10748	1.99454	239.014	7.11294
22.0	1.09182	9.11124	2.00447	240.192	7.10677
22.1	1.09227	9.11499	2.01441	241.383	7.10058
22.2	1.09272	9.11875	2.02436	242.575	7.09439
22.3	1.09317	9.12250	2.03432	243.768	7.08818
22.4	1.09362	9.12626	2.04428	244.962	7.08198
22.5	1.09408	9.13010	2.05427	246.159	7.07583
22.6	1.09453	9.13385	2.06425	247.355	7.06960
22.7	1.09498	9.13761	2.07424	248.552	7.06337
22.8	1.09543	9.14136	2.08423	249.749	7.05713
22.9	1.09588	9.14512	2.09423	250.947	7.05089
23.0	1.09634	9.14896	2.10426	252.149	7.04470
23.1	1.09679	9.15271	2.11428	253.350	7.03843
23.2	1.09725	9.15655	2.12432	254.553	7.03223
23.3	1.09770	9.16031	2.13435	255.755	7.02596
23.4	1.09816	9.16415	2.14441	256.960	7.01974
23.5	1.09862	9.16798	2.15448	258.167	7.01350
23.6	1.09907	9.17174	2.16453	259.371	7.00721

TABLE B.1.   *(Cont'd.)*

Degrees Brix	Specific Gravity	Lb Total Weight per U.S. Gallon	Lb Solids per U.S. Gallon	G Solids per Liter	Lb Water per U.S. Gallon
23.7	1.09953	9.17558	2.17461	260.579	7.00097
23.8	1.09998	9.17933	2.18468	261.786	6.99465
23.9	1.10044	9.18317	2.19478	262.996	6.98839
24.0	1.10090	9.18701	2.20488	264.206	6.98213
24.1	1.10135	9.19085	2.21499	265.418	6.97586
24.2	1.10182	9.19469	2.22511	266.630	6.96958
24.3	1.10228	9.19853	2.23524	267.844	6.96329
24.4	1.10274	9.20237	2.24538	269.059	6.95699
24.5	1.10320	9.20620	2.25552	270.236	6.95068
24.6	1.10366	9.21004	2.26567	271.491	6.94437
24.7	1.10412	9.21388	2.27583	272.708	6.93805
24.8	1.10458	9.21772	2.28599	273.926	6.93173
24.9	1.10504	9.22156	2.29617	275.145	6.92539
25.0	1.10550	9.22540	2.30635	276.365	6.91905
25.1	1.10596	9.22924	2.31654	277.586	6.91270
25.2	1.10642	9.23307	2.32673	278.807	6.90634
25.3	1.10688	9.23691	2.33694	280.031	6.89997
25.4	1.10734	9.24075	2.34715	281.254	6.89360
25.5	1.10781	9.24467	2.35739	282.481	6.88728
25.6	1.10827	9.24851	2.36762	283.707	6.88089
25.7	1.10873	9.25235	2.37785	284.933	6.87450
25.8	1.10919	9.25619	2.38810	286.161	6.86809
25.9	1.10965	9.26003	2.39835	287.389	6.86168
26.0	1.11012	9.26395	2.40863	288.621	6.85532
26.1	1.11058	9.26779	2.41889	289.851	6.84890
26.2	1.11105	9.27171	2.42919	291.085	6.84252
26.3	1.11151	9.27555	2.43947	292.317	6.83608
26.4	1.11198	9.27947	2.44978	293.552	6.82969
26.5	1.11245	9.28340	2.46010	294.789	6.82330
26.6	1.11291	9.28723	2.47040	296.023	6.81683
26.7	1.11338	9.29116	2.48074	297.262	6.81042
26.8	1.11384	9.29499	2.49106	298.499	6.80393
26.9	1.11431	9.29892	2.50141	299.739	6.79751
27.0	1.11478	9.30284	2.51177	300.980	6.79107
27.1	1.11525	9.30676	2.52213	302.222	6.78463
27.2	1.11572	9.31068	2.53250	303.464	6.77818
27.3	1.11619	9.31461	2.54289	304.709	6.77172
27.4	1.11666	9.31853	2.55328	305.954	6.76525
27.5	1.11713	9.32245	2.56367	307.199	6.75878
27.6	1.11760	9.32637	2.57408	308.447	6.75229
27.7	1.11807	9.33029	2.58449	309.694	6.74580
27.8	1.11854	9.33422	2.59491	310.943	6.73931
27.9	1.11901	9.33814	2.60534	312.193	6.73280
28.0	1.11948	9.34206	2.61578	313.444	6.72628
28.1	1.11995	9.34598	2.62622	314.695	6.71976
28.2	1.12042	9.34990	2.63667	315.947	6.71323
28.3	1.12089	9.35383	2.64713	317.200	6.70670
28.4	1.12137	9.35783	2.65762	318.457	6.70021
28.5	1.12184	9.36175	2.66810	319.713	6.69365
28.6	1.12231	9.36568	2.67858	320.969	6.68710
28.7	1.12279	9.36968	2.68910	322.229	6.68058
28.8	1.12326	9.37360	2.69960	323.488	6.67400
28.9	1.12373	9.37753	2.71011	324.747	6.66742
29.0	1.12421	9.38153	2.72064	326.009	6.66089
29.1	1.12468	9.38545	2.73117	327.271	6.65428
29.2	1.12516	9.38946	2.74172	328.535	6.64774
29.3	1.12564	9.39347	2.75229	329.801	6.64118
29.4	1.12611	9.39739	2.76283	331.064	6.63456
29.5	1.12659	9.40139	2.77341	332.332	6.62798

TABLE B.1.   *(Cont'd.)*

Degrees Brix	Specific Gravity	Lb Total Weight per U.S. Gallon	Lb Solids per U.S. Gallon	G Solids per Liter	Lb Water per U.S. Gallon
29.6	1.12707	9.40540	2.78400	333.601	6.62140
29.7	1.12754	9.40932	2.79457	334.868	6.61475
29.8	1.12802	9.41333	2.80517	336.138	6.60816
29.9	1.12850	9.41733	2.81578	337.409	6.60155
30.0	1.12898	9.42134	2.82640	338.682	6.59494
30.1	1.12945	9.42526	2.83700	339.952	6.58826
30.2	1.12993	9.42927	2.84764	341.227	6.58163
30.3	1.13041	9.43327	2.85828	342.502	6.57499
30.4	1.13089	9.43728	2.86893	343.778	6.56835
30.5	1.13137	9.44128	2.87959	345.056	6.56169
30.6	1.13185	9.44529	2.89026	346.334	6.55503
30.7	1.13223	9.44929	2.90093	347.613	6.54836
30.8	1.13281	9.45330	2.91162	348.894	6.54168
30.9	1.13329	9.45731	2.92231	350.175	6.53500
31.0	1.13377	9.46131	2.93301	351.457	6.52830
31.1	1.13425	9.46532	2.94371	352.739	6.52161
31.2	1.13473	9.46932	2.95443	354.023	6.51489
31.3	1.13521	9.47333	2.96515	355.308	6.50818
31.4	1.13570	9.47742	2.97591	356.597	6.50151
31.5	1.13618	9.48142	2.98665	357.884	6.49477
31.6	1.13666	9.48543	2.99740	359.172	6.48803
31.7	1.13715	9.48952	3.00818	360.464	6.48134
31.8	1.13763	9.49352	3.01894	361.754	6.47458
31.9	1.13811	9.49753	3.02971	363.044	6.46782
32.0	1.13860	9.50162	3.04052	364.339	6.46110
32.1	1.13908	9.50562	3.05130	365.631	6.45432
32.2	1.13957	9.50971	3.06213	366.929	6.44758
32.3	1.14005	9.51372	3.07293	368.223	6.44079
32.4	1.14054	9.51781	3.08377	369.522	6.43404
32.5	1.14103	9.52190	3.09462	370.822	6.42728
32.6	1.14151	9.52590	3.10544	372.119	6.42046
32.7	1.14200	9.52999	3.11631	373.421	6.41368
32.8	1.14248	9.53400	3.12715	374.720	6.40685
32.9	1.14297	9.53808	3.13803	376.024	6.40005
33.0	1.14346	9.54217	3.14892	377.329	6.39325
33.1	1.14395	9.54626	3.15981	378.634	6.38645
33.2	1.14444	9.55035	3.17072	379.941	6.37963
33.3	1.14493	9.55444	3.18163	381.248	6.37281
33.4	1.14542	9.55853	3.19255	382.557	6.36598
33.5	1.14591	9.56262	3.20248	383.867	6.35914
33.6	1.14640	9.56671	3.21441	385.176	6.35230
33.7	1.14689	9.57080	3.22536	386.488	6.34544
33.8	1.14738	9.57489	3.23631	387.801	6.33858
33.9	1.14787	9.57898	3.24727	389.114	6.33171
34.0	1.14836	9.58306	3.25824	390.428	6.32482
34.1	1.14885	9.58715	3.26922	391.744	6.31793
34.2	1.14934	9.59124	3.28020	393.060	6.31104
34.3	1.14984	9.59541	3.29123	394.382	6.30418
34.4	1.15033	9.59550	3.30223	395.700	6.29727
34.5	1.15083	9.60368	3.31327	397.023	6.29041
34.6	1.15132	9.60777	3.32429	398.343	6.28348
34.7	1.15181	9.61185	3.33531	399.664	6.27654
34.8	1.15231	9.61603	3.34638	400.990	6.26965
34.9	1.15280	9.62012	3.35742	402.313	6.26270
35.0	1.15330	9.62429	3.36850	403.641	6.25579
35.1	1.15379	9.62838	3.37956	404.966	6.24882
35.2	1.15429	9.63255	3.39066	406.296	6.24189
35.3	1.15479	9.63672	3.40176	407.626	6.23496
35.4	1.15528	9.64081	3.41285	408.955	6.22796

TABLE B.1.   *(Cont'd.)*

Degrees Brix	Specific Gravity	Lb Total Weight per U.S. Gallon	Lb Solids per U.S. Gallon	G Solids per Liter	Lb Water per U.S. Gallon
35.5	1.15578	9.64498	3.42397	410.287	6.22101
35.6	1.15628	9.64916	3.43510	411.621	6.21406
35.7	1.15677	9.65325	3.44621	412.952	6.20704
35.8	1.15727	9.65742	3.45736	414.289	6.20006
35.9	1.15777	9.66159	3.46851	415.625	6.19308
36.0	1.15827	9.66576	3.47967	416.962	6.18609
36.1	1.15877	9.66994	3.49085	418.302	6.17909
36.2	1.15927	9.67411	3.50203	419.641	6.17208
36.3	1.15977	9.67828	3.51322	420.982	6.16506
36.4	1.16027	9.68245	3.52441	422.323	6.15804
36.5	1.16078	9.68671	3.53565	423.670	6.15106
36.6	1.16128	9.69088	3.54686	425.013	6.14402
36.7	1.16178	9.69505	3.55808	426.358	6.13697
36.8	1.16228	9.69923	3.56932	427.704	6.12991
36.9	1.16278	9.70340	3.58055	429.050	6.12285
37.0	1.16329	9.70766	3.59183	430.402	6.11583
37.1	1.16379	9.71183	3.60309	431.751	6.10874
37.2	1.16429	9.71600	3.61435	433.100	6.10165
37.3	1.16480	9.72026	3.62566	434.456	6.09460
37.4	1.16530	9.72443	3.63694	435.807	6.08749
37.5	1.16581	9.72868	3.64826	437.164	6.08042
37.6	1.16631	9.73286	3.65956	438.518	6.07330
37.7	1.16681	9.73703	3.67086	439.872	6.06617
37.8	1.16732	9.74129	3.68221	441.232	6.05908
37.9	1.16782	9.74546	3.69353	442.588	6.05193
38.0	1.16833	9.74971	3.70489	443.950	6.04482
38.1	1.16883	9.75389	3.71623	445.308	6.03766
38.2	1.16934	9.75814	3.72761	446.672	6.03053
38.3	1.16985	9.76240	3.73900	448.037	6.02340
38.4	1.17036	9.76665	3.75039	449.402	6.01626
38.5	1.17087	9.77091	3.76180	450.769	6.00911
38.6	1.17138	9.77517	3.77322	452.137	6.00195
38.7	1.17189	9.77942	3.78464	453.506	5.99478
38.8	1.17240	9.78368	3.79607	454.875	5.98761
38.9	1.17291	9.78793	3.80750	456.245	5.98043
39.0	1.17342	9.79219	3.81895	457.617	5.97324
39.1	1.17393	9.79645	3.83041	458.990	5.96604
39.2	1.17444	9.80070	3.84187	460.364	5.95883
39.3	1.17495	9.80496	3.85335	461.739	5.95161
39.4	1.17546	9.80921	3.86483	463.115	5.94438
39.5	1.17598	9.81355	3.87635	464.495	5.93720
39.6	1.17649	9.81781	3.88785	465.873	5.92996
39.7	1.17700	9.82207	3.89936	467.253	5.92271
39.8	1.17751	9.82632	3.91188	468.633	5.91544
39.9	1.17802	9.83058	3.92240	470.013	5.90818
40.0	1.17854	9.83492	3.93397	471.340	5.90095
40.1	1.17905	9.83917	3.94551	472.783	5.89366
40.2	1.17957	9.84351	3.95709	474.170	5.88642
40.3	1.18008	9.84777	3.96865	475.555	5.87912
40.4	1.18060	9.85211	3.98025	476.945	5.87186
40.5	1.18111	9.85636	3.99183	478.333	5.86453
40.6	1.18163	9.86070	4.00344	479.724	5.85726
40.7	1.18214	9.86496	4.01504	481.114	5.84992
40.8	1.18266	9.86930	4.02667	482.508	5.84263
40.9	1.18317	9.87355	4.03828	483.899	5.83527
41.0	1.18369	9.87789	4.04993	485.295	5.82796
41.1	1.18420	9.88215	4.06156	486.689	5.82059
41.2	1.18472	9.88649	4.07323	488.087	5.81326
41.3	1.18524	9.89083	4.08491	489.487	5.80592

TABLE B.1.   *(Cont'd.)*

Degrees Brix	Specific Gravity	Lb Total Weight per U.S. Gallon	Lb Solids per U.S. Gallon	G Solids per Liter	Lb Water per U.S. Gallon
41.4	1.18576	9.89517	4.09660	490.887	5.79857
41.5	1.18628	9.89951	4.10830	492.289	5.79121
41.6	1.18679	9.90376	4.11996	493.687	5.78380
41.7	1.18731	9.90810	4.13168	495.091	5.77642
41.8	1.18783	9.91244	4.14340	496.495	5.76904
41.9	1.18835	9.91678	4.15513	497.901	5.76165
42.0	1.18887	9.92112	4.16687	499.308	5.75425
42.1	1.18939	9.92546	4.17862	500.716	5.74684
42.2	1.18991	9.92980	4.19038	502.125	5.73942
42.3	1.19043	9.93414	4.20214	503.534	5.73200
42.4	1.19096	9.93856	4.21395	504.949	5.72461
42.5	1.19148	9.94290	4.22573	506.361	5.71717
42.6	1.19200	9.94724	4.23752	507.774	5.70972
42.7	1.19253	9.95166	4.24936	509.192	5.70230
42.8	1.19305	9.95600	4.26117	510.607	5.69483
42.9	1.19357	9.96034	4.27299	512.024	5.68735
43.0	1.19410	9.96476	4.28485	513.445	5.67991
43.1	1.19462	9.96910	4.29668	514.863	5.67242
43.2	1.19515	9.97353	4.30856	517.711	5.66497
43.3	1.19568	9.97795	4.32045	519.132	5.65750
43.4	1.19620	9.98229	4.33231	520.559	5.64998
43.5	1.19673	9.98671	4.34422	521.986	5.64249
43.6	1.19726	9.99113	4.35613	523.411	5.63500
43.7	1.19778	9.99547	4.36802	524.842	5.62745
43.8	1.19831	9.99990	4.37996	526.273	5.61994
43.9	1.19884	10.00432	4.39190	527.705	5.61242
44.0	1.19937	10.00874	4.40385	529.126	5.60489
44.1	1.19990	10.01317	4.41581	530.571	5.59736
44.2	1.20043	10.01759	4.42777	532.006	5.58982
44.3	1.20096	10.02201	4.43975	533.442	5.58226
44.4	1.20149	10.02643	4.45173	534.880	5.57470
44.5	1.20202	10.03086	4.46373	536.318	5.56713
44.6	1.20255	10.03528	4.47573	537.758	5.55955
44.7	1.20308	10.03970	4.48775	539.198	5.55195
44.8	1.20361	10.04413	4.49977	540.640	5.54436
44.9	1.20414	10.04855	4.51180	541.722	5.53675
45.0	1.20467	10.05297	4.52384	542.803	5.52913
45.1	1.20520	10.05739	4.53588	543.525	5.52151
45.2	1.20573	10.06182	4.54794	544.971	5.51388
45.3	1.20627	10.06632	4.56004	546.420	5.50628
45.4	1.20680	10.07075	4.57212	547.868	5.49863
45.5	1.20734	10.07525	4.58424	549.320	5.49101
45.6	1.20787	10.07968	4.59633	550.769	5.48335
45.7	1.20840	10.08410	4.60843	552.219	5.47567
45.8	1.20894	10.08860	4.62058	553.675	5.46802
45.9	1.20947	10.09303	4.63270	555.127	5.46033
46.0	1.21001	10.09753	4.64486	556.584	5.45267
46.1	1.21054	10.10196	4.65700	558.039	5.44496
46.2	1.21108	10.10646	4.66918	559.499	5.43728
46.3	1.21162	10.11097	4.68138	560.960	5.42959
46.4	1.21216	10.11548	4.69358	562.422	5.42190
46.5	1.21270	10.11998	4.70579	563.885	5.41419
46.6	1.21323	10.12440	4.71797	565.345	5.40643
46.7	1.21377	10.12891	4.73020	567.028	5.39871
46.8	1.21431	10.13342	4.74244	568.277	5.39098
46.9	1.21485	10.13792	4.75468	569.744	5.38324
47.0	1.21539	10.14243	4.76694	571.213	5.37549
47.1	1.21593	10.14694	4.77921	572.683	5.36773
47.2	1.21647	10.15144	4.79148	574.153	5.35996

TABLE B.1.   *(Cont'd.)*

Degrees Brix	Specific Gravity	Lb Total Weight per U.S. Gallon	Lb Solids per U.S. Gallon	G Solids per Liter	Lb Water per U.S. Gallon
47.3	1.21701	10.15595	4.80376	575.625	5.35219
47.4	1.21755	10.16045	4.81605	577.098	5.34440
47.5	1.21809	10.16496	4.82836	578.573	5.33660
47.6	1.21863	10.16947	4.84067	580.049	5.32880
47.7	1.21917	10.17397	4.85298	581.523	5.32099
47.8	1.21971	10.17848	4.86531	583.000	5.31317
47.9	1.22025	10.18299	4.87765	584.479	5.30534
48.0	1.22080	10.18758	4.89004	585.964	5.29754
48.1	1.22134	10.19208	4.90239	587.444	5.28969
48.2	1.22189	10.19667	4.91479	588.929	5.28188
48.3	1.22243	10.20118	4.92717	590.413	5.27401
48.4	1.22298	10.20577	4.93959	591.901	5.26618
48.5	1.22353	10.21036	4.95202	593.391	5.25834
48.6	1.22407	10.21486	4.96442	594.877	5.25044
48.7	1.22462	10.21945	4.97687	596.368	5.24258
48.8	1.22516	10.22396	4.98929	597.857	5.23467
48.9	1.22571	10.22855	5.00176	599.351	5.22679
49.0	1.22626	10.23314	5.01424	600.846	5.21890
49.1	1.22680	10.23765	5.02669	602.338	5.21096
49.2	1.22735	10.24224	5.03918	603.835	5.20406
49.3	1.22790	10.24683	5.05169	605.334	5.19514
49.4	1.22845	10.25142	5.06420	606.833	5.18722
49.5	1.22900	10.25601	5.07562	608.333	5.17929
49.6	1.22955	10.26059	5.08925	609.935	5.17134
49.7	1.23010	10.26518	5.10179	611.337	5.16339
49.8	1.23065	10.26977	5.11435	612.842	5.15542
49.9	1.23120	10.27436	5.12691	614.347	5.14745
50.0	1.23175	10.27895	5.13948	615.854	5.13957
50.1	1.23230	10.28354	5.15205	617.360	5.13149
50.2	1.23285	10.28813	5.16464	618.868	5.12349
50.3	1.23340	10.29272	5.17724	620.378	5.11548
50.4	1.23396	10.29740	5.18989	621.894	5.10751
50.5	1.23451	10.30199	5.20250	623.405	5.09949
50.6	1.23506	10.30658	5.21513	624.919	5.09145
50.7	1.23562	10.31125	5.22780	626.437	5.08345
50.8	1.23617	10.31584	5.24045	627.953	5.07539
50.9	1.23672	10.32043	5.25310	629.468	5.06733
51.0	1.23728	10.32510	5.26580	630.990	5.05930
51.1	1.23783	10.32969	5.27847	632.509	5.05122
51.2	1.23839	10.33436	5.29119	634.033	5.04317
51.3	1.23894	10.33895	5.30388	635.553	5.03507
51.4	1.23950	10.34363	5.31663	637.081	5.02700
51.5	1.24006	10.34830	5.32937	638.608	5.01893
51.6	1.24061	10.35289	5.34209	640.132	5.01080
51.7	1.24117	10.35756	5.35486	641.662	5.00270
51.8	1.24172	10.36215	5.36759	643.188	4.99456
51.9	1.24228	10.36683	5.38038	644.720	4.98645
52.0	1.24284	10.37150	5.39318	646.254	4.97832
52.1	1.24340	10.37617	5.40598	647.788	4.97019
52.2	1.24396	10.38085	5.41880	649.324	4.96205
52.3	1.24452	10.38552	5.43163	650.861	4.95389
52.4	1.24508	10.39019	5.44446	652.399	4.94573
52.5	1.24564	10.39487	5.45731	653.939	4.93756
52.6	1.24620	10.39954	5.47016	655.478	4.92938
52.7	1.24676	10.40421	5.48302	657.019	4.92119
52.8	1.24732	10.40889	5.49589	658.562	4.91300
52.9	1.24788	10.41356	5.50877	660.105	4.90479
53.0	1.24845	10.41832	5.52171	661.655	4.89661
53.1	1.24901	10.42299	5.53461	663.201	4.88838

TABLE B.1.   *(Cont'd.)*

Degrees Brix	Specific Gravity	Lb Total Weight per U.S. Gallon	Lb Solids per U.S. Gallon	G Solids per Liter	Lb Water per U.S. Gallon
53.2	1.24957	10.42766	5.54752	664.748	4.88014
53.3	1.25014	10.43242	5.56048	666.301	4.87194
53.4	1.25070	10.43709	5.57341	667.851	4.86368
53.5	1.25127	10.44185	5.58639	669.406	4.85546
53.6	1.25183	10.44652	5.59933	670.957	4.84719
53.7	1.25239	10.45119	5.61229	672.509	4.82890
53.8	1.25296	10.45595	5.62530	674.068	4.83065
53.9	1.25352	10.46062	5.63827	675.623	4.82235
54.0	1.25409	10.46538	5.65131	677.185	4.81407
54.1	1.25465	10.47005	5.66430	678.742	4.80575
54.2	1.25522	10.47481	5.67735	680.305	4.79746
54.3	1.25579	10.47957	5.69041	681.870	4.78916
54.4	1.25636	10.48432	5.70347	683.435	4.78085
54.5	1.25693	10.48908	5.71655	685.003	4.77253
54.6	1.25749	10.49375	5.72959	686.565	4.76416
54.7	1.25806	10.49851	5.74268	688.134	4.75583
54.8	1.25863	10.50327	5.75579	689.705	4.74748
54.9	1.25920	10.50802	5.76890	691.276	4.73912
55.0	1.25977	10.51278	5.78203	692.849	4.73075
55.1	1.26034	10.51754	5.79516	694.422	4.72238
55.2	1.26091	10.52229	5.80830	695.997	4.71399
55.3	1.26148	10.52705	5.82146	697.574	4.70559
55.4	1.26205	10.53181	5.83462	699.151	4.69719
55.5	1.26263	10.53665	5.84784	700.735	4.68881
55.6	1.26320	10.54140	5.86102	702.314	4.68038
55.7	1.26377	10.54616	5.87421	703.895	4.67195
55.8	1.26434	10.55092	5.88741	705.477	4.66351
55.9	1.26491	10.55567	5.90062	707.059	4.65505
56.0	1.26549	10.56051	5.91389	708.650	4.64662
56.1	1.26606	10.56527	5.92712	710.235	4.63815
56.2	1.26664	10.57011	5.94040	711.826	4.62971
56.3	1.26721	10.57487	5.95365	713.414	4.62122
56.4	1.26779	10.57971	5.96696	715.009	4.61275
56.5	1.26836	10.58446	5.98022	716.598	4.60424
56.6	1.26894	10.58930	5.99354	718.194	4.59576
56.7	1.26951	10.59406	6.00683	719.786	4.58723
56.8	1.27009	10.59890	6.02018	721.386	4.57872
56.9	1.27066	10.60366	6.03348	722.980	4.57018
57.0	1.27124	10.60850	6.04685	724.582	4.56165
57.1	1.27182	10.61334	6.06022	726.184	4.55312
57.2	1.27240	10.61818	6.07360	727.787	4.54458
57.3	1.27298	10.62302	6.08699	729.392	4.53603
57.4	1.27356	10.62786	6.10039	730.998	4.52747
57.5	1.27414	10.63270	6.11380	732.604	4.51890
57.6	1.27472	10.63754	6.12722	734.213	4.51032
57.7	1.27530	10.64238	6.14065	735.822	4.50173
57.8	1.27588	10.64722	6.15409	737.432	4.49313
57.9	1.27646	10.65206	6.16754	739.044	4.48452
58.0	1.27704	10.65690	6.18100	740.657	4.47590
58.1	1.27762	10.66174	6.19447	742.271	4.46727
58.2	1.27820	10.66658	6.20795	743.886	4.45863
58.3	1.27878	10.67142	6.22144	745.503	4.44998
58.4	1.27937	10.67634	6.23498	747.125	4.44136
58.5	1.27995	10.68118	6.24849	748.744	4.43269
58.6	1.28053	10.68602	6.26201	750.364	4.42401
58.7	1.28112	10.69095	6.27559	751.991	4.41536
58.8	1.28170	10.69579	6.28912	753.613	4.40667
58.9	1.28228	10.70063	6.30267	755.236	4.39796
59.0	1.28287	10.70555	6.31627	756.866	4.38928

TABLE B.1.   *(Cont'd.)*

Degrees Brix	Specific Gravity	Lb Total Weight per U.S. Gallon	Lb Solids per U.S. Gallon	G Solids per Liter	Lb Water per U.S. Gallon
59.1	1.28345	10.71039	6.32984	758.492	4.38055
59.2	1.28404	10.71531	6.34346	760.124	4.37185
59.3	1.28463	10.72024	6.35710	761.759	4.36314
59.4	1.28521	10.72508	6.37070	763.388	4.35438
59.5	1.28580	10.73000	6.38435	765.024	4.34565
59.6	1.28639	10.73492	6.39801	766.661	4.33691
59.7	1.28697	10.73976	6.41164	768.294	4.32812
59.8	1.28756	10.74469	6.42532	769.933	4.31937
59.9	1.28815	10.74961	6.43902	771.575	4.31059
60.0	1.28874	10.75454	6.45272	773.217	4.30182
60.1	1.28933	10.75946	6.46644	774.861	4.29302
60.2	1.28992	10.76437	6.48016	776.505	4.28422
60.3	1.29051	10.76931	6.49389	778.150	4.27542
60.4	1.29110	10.77423	6.50763	779.796	4.26660
60.5	1.29169	10.77915	6.52139	781.445	4.25776
60.6	1.29228	10.78408	6.53515	783.094	4.24893
60.7	1.29287	10.78900	6.54892	784.744	4.24008
60.8	1.29346	10.79392	6.56270	786.395	4.23122
60.9	1.29405	10.79885	6.57650	788.049	4.22235
61.0	1.29465	10.80385	6.59035	789.708	4.21350
61.1	1.29524	10.80878	6.60416	791.363	4.20462
61.2	1.29584	10.81378	6.61803	793.025	4.19575
61.3	1.29643	10.81871	6.63187	794.684	4.18684
61.4	1.29703	10.82372	6.64576	796.348	4.17796
61.5	1.29762	10.82864	6.65961	798.008	4.16903
61.6	1.29822	10.83365	6.67353	799.676	4.16012
61.7	1.29881	10.83857	6.68740	801.338	4.15117
61.8	1.29941	10.84358	6.70133	803.007	4.14225
61.9	1.30000	10.84850	6.71522	804.671	4.13328
62.0	1.30060	10.85351	6.72918	806.344	4.12433
62.1	1.30119	10.85843	6.74309	808.011	4.11534
62.2	1.30179	10.86344	6.74706	809.685	4.10638
62.3	1.30239	10.86844	6.77104	811.360	4.09740
62.4	1.30299	10.87345	6.78503	813.037	4.08842
62.5	1.30359	10.87846	6.79904	814.715	4.07942
62.6	1.30418	10.88338	6.81300	816.388	4.07038
62.7	1.30478	10.88839	6.82702	818.068	4.06137
62.8	1.30538	10.89340	6.84106	819.751	4.05234
62.9	1.30598	10.89840	6.85509	821.432	4.04331
63.0	1.30658	10.90341	6.86915	823.117	4.03426
63.1	1.30718	10.90842	6.88321	824.801	4.02521
63.2	1.30778	10.91342	6.89728	826.487	4.01614
63.3	1.30838	10.91843	6.91137	828.176	4.00706
63.4	1.30899	10.92352	6.92551	829.870	3.99802
63.5	1.30959	10.92853	6.93962	831.561	3.98891
63.6	1.31019	10.93354	6.95373	833.252	3.97981
63.7	1.31080	10.93863	6.96791	834.951	3.97072
63.8	1.31140	10.94363	6.98204	836.644	3.96159
63.9	1.31200	10.94864	6.99618	838.338	3.95246
64.0	1.31261	10.95373	7.01039	840.041	3.94334
64.1	1.31321	10.95874	7.02455	841.738	3.93419
64.2	1.31382	10.96383	7.03878	843.443	3.92505
64.3	1.31442	10.96883	7.05296	845.142	3.91587
64.4	1.31503	10.97393	7.06721	846.850	3.90672
64.5	1.31564	10.97902	7.08147	848.558	3.89755
64.6	1.31624	10.98402	7.09568	850.261	3.88834
64.7	1.31685	10.98911	7.10995	851.971	3.87916
64.8	1.31745	10.99412	7.12419	853.677	3.86993

TABLE B.1.   *(Cont'd.)*

Degrees Brix	Specific Gravity	Lb Total Weight per U.S. Gallon	Lb Solids per U.S. Gallon	G Solids per Liter	Lb Water per U.S. Gallon
64.9	1.31806	10.99921	7.13849	855.391	3.86072
65.0	1.31867	11.00430	7.15280	857.106	3.85150
65.1	1.31928	11.00939	7.16711	858.820	3.84228
65.2	1.31989	11.01448	7.18144	860.538	3.83304
65.3	1.32050	11.01957	7.19578	862.256	3.82379
65.4	1.32111	11.02466	7.21013	863.975	3.81453
65.5	1.32172	11.02975	7.22449	865.696	3.80526
65.6	1.32233	11.03484	7.23886	867.418	3.79598
65.7	1.32294	11.03993	7.25323	869.140	3.78670
65.8	1.32355	11.04502	7.26762	870.864	3.77740
65.9	1.32416	11.05012	7.28203	872.591	3.76809
66.0	1.32477	11.05521	7.29644	874.318	3.75877
66.1	1.32538	11.06030	7.31086	876.046	3.74944
66.2	1.32600	11.06547	7.32534	877.781	3.74013
66.3	1.32661	11.07056	7.33978	879.511	3.73078
66.4	1.32723	11.07573	7.35428	881.249	3.72145
66.5	1.32784	11.08082	7.36875	882.983	3.71207
66.6	1.32846	11.08600	7.38328	884.724	3.70272
66.7	1.32907	11.09109	7.39776	886.459	3.69333
66.8	1.32969	11.09626	7.41230	888.201	3.68396
66.9	1.33030	11.10135	7.42680	889.939	3.67455
67.0	1.33092	11.10653	7.44138	891.686	3.66515
67.1	1.33153	11.11162	7.45590	893.426	3.65572
67.2	1.33215	11.11679	7.47048	895.173	3.64631
67.3	1.33277	11.12197	7.48509	896.923	3.63688
67.4	1.33338	11.12706	7.49964	898.667	3.62742
67.5	1.33400	11.13223	7.51426	900.419	3.61797
67.6	1.33462	11.13740	7.52888	902.171	3.60852
67.7	1.33523	11.14249	7.54347	903.919	3.59902
67.8	1.33585	11.14767	7.55812	905.674	3.58955
67.9	1.33647	11.15284	7.57278	907.431	3.58006
68.0	1.33709	11.15802	7.58745	909.189	3.57057
68.1	1.33771	11.16319	7.60213	910.948	3.56106
68.2	1.33833	11.16836	7.61682	912.708	3.55154
68.3	1.33895	11.17354	7.63153	914.471	3.54201
68.4	1.33957	11.17871	7.64624	916.234	3.53247
68.5	1.34020	11.18397	7.66102	918.005	3.52295
68.6	1.34082	11.18914	7.67575	919.770	3.51339
68.7	1.34144	11.19432	7.69050	921.537	3.50382
68.8	1.34206	11.19949	7.70525	923.305	3.49424
68.9	1.34268	11.20466	7.72001	925.073	3.48465
69.0	1.34331	11.20992	7.73484	926.850	3.47508
69.1	1.34393	11.21510	7.74963	928.623	3.46547
69.2	1.34456	11.22035	7.76448	930.402	3.45587
69.3	1.34518	11.22553	7.77929	932.177	3.44624
69.4	1.34581	11.23078	7.79416	933.959	3.43662
69.5	1.34643	11.23596	7.80899	935.736	3.42697
69.6	1.34706	11.24122	7.82389	937.521	3.41733
69.7	1.34768	11.24639	7.83873	939.299	3.40766
69.8	1.34831	11.25165	7.85365	941.087	3.39800
69.9	1.34893	11.25682	7.86852	942.869	3.38830
70.0	1.34956	11.26208	7.88346	944.659	3.37862
70.1	1.35019	11.26734	7.89841	946.451	3.36893
70.2	1.35083	11.27268	7.91342	948.249	3.35926
70.3	1.35147	11.27802	7.92845	950.050	3.34957
70.4	1.35210	11.28327	7.94342	951.844	3.33985
70.5	1.35274	11.28862	7.95848	953.649	3.33014
70.6	1.35338	11.29396	7.97354	955.453	3.32042
70.7	1.35401	11.29921	7.98854	957.251	3.31067

TABLE B.1.   (Cont'd.)

Degrees Brix	Specific Gravity	Lb Total Weight per U.S. Gallon	Lb Solids per U.S. Gallon	G Solids per Liter	Lb Water per U.S. Gallon
70.8	1.35465	11.30455	8.00362	959.058	3.30093
70.9	1.35529	11.30990	8.01872	960.867	3.29118
71.0	1.35593	11.31524	8.03382	962.677	3.28142
71.1	1.35656	11.32049	8.04887	964.480	3.27162
71.2	1.35720	11.32583	8.06399	966.292	3.26184
71.3	1.35784	11.33117	8.07912	968.105	3.25205
71.4	1.35847	11.33643	8.09421	969.913	3.24222
71.5	1.35911	11.34177	8.10937	971.730	3.23240
71.6	1.35975	11.34711	8.12453	973.546	3.22258
71.7	1.36038	11.35237	8.13965	975.358	3.21272
71.8	1.36102	11.35771	8.15484	977.178	3.20287
71.9	1.36166	11.36305	8.17003	978.998	3.19302
72.0	1.36230	11.36839	8.18524	980.821	3.18315
72.1	1.36293	11.37365	8.20040	982.638	3.17325
72.2	1.36357	11.37899	8.21563	984.463	3.16336
72.3	1.36421	11.38433	8.23087	986.289	3.15346
72.4	1.36484	11.38959	8.24606	988.109	3.14353
72.5	1.36548	11.39493	8.26132	989.937	3.13361
72.6	1.36612	11.40027	8.27660	991.768	3.12367
72.7	1.36675	11.40553	8.29182	993.592	3.11371
72.8	1.36739	11.41087	8.30711	995.424	3.10376
72.9	1.36803	11.41621	8.32242	997.259	3.09379
73.0	1.36867	11.42155	8.33773	999.094	3.08382
73.1	1.36930	11.42681	8.35300	1000.923	3.07381
73.2	1.36994	11.43215	8.36833	1002.760	3.06382
73.3	1.37058	11.43749	8.38368	1004.600	3.05381
73.4	1.37121	11.44275	8.39898	1006.433	3.04377
73.5	1.37185	11.44809	8.41435	1008.275	3.03374
73.6	1.37249	11.45343	8.42972	1010.116	3.02371
73.7	1.37312	11.45869	8.44505	1011.953	3.01364
73.8	1.37376	11.46403	8.46045	1013.799	3.00358
73.9	1.37440	11.46937	8.47586	1015.645	2.99351
74.0	1.37504	11.47471	8.49129	1017.494	2.98342
74.1	1.37567	11.47997	8.50666	1019.336	2.97331
74.2	1.37631	11.48531	8.52210	1021.186	2.96321
74.3	1.37695	11.49065	8.53755	1023.038	2.95310
74.4	1.37758	11.49591	8.55296	1024.884	2.94295
74.5	1.37822	11.50125	8.56843	1026.738	2.93282
74.6	1.37886	11.50659	8.58392	1028.594	2.92267
74.7	1.37949	11.51184	8.59934	1030.442	2.91250
74.8	1.38013	11.51718	8.61485	1032.300	2.90233
74.9	1.38077	11.52253	8.63037	1034.160	2.89216
75.0	1.38142	11.52795	8.64596	1036.028	2.88199
75.1	1.38207	11.53337	8.66156	1037.897	2.87181
75.2	1.38273	11.53888	8.67724	1038.776	2.86164
75.3	1.38338	11.54431	8.69287	1041.649	2.85144
75.4	1.38404	11.54981	8.70856	1043.529	2.84125
75.5	1.38470	11.55532	8.72427	1045.412	2.83105
75.6	1.38535	11.56075	8.73993	1047.288	2.82082
75.7	1.38601	11.56625	8.75565	1049.172	2.81060
75.8	1.38666	11.57168	8.77133	1051.051	2.80035
75.9	1.38732	11.57719	8.78709	1052.939	2.79010
76.0	1.38798	11.58269	8.80284	1054.827	2.77985
76.1	1.38863	11.58812	8.81856	1056.710	2.76956
76.2	1.38929	11.59363	8.83435	1058.602	2.75928
76.3	1.38994	11.59905	8.85008	1060.487	2.74897
76.4	1.39060	11.60456	8.86588	1062.381	2.73868
76.5	1.39126	11.61006	8.88170	1064.276	2.72836
76.6	1.39191	11.61549	8.89747	1066.166	2.71802

TABLE B.1. *(Cont'd.)*

Degrees Brix	Specific Gravity	Lb Total Weight per U.S. Gallon	Lb Solids per U.S. Gallon	G Solids per Liter	Lb Water per U.S. Gallon
76.7	1.39257	11.62100	8.91331	1068.064	2.70769
76.8	1.39322	11.62642	8.92909	1069.955	2.69733
76.9	1.39388	11.63193	8.94495	1071.855	2.68698
77.0	1.39454	11.63744	8.96083	1073.758	2.67661
77.1	1.39519	11.64286	8.97665	1075.654	2.66621
77.2	1.39585	11.64837	8.99254	1077.558	2.65583
77.3	1.39650	11.65379	9.00838	1079.456	2.64541
77.4	1.39716	11.65930	9.02430	1081.364	2.63500
77.5	1.39782	11.66481	9.04023	1083.273	2.62458
77.6	1.39847	11.67023	9.05610	1085.174	2.61413
77.7	1.39913	11.67574	9.07205	1087.086	2.60369
77.8	1.39978	11.68116	9.08794	1088.990	2.59322
77.9	1.40044	11.68667	9.10392	1090.905	2.58275
78.0	1.40110	11.69218	9.11990	1092.819	2.57228
78.1	1.40175	11.69760	9.13583	1094.728	2.56177
78.2	1.40241	11.70311	9.15183	1096.645	2.55128
78.3	1.40306	11.70854	9.16779	1098.558	2.54075
78.4	1.40372	11.71404	9.18381	1100.478	2.53023
78.5	1.40438	11.71955	9.19985	1102.400	2.51970
78.6	1.40503	11.72498	9.21583	1104.314	2.50915
78.7	1.40569	11.73048	9.23189	1106.239	2.49859
78.8	1.40634	11.73591	9.24790	1108.157	2.48801
78.9	1.40700	11.74142	9.26398	1110.084	2.47744
79.0	1.40766	11.74692	9.28007	1112.012	2.46685
79.1	1.40831	11.75235	9.29611	1113.934	2.45624
79.2	1.40897	11.75785	9.31222	1115.865	2.44563
79.3	1.40962	11.76328	9.32828	1117.789	2.43500
79.4	1.41028	11.76879	9.34442	1119.723	2.42437
79.5	1.41094	11.77429	9.36056	1121.657	2.41373
79.6	1.41159	11.77972	9.37666	1123.586	2.40306
79.7	1.41225	11.78523	9.39283	1125.524	2.39240
79.8	1.41290	11.79065	9.40894	1127.454	2.38171
79.9	1.41356	11.79616	9.42513	1129.394	2.37103

TABLE B.2.   TRADITIONAL BULK WINE VESSEL NAMES

Vessel	Origin	Capacity in U.S. Gal.	Liters
Aroba	Spain	3–5	12–18
Aum	Germany	42.27	160.00
Baril	Lisbon	4.42	16.74
Baril	Malaga	7.93	30.00
Barile	Rome	15.41	58.34
Barrel	U.S.	31.50	119.24
Barril	Madeira	4.08	15.44
Barrique	Algeria	58.12	220.00
Barrique	Beaujolais	57.07	216.00
Barrique	Bordeaux	59.45	225.00
Barrique	Champagne	52.84	200.00
Barrique	Côte d'Or	60.24	228.00
Barrique	Côtes du Rhone	59.45	225.00
Barrique	Macon	56.80	215.00
Barrique	Yonne	65.52	248.00
Bocoy	Spain	162.00	613.17
Botte (small)	Sardinia	11.77	44.54
Botte (large)	Sardinia	132.10	500.00
Brente	Switzerland	13.21	50.00
Butt	Spain	132.00	499.62
Butte (Buttig)	Hungary	3.59	13.6
Demi-queue	Burgundy	60.24	228.00
Double-Aum	Germany	84.54	320.00
Dreiling	Vienna	358.78	1358.00
Eimer	Switzerland	9.91	37.50
Eimer	Vienna	14.95	56.58
Eimer	Württemberg	77.65	293.92
Fuder	Mosel	253.63	960.00
Fuder	Germany (Prussian)	217.81	824.42
Fuder	Württemberg	465.94	1763.57
Gallon	England	1.20	5.54
Gallon	U.S.	1.00	3.79
Halb-Fuder	Mosel	126.82	480.00
Halb-Stuck	Rhine and Palatinate	169.09	640.00
Hogshead	England	66.00	249.81
Octavilla	England and Spain	16.50	62.45
Ohm	Alsace	13.21	50.00
Ohm	Baden	39.63	150.00
Ohm	Bavaria	33.82	128.00
Ohm	Saar	38.04	144.00
Ohm	Switzerland	10.57	40.00
Oka	Balkans	.34	1.28
Oxhoft	Hamburg	59.71	226.00
Piece	Burgundy	60.24	228.00
Piece	Champagne	52.84	200.00
Piece	Saumur	58.12	220.00
Piece	Vouvray	66.05	250.00
Pipe	England	126.02	477.00
Pipe	Lisbon	132.63	502.00
Pipe	Madeira	109.91	416.00
Pipe	Oporto	141.35	535.00
Pipe	Tarragona	134.74	510.00
Pipe	Valencia	19.92	75.39
Puncheon	England	84.00	317.94
Queue	Burgundy	120.48	456.00
Stuck	Rhineland	317.04	1200.00
Tierce	England	42.00	158.97
Tonneau	Bordeaux	237.78	900.00
Tun	England	252.05	954.00
Vedro	Russia	3.27	12.39

TABLE B.3.   ALCOHOL MEASUREMENTS[1]

Specific Gravity[2] $\frac{20°C}{4°C}$	Percent by Volume at 20°C	Percent by Weight	G per 100 ml
0.99823	0.00	0.00	0.00
0.99675	1.00	0.79	0.79
0.99528	2.00	1.59	1.58
0.99384	3.00	2.38	2.37
0.99243	4.00	3.18	3.16
0.99106	5.00	3.98	3.95
0.98973	6.00	4.78	4.74
0.98845	7.00	5.59	5.53
0.98718	8.00	6.40	6.32
0.98596	9.00	7.20	7.10
0.98476	10.00	8.02	7.89
0.98416	11.00	8.83	8.68
0.98296	11.50	9.23	9.08
0.98238	12.00	9.64	9.47
0.98180	12.50	10.05	9.87
0.98122	13.00	10.40	10.26
0.98066	13.50	10.86	10.66
0.98009	14.00	11.28	11.05
0.97953	14.50	11.68	11.44
0.97897	15.00	12.09	11.84
0.97841	15.50	12.50	12.23
0.97786	16.00	12.92	12.63
0.97732	16.50	13.33	13.02
0.97678	17.00	13.74	13.42
0.97624	17.50	14.15	13.81
0.97570	18.00	14.56	14.21
0.97517	18.50	14.97	14.60
0.97464	19.00	15.39	15.00
0.97412	19.50	15.80	15.39
0.97359	20.00	16.21	15.79
0.97306	20.50	16.63	16.18
0.97252	21.00	17.04	16.58
0.97199	21.50	17.46	16.97
0.97145	22.00	17.88	17.37
0.97091	22.50	18.29	17.76
0.97036	23.00	18.71	18.16
0.96980	23.50	19.13	18.55
0.96925	24.00	19.55	18.94
0.96869	24.50	19.96	19.34
0.96812	25.00	20.38	19.73
0.96755	25.50	20.80	20.13
0.96699	26.00	21.22	20.52
0.96641	26.50	21.64	20.92
0.96583	27.00	22.07	21.31
0.96525	27.50	22.49	21.71
0.96465	28.00	22.91	22.10
0.96406	28.50	23.33	22.50
0.96346	29.00	23.76	22.89
0.96285	29.50	24.18	23.29
0.96224	30.00	24.61	23.68
0.96163	30.50	25.04	24.08
0.96100	31.00	25.46	24.47
0.96036	31.50	25.89	24.86
0.95972	32.00	26.32	25.26

TABLE B.3.   *(Cont'd.)*

Specific Gravity[2] 20°C/4°C	Percent by Volume at 20°C	Percent by Weight	G per 100 ml
0.95906	32.50	26.75	25.64
0.95839	33.00	27.18	26.05
0.95771	33.50	27.61	26.44
0.95703	34.00	28.04	26.84
0.95634	34.50	28.48	27.23
0.95563	35.00	28.91	27.63
0.95492	35.50	29.34	28.02
0.95419	36.00	29.78	28.42
0.95346	36.50	30.22	28.81
0.95272	37.00	30.66	29.21
0.95196	37.50	31.09	29.60
0.95120	38.00	31.53	29.99
0.95043	38.50	31.97	30.39
0.94964	39.00	32.42	30.79
0.94885	39.50	32.86	31.18
0.94805	40.00	33.30	31.57
0.94724	40.50	33.75	31.97
0.94643	41.00	34.19	32.36
0.94560	41.50	34.64	32.76
0.94477	42.00	35.09	33.15
0.94393	42.50	35.54	33.55
0.94308	43.00	35.99	33.94
0.94222	43.50	36.44	34.34
0.94135	44.00	36.89	34.73
0.94046	44.50	37.35	35.13
0.93957	45.00	37.80	35.52
0.93867	45.50	38.26	35.92
0.93776	46.00	38.72	36.31
0.93684	46.50	39.18	36.70
0.93591	47.00	39.64	37.10
0.93498	47.50	40.10	37.49
0.93404	48.00	40.56	37.89
0.93308	48.50	41.03	38.29
0.93213	49.00	41.49	38.68
0.93116	49.50	41.96	39.07
0.93017	50.00	42.43	39.47

[1] Calculated by U.S. Bureau of Standards.
[2] This table is valid for specific gravity determinations at temperatures between an upper limit of 20°C and a lower limit of 4°C.

TABLE B.4.   APPROXIMATE DILUTION OF EXTRACT IN WINES FROM THE ADDITION OF HIGH-PROOF BRANDY

Percent Brandy: By Volume Added to Resulting Product	Degrees Extract of Wine Resulting from the Brandy Added					
	4	6	8	10	12	14
0	4	6	8	10	12	14
1	3.95	5.95	7.9	9.9	11.9	13.9
2	3.9	5.9	7.85	9.8	11.8	13.75
3	3.9	5.8	7.8	9.7	11.7	13.6
4	3.85	5.75	7.7	9.65	11.55	13.5
5	3.8	5.7	7.65	9.55	11.5	13.4
6	3.8	5.7	7.55	9.45	11.35	13.25
7	3.75	5.6	7.5	9.4	11.25	13.15
8	3.7	5.55	7.4	9.3	11.15	13.5
9	3.7	5.5	7.35	9.2	11.05	12.9
10	3.65	5.5	7.3	9.15	10.95	12.8
11	3.6	5.4	7.25	9.05	10.85	12.7
12	3.6	5.4	7.2	9.0	10.75	12.6
13	3.55	5.3	7.1	8.9	10.7	12.45
14	3.5	5.3	7.05	8.8	10.6	12.35
15	3.5	5.25	7.0	8.75	10.5	12.25
16	3.45	5.2	6.95	8.7	10.4	12.15
17	3.4	5.15	6.9	8.6	10.3	12.05
18	3.4	5.1	6.8	8.5	10.25	11.95
19	3.35	5.05	6.75	8.45	10.15	11.85
20	3.35	5.0	6.7	8.4	10.1	11.75

TABLE B.5. TONS OF GRAPES PER ACRE

(Acre = 202 ft × 202 ft = 20 rows at 200 ft spaced 10 ft apart)

Lb Grapes per Vine	Vines at 5 ft (800 Vines)	Vines at 6 ft (667 Vines)	Vines at 7 ft (571 Vines)	Vines at 8 ft (500 Vines)	Vines at 15 ft (267 Vines)	Vines at 20 ft (200 Vines)
2	.80	.67	.57	.50	.27	.20
4	1.60	1.33	1.14	1.00	.54	.40
6	2.40	2.00	1.71	1.50	.81	.60
8	3.20	2.67	2.28	2.00	1.08	.80
10	4.00	3.34	2.85	2.50	1.35	1.00
12	4.80	4.00	3.42	3.00	1.62	1.20
14	5.60	4.67	3.99	3.50	1.89	1.40
16	6.40	5.34	4.56	4.00	2.16	1.60
18	7.20	6.00	5.13	4.50	2.43	1.80
20	8.00	6.67	5.70	5.00	2.70	2.00
22	8.80	7.34	6.27	5.50	2.97	2.20
24	9.60	8.00	6.84	6.00	3.24	2.40
26	10.40	8.67	7.41	6.50	3.51	2.60
28	11.20	9.34	7.98	7.00	3.78	2.80
30	12.00	10.01	8.55	7.50	4.05	3.00
32	12.80	10.67	9.12	8.00	4.32	3.20
34	13.60	11.34	9.69	8.50	4.59	3.40
36	14.40	12.01	10.26	9.00	4.86	3.60
38	15.20	12.67	10.83	9.50	5.13	3.80
40	16.00	13.34	11.40	10.00	5.40	4.00
42	16.80	14.07	11.97	10.50	5.67	4.20
44	17.60	14.74	12.54	11.00	5.94	4.40
46	18.40	15.41	13.11	11.50	6.21	4.60
48	19.20	16.08	13.68	12.00	6.48	4.80
50	20.00	16.75	14.25	12.50	6.75	5.00
60	24.00	20.04	17.10	15.00	8.10	6.00
70	28.00	23.38	19.95	17.50	9.45	7.00
80	32.00	26.72	22.80	20.00	10.80	8.00
90	36.00	30.06	25.65	22.50	12.15	9.00
100	40.00	33.40	28.50	25.00	13.50	10.00
110	44.00	36.74	31.35	27.50	14.85	11.00
120	48.00	40.08	34.20	30.00	16.20	12.00
130	52.00	43.42	37.05	32.50	17.55	13.00
140	56.00	46.76	39.90	35.00	18.90	14.00
150	60.00	50.10	42.75	37.50	20.25	15.00

TABLE B.5. (Cont'd.)

(Acre = 202 ft × 202 ft = 20 rows at 200 ft spaced 10 ft apart)

Lb Grapes per Vine	Vines at 5 ft (800 Vines)	Vines at 6 ft (667 Vines)	Vines at 7 ft (571 Vines)	Vines at 8 ft (500 Vines)	Vines at 15 ft (267 Vines)	Vines at 20 ft (200 Vines)
160		53.44	45.60	40.00	21.60	16.00
170		56.78	48.45	42.50	22.95	17.00
180		60.12	51.30	45.00	24.30	18.00
190			54.15	47.50	25.65	19.00
200			57.00	50.00	27.00	20.00
210			59.85	52.50	28.35	21.00
220				55.00	29.70	22.00
230				57.50	31.05	23.00
240				60.00	32.40	24.00
250					33.75	25.00
260					35.10	26.00
270					36.45	27.00
280					37.80	28.00
290					39.15	29.00
300					40.50	30.00

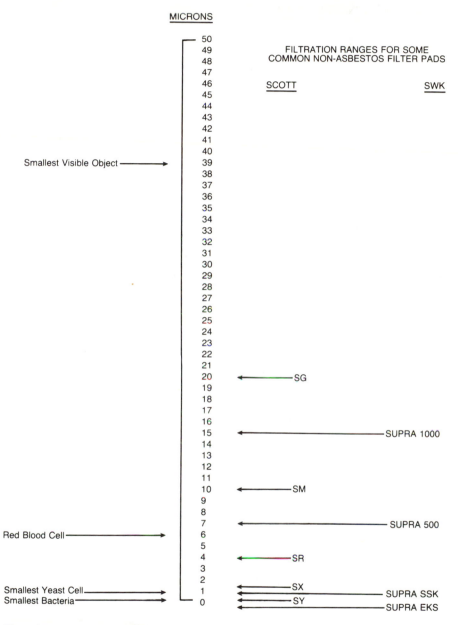

FIG. B.1.   FILTRATION POROSITY COMPARISONS

TABLE B.6.   WINERY HOSE DATA

Hose Inside Diameter (in.)	Cross-section (Sq. in.)	Volume per 25-ft Length	
		(In.^{3})	(Gal.)
1	0.785	235.5	1.0
1½	1.87	561.0	2.4
2	3.14	942.0	4.1
2½	4.91	1473.0	6.4
3	7.07	2121.0	9.2
4	12.60	3768.0	16.3
5	19.60	5892.0	25.5

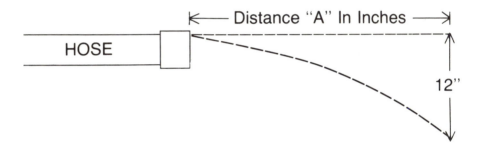

FIG. B.2.   FLOW ESTIMATION FROM A HORIZONTAL 2-INCH HOSE

TABLE B.7.   FLOW ESTIMATION FROM A HORIZONTAL 2-INCH HOSE

Distance "A" in Inches	Flow (Gal. per Hour)	Distance "A" in Inches	Flow (Gal. per Hour)
2	396	20	3960
4	780	22	4380
6	1200	24	4740
8	1560	26	5160
10	1980	28	5520
12	2400	30	5940
14	2760	32	6360
16	3180	34	6720
18	3540	36	7140

TABLE B.8.   SULFUR DIOXIDE ADDITIONS

Ppm SO$_2$ to be Added	SO$_2$ Gas Cylinder		Potassium Metabisulfite[1]		Sodium Bisulfite[1]	
	Lb	Oz	Lb	Oz	Lb	Oz
Addition required per 1000 gal. of wine						
10		1.3		2.3		2.1
20		2.7		4.5		4.3
30		4.0		6.8		6.4
40		5.3		9.1		8.5
50		6.7		11.3		10.7
60		8.0		13.6		12.8
70		9.3		15.9		14.9
80		10.7	1	2.1	1	1.1
90		12.0	1	4.4	1	3.2
100		13.3	1	6.7	1	5.4
110		14.7	1	8.9	1	7.5
120	1	0.0	1	11.2	1	9.6
130	1	1.4	1	13.5	1	11.8
140	1	2.7	1	15.8	1	13.9
150	1	4.0	2	2.0	2	0.0
160	1	5.4	2	4.2	2	2.2
170	1	6.7	2	6.5	2	4.3
180	1	8.0	2	8.8	2	6.4
190	1	9.4	2	11.0	2	8.6
200	1	10.7	2	13.3	2	10.7
Addition required for ton of crushed grapes						
10		0.3		0.5		0.5
20		0.6		1.1		1.0
30		1.0		1.6		1.5
40		1.3		2.2		2.0
50		1.6		2.7		2.6
60		1.9		3.3		3.1
70		2.2		3.8		3.6
80		2.6		4.3		4.1
90		2.9		4.9		4.6
100		3.2		5.4		5.1
110		3.5		6.0		5.6
120		3.8		6.5		6.1
130		4.2		7.0		6.7
140		4.5		7.6		7.2
150		4.8		8.2		7.7

[1] Providing compound has not been allowed to weaken.

TABLE B.9.   LEGAL U.S. WINE BOTTLE SIZES

Bottle Size	Equivalent Fluid Oz	Bottles per Case	Liters per Case	Gal. per Case
100 ml	3.4	60	6.00	1.58502
187 ml	6.3	48	8.976	2.37119
375 ml	12.7	24	9.00	2.37753
750 ml	25.4	12	9.00	2.37753
1 liter	33.8	12	12.00	3.17004
1.5 liters	50.7	6	9.00	2.37753
3 liters	101.0	4	12.00	3.17004

PERIODIC TABLE

Table of Radioactive Isotopes

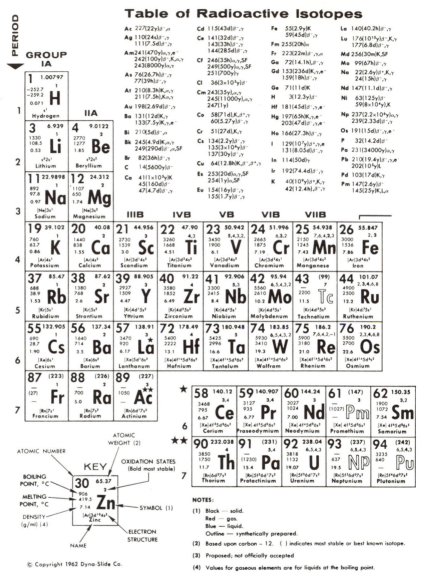

FIG. B.3.   PERIODIC TABLE OF THE ELEMENTS

OF THE ELEMENTS

Po 210(138.4d)α,γ
209(100y)α,K,γ
Pr 143(13.8d)β-
Pt 197(18h)β-,γ
Pu 242(3.8 × 10⁵y)α,SF
241(13y)β-,α,γ
239(24300y)α,γ,SF
Ra 226(1620y)α,γ
Rb 86(18.6h)β-,γ
Re 188(16.7h)β-,γ
186(3.7d)β-,γ
Rn 222(3.82d)α
Ru 103(40d)β-,γ
97(2.9d),K,γ,e-
S 35(87d)β-
Sb 122(2.8d)β-,K,β+,γ
124(60d)β-,γ
Sc 46(84d)β-,γ
Se 75(121d)K,γ
Sm 153(47h)β-,γ
145(340d)K,γ
Sn 113(119d)K,L,γ,e-

Sr 90(28y)β-
89(51d)β-,γ
85(64d)K,γ
Ta 182(115d)β-,γ
Tb 160(73d)β-,γ
Tc 99(2×10⁵y)β-
97(10⁶y)K
Te 127(9.3h)β-
Th 232(1.4×10¹⁰y)α,γ,SF
228(1.91y)β-
Tl 204(3.56y)β-,K
Tm 170(127d)β-,γ,e-
U 238(4.5×10⁹y)α,γ,SF
234(2.5×10⁵y)α,γ,SF
235(7.1×10⁸y)α,γ,SF
233(1.6×10⁵y)α,γ
W 185(73d)β-
Y 90(64h)β-,e-
Yb 175(4.2d)β-,γ
169(31d)K,γ,e-
Zn 65(245d)K,β+,γ
Zr 95(65d)β-,γ,e-
93(9×10⁵y)β-,γ

Naturally occurring radioactive isotopes are indicated by a blue mass number. Half lives are in parentheses where s, m, h, d and y stand for seconds, minutes, hours, days and years respectively. The symbols describing the mode of decay and resulting radiation are defined as follows:

α	alpha particle	L	L-electron capture
β-	beta particle	SF	spontaneous fission
β+	positron	γ	gamma ray
K	K-electron capture	e-	internal electron conversion

Periodic Table of the Elements

INERT GASES

2 4.0026 — He (Helium); 0; -268.9; -269.7; 0.126

Group IIIA – VIIA

IIIA	IVA	VA	VIA	VIIA
5 10.811 B (Boron); 3; (2030); 2.34; s²2s²2p¹	**6 12.0111** C (Carbon); ±4,2; 4830; 3727g; 2.26; s²2s²2p²	**7 14.0067** N (Nitrogen); ±3,5,4,2; -195.8; -210; 0.81; s²2s²2p³	**8 15.9994** O (Oxygen); -2; -183; -218.8; 1.14; s²2s²2p⁴	**9 18.9984** F (Fluorine); -1; -188.2; -219.6; 1.11; s²2s²2p⁵
13 26.9815 Al (Aluminum); 3; 2450; 660; 2.70; [Ne]3s²3p¹	**14 28.086** Si (Silicon); 4; 2680; 1410; 2.33; [Ne]3s²3p²	**15 30.9738** P (Phosphorus); ±3,5,4; 280w; 44.2w; 1.82w; [Ne]3s²3p³	**16 32.064** S (Sulfur); 6,3,4,-2; 444.6; 119.0; 2.07; [Ne]3s²3p⁴	**17 35.453** Cl (Chlorine); ±1,4,5,6,7; -34.7; -101.0; 1.56; [Ne]3s²3p⁵

10 20.183 Ne (Neon); 0; -246; -248.6; 1.20; s²2s²2p⁶

18 39.948 Ar (Argon); 0; -185.8; -189.4; 1.40; [Ne]3s²3p⁶

Groups VIII, IB, IIB

VIII			IB	IIB
27 58.933 Co (Cobalt); 2,3; 2900; 1495; 8.9; [Ar]3d⁷4s²			**29 63.54** Cu (Copper); 2,1; 2595; 1083; 8.96; [Ar]3d¹⁰4s¹	**30 65.37** Zn (Zinc); 2; 906; 419.5; 7.14; [Ar]3d¹⁰4s²
28 58.71 Ni (Nickel); 2,1; 2730; 1453; 8.9; [Ar]3d⁸4s²				

31 69.72 Ga (Gallium); 3; 2237; 29.8; 5.91; [Ar]3d¹⁰4s²4p¹
32 72.59 Ge (Germanium); 4; 2830; 937.4; 5.72; [Ar]3d¹⁰4s²4p²
33 74.922 As (Arsenic); ±3,5; 613*; 817; 5.72; [Ar]3d¹⁰4s²4p³
34 78.96 Se (Selenium); 6,4,-2; 685; 217; 4.79; [Ar]3d¹⁰4s²4p⁴
35 79.909 Br (Bromine); ±1,5; 58; -7.2; 3.12; [Ar]3d¹⁰4s²4p⁵
36 83.80 Kr (Krypton); 0; -152; -157.3; 2.6; [Ar]3d¹⁰4s²4p⁶

45 102.905 Rh (Rhodium); 2,3,4; 4500; 1966; 12.4; [Kr]4d⁸5s¹
46 106.4 Pd (Palladium); 2,4; 3980; 1552; 12.0; [Kr]4d¹⁰5s⁰
47 107.870 Ag (Silver); 1; 2210; 960.8; 10.5; [Kr]4d¹⁰5s¹
48 112.40 Cd (Cadmium); 2; 765; 320.9; 8.65; [Kr]4d¹⁰5s²
49 114.82 In (Indium); 3; 2000; 156.2; 7.31; [Kr]4d¹⁰5s²5p¹
50 118.69 Sn (Tin); 4,2; 2270; 231.9; 7.30; [Kr]4d¹⁰5s²5p²
51 121.75 Sb (Antimony); ±3,5; 1380; 630.5; 6.62; [Kr]4d¹⁰5s²5p³
52 127.60 Te (Tellurium); 6,4,-2; 989.8; 449.5; 6.24; [Kr]4d¹⁰5s²5p⁴
53 126.904 I (Iodine); ±1,4,5,7; 183; 113.7; 4.94; [Kr]4d¹⁰5s²5p⁵
54 131.30 Xe (Xenon); 0; -108.0; -111.9; 3.06; [Kr]4d¹⁰5s²5p⁶

77 192.2 Ir (Iridium); 2,3,4,6; 5300; 2454; 22.5; [Xe]4f¹⁴5d⁷6s²
78 195.09 Pt (Platinum); 2,4; 4530; 1769; 21.4; [Xe]4f¹⁴5d⁹6s¹
79 196.967 Au (Gold); 3,1; 2970; 1063; 19.3; [Xe]4f¹⁴5d¹⁰6s¹
80 200.59 Hg (Mercury); 2,1; 357; -38.4; 13.6; [Xe]4f¹⁴5d¹⁰6s²
81 204.37 Tl (Thallium); 3,1; 1457; 303; 11.85; [Xe]4f¹⁴5d¹⁰6s²6p¹
82 207.19 Pb (Lead); 4,2; 1725; 327.4; 11.4; [Xe]4f¹⁴5d¹⁰6s²6p²
83 208.980 Bi (Bismuth); 3,5; 1560; 271.3; 9.8; [Xe]4f¹⁴5d¹⁰6s²6p³
84 (210) Po (Polonium); 4,2; 254; (9.2); [Xe]4f¹⁴5d¹⁰6s²6p⁴
85 (210) At (Astatine); --; (302); --; [Xe]4f¹⁴5d¹⁰6s²6p⁵
86 (222) Rn (Radon); 0; (-61.8); (-71); [Xe]4f¹⁴5d¹⁰6s²6p⁶

Lanthanides

63 151.96	64 157.25	65 158.924	66 162.50	67 164.930	68 167.26	69 168.934	70 173.04	71 174.97
Eu	Gd	Tb	Dy	Ho	Er	Tm	Yb	Lu
3,2	3	3,4	3	3	3	3,2	3,2	3
1439 / 826	2800 / 1312	2800 / 1356	2600 / 1407	2600 / 1461	2900 / 1497	1727 / 1545	1427 / 824	3327 / 1652
5.26	7.89	8.27	8.54	8.80	9.05	9.33	6.98	9.84
[Xe]4f⁷5d⁰6s²	[Xe]4f⁷5d¹6s²	[Xe]4f⁸5d⁰6s²	[Xe]4f¹⁰5d⁰6s²	[Xe]4f¹¹5d⁰6s²	[Xe]4f¹²5d⁰6s²	[Xe]4f¹³5d⁰6s²	[Xe]4f¹⁴5d⁰6s²	[Xe]4f¹⁴5d¹6s²
Europium	Gadolinium	Terbium	Dysprosium	Holmium	Erbium	Thulium	Ytterbium	Lutetium

Actinides

95 (243)	96 (247)	97 (247)	98 (251)	99 (254)	100 (253)	101 (256)	102 (254)	103 (257)
6,5,4,3			3					
Am	Cm	Bk	Cf	Es	Fm	Md	No	(Lw)
11.7								See note 3
[Rn]5f⁷6d⁰7s²	[Rn]5f⁷6d¹7s²	[Rn]5f⁸6d¹7s²	[Rn]5f¹⁰6d⁰7s²					
Americium	Curium	Berkelium	Californium	Einsteinium	Fermium	Mendelevium	Nobelium	(Lawrencium)

Courtesy of E.H. Sargent & Co.

TABLE OF PERIODIC PROPERTIES

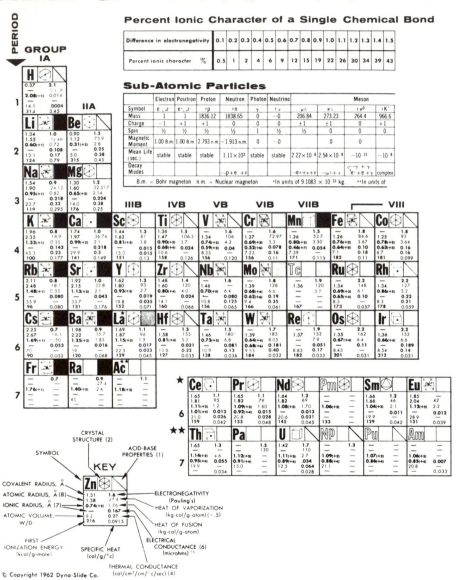

FIG. B.4.   PERIODIC PROPERTIES OF THE ELEMENTS

OF THE ELEMENTS

1.6	1.7	1.8	1.9	2.0	2.1	2.2	2.3	2.4	2.5	2.6	2.7	2.8	2.9	3.0	3.1	3.2
47	51	55	59	63	67	70	74	76	79	82	84	86	88	89	91	92

INERT GASES

Hyperon

	· Λ^0	· Σ·	· Σ	· Ξ
· K^0	2181.4	2327.7	2343.2	2584
974.4	0	+1	1	-1
0	½	½-integral	½-integral	½-integral
0				
10^{-10} 10^{-7}	2.8×10^{-10}	5×10^{-11}	10^{-10}	10^{-10}
	-p + π	-p + π⁰		
complex	·n + π⁰	·n + π·	-n + π	-Λ^0+π

4.80286×10^{-10} esu.   ·Exists as an antiparticle not listed.

He
0.93	
—	0.020
—	0.005
318	.0003
567	1.25

IIIA

B
0.82	2.0
0.98	7.5
0.20(+3)	5.3
—	10^{-13}
4.6	—
191	0.309

Al
1.18	1.5
1.43	67.9
0.50(+3)	2.55
—	0.382
10.0	0.50
138	0.215

IVA

C
0.77	2.5
0.914	171.7·
2.60(-4)	—
0.15(+4)	.0007
5.3	0.057
260	0.165

Si
1.11	1.8
1.32	(40.6)
2.71(-4)	11.1
0.41(+4)	0.10
12.1	0.20
188	0.162

VA

N
0.75	3.0
0.92	0.666
1.71(-3)	0.086
0.11(+5)	—
17.3	.00006
336	0.247

P
1.06	2.1
1.28	2.97
2.12(-3)	0.15
0.34(+5)	10^{-17}
17.0	—
254	0.177

VIA

O
0.73	3.5
0.92	0.815
1.40(-2)	0.053
0.09(+6)	—
14.0	.00006
314	0.218

S
1.02	2.5
1.27	3.01·
1.84(-2)	0.34
0.29(+6)	10^{-23}
15.5	.0007
239	0.175

VIIA

F
0.72	4.0
—	0.755
1.36(-1)	0.061
0.07(+7)	—
17.1	—
402	0.18

Cl
0.99	3.0
1.27	2.44
1.81(-1)	0.77
0.26(+7)	—
22.2	.00002
300	0.116

Ne
1.31	
—	0.431
—	0.080
16.8	.0001
497	—

Ar
1.74	
—	1.56
—	0.281
24.2	.00004
363	0.125

IB

Ni
1.24	1.8
0.78(+2)	4.21
0.62(+3)	0.145
6.6	0.22
176	0.105

Pd
1.37	2.2
0.50(+3)	9.0
—	0.093
8.9	0.17
192	0.058

Pt
1.38	2.2
0.52(+2)	5.2
—	0.095
9.10	0.17
207	0.032

Cu
1.28	1.9
0.96(+1)	72.8
0.69(+2)	3.11
7.1	0.593
178	0.092

Ag
1.44	1.9
1.26(+1)	60.7
—	2.70
10.3	0.616
175	0.98

Au
1.44	2.4
1.37(+1)	81.8
—	3.03
10.2	0.42
213	0.71

IIB

Zn
1.31	1.6
0.74(+2)	27.4
—	0.167
9.2	0.27
216	0.0915

Cd
1.48	1.7
0.97(+2)	23.9
—	1.45
13.1	0.146
207	0.22

Hg
1.49	1.9
1.57	13.9
1.10(+2)	0.56
14.8	0.011
241	0.02

Ga
1.26	1.6
1.41	—
1.48(+1)	1.34
0.62(+3)	.058
11.8	—
138	0.079

Ge
1.22	1.8
1.37	68
0.93(+2)	7.6
0.53(+4)	0.022
13.6	0.14
187	0.073

As
1.19	2.0
1.39	7.75·
2.22(-3)	6.62
0.47(+5)	0.029
13.1	—
231	0.082

Se
1.16	2.4
1.40	3.34
1.98(-2)	1.25
0.42(+6)	0.08
16.5	.00001
225	0.084

Br
1.14	2.8
—	3.58
1.95(-1)	1.26
0.39(+7)	10^{-18}
23.5	—
273	0.070

Kr
1.89	
—	2.16
—	0.39
32.2	.00002
323	—

In
1.44	1.7
1.66	53.7
1.32(+1)	0.78
0.81(+3)	0.111
15.7	0.057
133	0.057

Sn
1.41	1.8
1.62	70
1.12(+2)	1.72
0.71(+4)	0.088
16.3	0.16
169	0.054

Sb
1.38	1.9
1.59	46.6
2.45(-3)	4.74
0.62(+5)	0.026
18.4	0.05
199	0.049

Te
1.35	2.1
1.60	11.9
2.21(-2)	4.28
0.56(+6)	10^{-6}
20.5	0.014
208	0.047

I
1.33	2.5
—	13.4
2.16(-1)	1.87
0.50(+7)	10^{-13}
25.7	.001
241	0.052

Xe
2.09	
—	3.02
—	0.55
42.9	.0001
280	—

Tl
1.48	1.8
1.71	38.8
1.40(+1)	1.02
0.95(+3)	0.055
17.2	0.093
141	0.031

Pb
1.47	1.8
1.75	42.4
1.20(+2)	1.72
0.84(+4)	0.046
18.3	0.083
171	0.031

Bi
1.46	1.9
1.59	42.7
1.20(+3)	—
0.74(+5)	0.009
213	0.02
185	0.034

Po
—	2.0
1.76	29
—	—
—	0.02
22.7	—
—	—

At
—	2.2
—	8
—	—
—	—
—	—
248	—

Rn
2.14	
—	3.92
—	0.69
—	—
—	—

Gd
1.61	1.1
1.79	72
1.02(+3)	3.70
—	0.007
19.9	0.021
142	0.071

Tb
1.59	1.2
1.77	70
1.00(+3)	3.9
—	0.009
19.2	—
155	0.044

Dy
1.59	—
1.77	67
0.99(+3)	4.1
—	0.011
19.0	0.024
157	0.041

Ho
1.58	1.2
1.76	67
0.97(+3)	4.1
—	0.011
18.7	—
—	0.039

Er
1.57	1.2
1.75	67
0.96(+3)	4.1
—	0.012
18.4	0.023
—	0.040

Tm
1.56	1.2
1.74	59
0.95(+3)	4.4
—	0.011
18.1	—
—	0.038

Yb
1.70	1.1
1.92	38
1.13(+3)	1.8
—	0.035
24.8	—
143	0.035

Lu
1.56	1.2
1.74	90
0.93(+3)	4.6
—	0.015
17.8	—
115	0.037

Cm · **Bk** · **Cf** · **Es** · **Fm** · **Md** · **No** · **(Lw)**

See note 3

NOTES: · Element sublimes

(1) For representative oxides (higher valence) of group. Oxide is acidic if color is red, basic if color is blue and amphoteric if both colors are shown. Intensity of color indicates relative strength.

(2) Cubic, face centered; cubic, body centered; diamond; cubic; hexagonal; rhombohedral; tetragonal; orthorhombic; monoclinic.

(3) Proposed; not officially accepted   (4) At room temperature.   (5) At boiling point.

(6) From 0° to 20°C.   (7) Ionic (crystal) radii for coordination number 6.

(8) Metallic radii for coordination number of 12.

Courtesy of E.H. Sargent & Co.

TABLE B.10.   TRADITIONAL U.S. WINE BOTTLE SIZES

Common Name	Gal.	Contents in Oz	Ml
Split	0.05	6.4	189.3
Tenth	0.10	12.8	378.5
Fifth	0.20	25.6	757.1
Magnum	0.40	51.2	1514.1
Jeroboam	0.80	102.4	3028.3
Rehoboam	1.20	153.6	4542.4
Methuselah	1.60	204.8	6056.5
Salmanazar	2.40	307.2	9084.7
Balthazar	3.20	409.6	12,113.1
Nebuchadnezzer	4.00	512.0	15,141.3

TABLE B.11.   CORRECTION OF WINE SPIRITS VOLUME TO 60°F

Temp. °F	Degrees Proof							
	165	170	175	180	185	190	195	200
40	1.011	1.011	1.011	1.011	1.011	1.012	1.012	1.012
42	1.010	1.010	1.010	1.010	1.010	1.010	1.011	1.011
44	1.009	1.009	1.009	1.009	1.009	1.009	1.009	1.010
46	1.008	1.008	1.008	1.008	1.008	1.008	1.008	1.008
48	1.007	1.007	1.007	1.007	1.007	1.007	1.007	1.007
50	1.005	1.006	1.006	1.006	1.006	1.006	1.006	1.006
52	1.004	1.004	1.004	1.005	1.005	1.005	1.005	1.005
54	1.003	1.003	1.003	1.003	1.003	1.004	1.004	1.004
56	1.002	1.002	1.002	1.002	1.002	1.002	1.002	1.002
58	1.001	1.001	1.001	1.001	1.001	1.001	1.001	1.001
60	1.000	1.000	1.000	1.000	1.000	1.000	1.000	1.000
62	.999	.999	.999	.999	.999	.999	.999	.999
64	.998	.998	.998	.998	.998	.998	.998	.998
66	.997	.997	.997	.997	.997	.996	.996	.996
68	.996	.995	.995	.995	.995	.995	.995	.995
70	.994	.994	.994	.994	.994	.994	.994	.994
72	.993	.993	.993	.993	.993	.993	.993	.993
74	.992	.992	.992	.992	.992	.992	.992	.992
76	.991	.991	.991	.991	.991	.991	.990	.990
78	.990	.990	.990	.990	.989	.989	.989	.989
80	.989	.989	.989	.988	.988	.988	.988	.988

Based on information published in *The Bulletin of The Bureau of Standards*, Volume 9, No. 3, October 15, 1913.

TABLE B.12.   ESTIMATION OF PARTIALLY FILLED HORIZONTAL TANKS (CIRCULAR, STRAIGHT-SIDED TANKS ONLY; WHERE TOTAL CAPACITY OF TANK IS KNOWN)

Procedure:

(1) Determine inside diameter of tank.
(2) Determine depth of wine in tank.
(3) Compute percent depth of wine (line 2 less line 1).
(4) From table below find percent total volume that corresponds to this percent depth
(5) Multiply percent found by total volume = amount of wine in tank

Percent Depth Wine	Percent Volume	Percent Depth Wine	Percent Volume	Percent Depth Wine	Percent Volume
99	99.6	64	67.8	29	23.8
98	99.2	63	66.6	28	22.7
97	98.8	62	65.4	27	21.6
96	98.4	61	64.2	26	20.5
95	98.0	60	63.0	25	19.4
94	97.4	59	61.6	24	18.3
93	96.8	58	60.2	23	17.2
92	96.2	57	58.8	22	16.1
91	95.6	56	57.4	21	15.0
90	95.0	55	56.0	20	14.0
89	94.2	54	54.8	19	13.0
88	93.4	53	53.6	18	12.0
87	92.6	52	52.4	17	11.0
86	91.8	51	51.2	16	10.0
85	91.0	50	50.0	15	9.0
84	90.0	49	48.8	14	8.2
83	89.0	48	47.6	13	7.4
82	88.0	47	46.4	12	6.6
81	87.0	46	45.2	11	5.8
80	86.0	45	44.0	10	5.0
79	85.0	44	42.6	9	4.4
78	83.9	43	41.2	8	3.8
77	82.8	42	39.8	7	3.2
76	81.7	41	38.4	6	2.6
75	80.5	40	37.0	5	2.0
74	79.5	39	35.8	4	1.6
73	78.4	38	34.6	3	1.2
72	77.3	37	33.4	2	.8
71	76.2	36	32.2	1	.4
70	75.0	35	31.0	0	.0
69	73.8	34	29.8		
68	72.6	33	28.6		
67	71.4	32	27.4		
66	70.2	31	26.2		
65	69.0	30	25.0		

TABLE B.13.   WINE TANK CAPACITIES (FOR CYLINDRICAL STRAIGHT-SIDED TANKS ONLY MEASURED AT THE WAIST[1]

Calculation of total capacity in U.S. gal.:
Diam. (in.) × diam. (in.) × height (in.) × 0.0034 = U.S. gal capacity of tank.
Example:
72 in. diam. × 72 in. diam × 96 in. height (or length) × 0.0034 = 1692.1 U.S. gal. capacity of tank.

Calculation of U.S. gal. per in.:
Total U.S. gal. capacity divided by total number of in. = U.S. gal. per in.
Example:
1692.1 U.S. gal. capacity divided by 96 in. = 17.626 U.S. gal. per in.

Calculation of total capacity in liters:
Total U.S. gal. capacity × 3.785 = liters capacity of tank.
Example 1692.1 U.S. gal. capacity × 3.785 = 6404.6 liters capacity of tank.

Calculation of liters per cm:
U.S. gal. per in. × 1.49016 = liters per cm.
Example:
17.626 gal. per in. × 1.49016 = 26.266 liters per cm.

More Examples:

Diameter In. I.D.	Length In. I.D.	Capacity Gal. U.S.	Gal. per In. U.S.	Capacity Liters	Liters per Cm
36	36	158.6	4.41	600.3	6.44
36	48	211.5	4.41	800.5	6.44
36	60	264.4	4.41	1000.8	6.44
48	48	376.0	7.83	1423.2	11.67
48	60	470.0	7.83	1779.0	11.67
48	72	564.0	7.83	2134.7	11.67
60	60	734.4	12.24	2779.7	18.24
60	72	881.3	12.24	3335.7	18.24
60	84	1028.2	12.24	3891.7	18.24
60	96	1175.0	12.24	4447.4	18.24
72	72	1269.0	17.63	4803.2	26.27
72	84	1480.6	17.63	5604.1	26.27
72	96	1692.1	17.63	6404.6	26.27
72	108	1903.6	17.63	7205.1	26.27
72	120	2115.1	17.63	8005.7	26.27
84	84	2015.2	23.99	7627.5	35.75
84	96	2303.1	23.99	8717.2	35.75
84	108	2591.0	23.99	9806.9	35.75
84	120	2878.8	23.99	10,896.3	35.75
84	132	3166.7	23.99	11,986.0	35.75
84	144	3454.6	23.99	13,075.7	35.75
96	96	3008.1	31.33	11,385.7	46.69
96	108	3384.1	31.33	12,810.0	46.69
96	120	3760.1	31.33	14,232.0	46.69
96	132	4136.1	31.33	15,655.1	46.69
96	144	4512.2	31.33	17,078.7	46.69

[1] Whether the tank is horizontal or vertical, the diameter must be measured at the center of the tank between the two parallel, flat ends.

TABLE B.14.   VAT CAPACITY

(Per 1 Ft of Depth—Inside Measure)

Diameter (Ft)	(In.)	Area (Ft)	U.S. Gal.
1	0	.7854	5.8735
1	2	1.0690	7.9944
1	4	1.3962	10.4413
1	6	1.7671	13.2150
1	8	2.1816	16.3148
1	10	2.6398	19.7414
2	0	3.1416	23.4940
2	2	3.6869	27.5720
2	4	4.2760	32.6976
2	6	4.9087	36.7092
2	8	5.5850	41.7668
2	10	6.3049	47.1505
3	0	7.0686	52.8618
3	2	7.8757	58.8976
3	4	8.7265	65.2602
3	6	9.6211	73.1504
3	8	10.5591	78.9652
3	10	11.5409	86.3074
4	0	12.5664	93.9754
4	2	13.6353	101.9701
4	4	14.7479	110.2907
4	6	15.9043	118.9386
4	8	17.1041	127.9112
4	10	18.3476	137.2105
5	0	19.6350	146.8384
5	2	20.9656	156.7891
5	4	22.3400	167.0674
5	6	23.7583	177.6740
5	8	25.2199	188.6045
5	10	26.7251	199.8610
6	0	28.2744	211.4472
6	6	33.1831	248.1564
7	0	38.4846	287.8230
7	6	44.1787	330.3859
8	0	50.2656	375.9062
8	6	56.7451	424.3625
9	0	63.6174	475.7563
9	6	70.8823	530.0861
10	0	78.5400	587.3534
10	6	86.5903	647.5568
11	0	95.0034	710.6977
11	6	103.8691	776.7746
12	0	113.0976	848.1890

TABLE B.15.   EXAMPLE OF ENGLISH MEASURE TANK CHART FOR STRAIGHT-SIDED HORIZONTAL TANK[1]

Tank No. S-41		Gal. per In.		Wet In.	
In.	Gal.	In.	Gal.	In.	Gal.
1	10	37	1855	73	3896
2	23	38	1920	74	3928
3	41	39	1987	75	3955
4	66	40	2052	76	3980
5	91	41	2118	77	3998
6	120	42	2183	78	4008
7	153	43	2248		
8	189	44	2314		
9	228	45	2379		
10	271	46	2445		
11	317	47	2508		
1 ft	362	4 ft	2573		
13	409	49	2638		
14	459	50	2703		
15	512	51	2765		
16	568	52	2827		
17	622	53	2888		
18	677	54	2949		
19	733	55	3010		
20	791	56	3070		
21	842	57	3130		
22	903	58	3188		
23	962	59	3245		
2 ft	1023	5 ft	3300		
25	1084	61	3356		
26	1147	62	3411		
27	1209	63	3463		
28	1271	64	3515		
29	1336	65	3565		
30	1399	66	3612		
31	1464	67	3658		
32	1528	68	3703		
33	1593	69	3745		
34	1659	70	3786		
35	1724	71	3825		
3 ft	1789	6 ft	3862		

[1] Tank has an inside diameter of 122.41 in. at the waist (39 in.).

TABLE B.16.   EXAMPLE OF METRIC MEASURE TANK CHART FOR STRAIGHT-SIDED VERTICAL TANK[1]

	Tank No. S-56 Liters per Cm			Measured at Center of Tank Wet Cm	
Cm	Liters	Cm	Liters	Cm	Liters
0	0	59	1040.64	118	2081.28
1	17.64	60	1058.28	119	2098.92
2	35.28	61	1075.92	120	2116.56
3	52.91	62	1093.56	121	2134.20
4	70.55	63	1111.19	122	2151.84
5	88.19	64	1128.83	123	2169.47
6	105.83	65	1146.47	124	2187.11
7	123.47	66	1164.11	125	2204.75
8	141.10	67	1181.75	126	2222.39
9	158.74	68	1199.38	127	2240.03
10	176.38	69	1217.02	128	2257.66
11	194.02	70	1234.66	129	2275.30
12	211.66	71	1252.30	130	2292.94
13	229.29	72	1269.94	131	2310.58
14	246.93	73	1287.57	132	2328.22
15	264.57	74	1305.21	133	2345.85
16	282.21	75	1322.85	134	2363.49
17	299.85	76	1340.49	135	2381.13
18	317.48	77	1358.13	136	2398.77
19	335.12	78	1375.76	137	2416.41
20	352.76	79	1393.40	138	2434.04
21	370.40	80	1411.04	139	2451.68
22	388.04	81	1428.68	140	2469.32
23	405.67	82	1446.32	141	2486.96
24	423.31	83	1463.95	142	2504.60
25	440.95	84	1481.59	143	2522.23
26	458.59	85	1499.23	144	2539.87
27	476.23	86	1516.87	145	2557.51
28	493.86	87	1534.51	146	2575.15
29	511.50	88	1552.14	147	2592.79
30	529.14	89	1569.78	148	2610.42
31	546.78	90	1587.42	149	2628.06
32	564.42	91	1605.06	150	2645.70
33	582.05	92	1622.70	151	2663.34
34	599.69	93	1640.33	152	2680.98
35	617.33	94	1657.97	153	2698.61
36	634.97	95	1675.61	154	2716.25
37	652.61	96	1693.25	155	2733.89
38	670.24	97	1710.89	156	2751.53
39	687.88	98	1728.52	157	2769.17
40	705.52	99	1746.16	158	2786.80
41	723.16	100	1763.80	159	2804.44
42	740.80	101	1781.44	160	2822.08
43	758.43	102	1799.08	161	2839.72
44	776.07	103	1816.71	162	2857.36
45	793.71	104	1834.35	163	2874.99
46	811.35	105	1851.99	164	2892.63
47	828.99	106	1869.63	165	2910.27
48	846.62	107	1887.27	166	2927.91
49	864.26	108	1904.90	167	2945.55
50	881.90	109	1922.54	168	2963.18
51	899.54	110	1940.18	169	2980.82
52	917.18	111	1957.82	170	2998.46
53	934.81	112	1975.46	171	3016.10
54	952.45	113	1993.09	172	3033.74
55	970.09	114	2010.73	173	3051.37
56	987.73	115	2028.37	174	3069.01
57	1005.37	116	2046.01	175	3086.65
58	1023.00	117	2063.65	176	3104.29

[1] Tank has an inside diameter of 149.86 cm. at the waist (137 cm.).

TABLE B.16.   *(Cont'd)*

	Tank No. S-56 Liters per Cm			Measured at Center of Tank Wet Cm	
Cm	Liters	Cm	Liters	Cm	Liters
177	3121.93	210	3703.98	242	4268.40
178	3139.56	211	3721.62	243	4286.03
179	3157.20	212	3739.26	244	4303.67
180	3174.84	213	3756.89	245	4321.31
181	3192.48	214	3774.53	246	4338.95
182	3210.12	215	3792.17	247	4356.59
183	3227.75	216	3809.81	248	4374.22
184	3245.39	217	3827.45	249	4391.86
185	3263.03	218	3845.08	250	4409.50
186	3280.67	219	3862.72	251	4427.14
187	3298.31	220	3880.36	252	4444.78
188	3315.94	221	3898.00	253	4462.41
189	3333.58	222	3915.64	254	4480.05
190	3351.22	223	3933.27	255	4497.69
191	3368.86	224	3950.91	256	4515.33
192	3386.50	225	3968.56	257	4532.97
193	3404.13	226	3986.19	258	4550.60
194	3421.77	227	4003.83	259	4568.24
195	3439.41	228	4021.46	260	4585.88
196	3457.05	229	4039.10	261	4603.52
197	3474.69	230	4056.74	262	4621.16
198	3492.32	231	4074.38	263	4638.79
199	3509.96	232	4092.02	264	4656.43
200	3527.60	233	4109.65	265	4674.07
201	3545.24	234	4127.29	266	4691.71
202	3562.88	235	4144.93	267	4709.35
203	3580.51	236	4162.57	268	4726.98
204	3598.15	237	4180.21	269	4744.62
205	3615.80	238	4197.84	270	4762.26
206	3633.43	239	4215.48	271	4779.90
207	3651.07	240	4233.12	272	4797.54
208	3668.70	241	4250.76	273	4815.70
209	3686.34				

[1] Tank has an inside diameter of 149.86 cm. at the waist (137 cm.).

Appendix C

Conversion Tables

Units of Areas

Units of Capacity and Volume

Units of Length

Units of Power and Pressure

Units of Temperature

Units of Weight

TABLE C.1.   BASIC CONVERSIONS: UNITS OF AREA

1 square inch	645.2 square millimeters
1 square inch	6.452 square centimeters
1 square inch	.000645 square meters
1 square inch	.006944 square feet
1 square inch	.000772 square yards
1 square foot	92,903 square millimeters
1 square foot	929.034 square centimeters
1 square foot	.092903 square meters
1 square foot	144 square inches
1 square foot	0.11111 square yards
1 square yard	8361.31 square centimeters
1 square yard	.836131 square meters
1 square yard	1296 square inches
1 square yard	9 square feet
1 square rod	30.25 square feet
1 acre	40.47 ares
1 acre	0.4047 hectares
1 acre	4480 square yards
1 square mile	2.58998 kilometers
1 square mile	640 acres
1 square centimeter	100 square millimeters
1 square centimeter	0.1550 square inches
1 square centimeter	.001076 square feet
1 square centimeter	.00012 square yards
1 square centimeter	.00010 square meters
1 square meter	10000 square centimeters
1 square meter	1550 square inches
1 square meter	10.7639 square feet
1 square meter	1.19598 square yards
1 square kilometer	247.1 acres
1 square kilometer	0.3861 square miles

TABLE C.2.   CONVERSION FORMULAS

To Convert	into	Multiply by
square inches	square centimeters	6.451600
square feet	square meters	0.092903
square yards	square meters	0.836127
acres	square meters	4046.856422
square miles	square kilometers	2.589980
square centimeters	square inches	0.155000
square meters	square feet	10.763910
square meters	square yards	1.195990
square kilometers	square miles	0.386102

TABLE C.3.   SQUARE INCHES INTO SQUARE CENTIMETERS

Sq. In	0	1	2	3	4	5	6	7	8	9
					Sq. Cm					
0	—	6.45	12.90	19.36	25.81	32.26	38.71	45.16	51.61	58.06
10	64.52	70.97	77.42	83.87	90.32	96.77	103.23	109.68	116.13	122.58
20	129.03	135.48	141.94	148.39	154.84	161.29	167.74	174.19	180.65	187.10
30	193.55	200.00	206.45	212.90	219.35	225.81	232.26	238.71	245.16	251.61
40	258.06	264.52	270.97	277.42	283.87	290.32	296.77	303.23	309.68	316.13
50	322.58	329.03	335.48	341.94	348.39	354.84	361.29	367.74	374.19	380.64
60	387.10	393.55	400.00	406.45	412.90	419.35	425.81	432.26	438.71	455.16
70	451.61	458.06	464.52	470.97	477.42	483.87	490.32	496.77	503.23	509.68
80	516.13	522.58	529.03	535.48	541.93	548.39	554.84	561.29	567.74	574.19
90	580.64	587.10	593.55	600.00	606.45	612.90	619.35	625.81	632.26	638.71
100	645.16	651.61	658.06	664.52	670.97	677.42	683.87	690.32	696.77	703.22

TABLE C.4.   SQUARE CENTIMETERS INTO SQUARE INCHES

Sq. cm	0	1	2	3	4	5	6	7	8	9
					Sq. In					
0	—	0.155	0.310	0.465	0.620	0.775	0.930	1.085	1.240	1.395
10	1.550	1.705	1.860	2.015	2.170	2.325	2.480	2.635	2.790	2.945
20	3.100	3.255	3.410	3.565	3.720	3.875	4.030	4.185	4.340	4.495
30	4.650	4.805	4.960	5.115	5.270	5.425	5.580	5.735	5.890	6.045
40	6.200	6.355	6.510	6.665	6.820	6.975	7.130	7.285	7.440	7.595
50	7.750	7.905	8.060	8.215	8.370	8.525	8.680	8.835	8.990	9.145
60	9.300	9.455	9.610	9.765	9.920	10.075	10.230	10.385	10.540	10.695
70	10.850	11.005	11.160	11.315	11.470	11.625	11.780	11.935	12.090	12.245
80	12.400	12.555	12.710	12.865	13.020	13.175	13.330	13.485	13.640	13.795
90	13.950	14.105	14.260	14.415	14.570	14.725	14.880	15.035	15.190	15.345
100	15.500	15.655	15.810	15.965	16.120	16.275	16.430	16.585	16.740	16.895

TABLE C.5.   SQUARE FEET INTO SQUARE METERS

Sq. Ft	0	1	2	3	4	5	6	7	8	9
					Sq. M					
0	—	0.0929	0.1858	0.2787	0.3716	0.4645	0.5574	0.6503	0.7432	0.8361
10	0.9290	1.0219	1.1148	1.2077	1.3006	1.3936	1.4865	1.5794	1.6723	1.7652
20	1.8581	1.9510	2.0439	2.1368	2.2297	2.3226	2.4155	2.5084	2.6013	2.6942
30	2.7871	2.8800	2.9729	3.0658	3.1587	3.2516	3.3445	3.4374	3.5303	3.6232
40	3.7161	3.8090	3.9019	3.9948	4.0877	4.1806	4.2735	4.3664	4.4594	4.5523
50	4.6452	4.7381	4.8310	4.9239	5.0168	5.1097	5.2026	5.2955	5.3884	5.4813
60	5.5742	5.6671	5.7600	5.8529	5.9458	6.0387	6.1316	6.2245	6.3174	6.4103
70	6.5032	6.5961	6.6890	6.7819	6.8748	6.9677	7.0606	7.1535	7.2464	7.3393
80	7.4322	7.5252	7.6181	7.7110	7.8039	7.8968	7.9897	8.0826	8.1755	8.2684
90	8.3613	8.4542	8.5471	8.6400	8.7329	8.8258	8.9187	9.0116	9.1045	9.1974
100	9.2903	9.3832	9.4761	9.5690	9.6619	9.7548	9.8477	9.9406	10.0335	10.1264

TABLE C.6.   SQUARE METERS INTO SQUARE FEET

Sq. M	0	1	2	3	4	5	6	7	8	9
					Sq. Ft					
0	—	10.76	21.53	32.29	43.06	53.82	64.58	75.35	86.11	96.88
10	107.64	118.40	129.17	139.93	150.70	161.46	172.22	182.99	193.75	204.51
20	215.28	226.04	236.81	247.57	258.33	269.10	279.86	290.63	301.39	312.15
30	322.92	333.68	344.45	355.21	365.97	376.74	387.50	398.27	409.03	419.79
40	430.56	441.32	452.08	462.85	473.61	484.38	495.14	505.90	516.67	527.43
50	538.20	548.96	559.72	570.49	581.25	592.02	602.78	613.54	624.31	635.07
60	645.84	656.60	667.36	678.13	688.89	699.65	710.42	721.18	731.95	742.71
70	753.47	764.24	775.00	785.77	796.53	807.29	818.06	828.82	839.59	850.35
80	861.11	871.88	882.64	893.41	904.17	914.93	925.70	936.46	947.22	957.99
90	968.75	979.52	990.28	1001.04	1011.81	1022.57	1033.34	1044.10	1054.86	1065.63
100	1076.39	1087.15	1097.92	1108.68	1119.45	1130.21	1140.97	1151.74	1162.50	1173.27

TABLE C.7. ACRES TO HECTARES

Acres	0	10	20	30	40	50	60	70	80	90
					Hectares					
0	—	4.047	8.094	12.141	16.187	20.234	24.281	28.328	32.375	36.422
100	40.469	44.515	48.562	52.609	56.656	60.703	64.750	68.797	72.843	76.890
200	80.937	84.984	89.031	93.078	97.125	101.171	105.218	109.265	113.312	117.359
300	121.406	125.453	129.499	133.546	137.593	141.640	145.687	149.734	153.781	157.827
400	161.874	165.921	169.968	174.015	178.062	182.109	186.155	190.202	194.249	198.296
500	202.343	206.390	210.437	214.483	218.530	222.577	226.624	230.671	234.718	238.765
600	242.811	246.858	250.905	254.952	258.999	263.046	267.093	271.139	275.186	279.233
700	283.280	287.327	291.374	295.421	299.467	303.514	307.561	311.608	315.655	319.702
800	323.749	327.795	331.842	335.889	339.936	343.983	348.030	352.077	356.123	360.170
900	364.217	368.264	372.311	376.358	380.405	384.451	388.498	392.545	396.592	400.639
1000	404.686	—								

TABLE C.8. HECTARES TO ACRES

Hectares	0	1	2	3	4	5	6	7	8	9
					Acres					
0	—	24.71	49.42	74.13	98.84	123.55	148.26	172.97	197.68	222.40
100	247.11	271.82	296.53	321.24	345.95	370.66	395.37	420.08	444.79	469.50
200	484.21	518.92	543.63	568.34	593.05	617.76	642.47	667.19	691.90	716.61
300	741.32	766.03	790.74	815.45	840.16	864.87	889.58	914.29	939.00	963.71
400	988.42	1013.13	1037.84	1062.55	1087.26	1111.97	1136.68	1161.40	1186.11	1210.82
500	1235.53	1260.24	1284.95	1309.66	1334.37	1359.08	1383.79	1408.50	1433.21	1457.92
600	1482.63	1507.34	1532.05	1556.76	1581.47	1606.18	1630.90	1655.61	1680.32	1705.03
700	1729.74	1754.45	1779.16	1803.87	1828.58	1853.29	1878.00	1902.71	1927.42	1952.13
800	1976.84	2001.55	2026.26	2050.97	2075.69	2100.40	2125.11	2149.82	2174.53	2199.24
900	2223.95	2248.66	2273.37	2298.08	2322.79	2347.50	2372.21	2396.92	2421.63	2446.34
1000	2471.05	—								

UNITS OF CAPACITY AND VOLUME

TABLE C.9. BASIC CONVERSIONS: UNITS OF CAPACITY AND VOLUME

1 cubic inch	16.38706 cubic centimeters
1 cubic inch	.01638706 liters
1 cubic inch	0.5541 fluid ounces
1 cubic inch	.017316 U.S. quarts
1 cubic inch	.004329 U.S. gallons
1 cubic foot	0.02832 cubic meters
1 cubic foot	1728 cubic inches
1 cubic foot	28.317 liters
1 cubic foot	.03704 cubic yards
1 cubic foot	7.48052 U.S. gallons
1 cubic yard	764,559 cubic centimeters
1 cubic yard	46,656 cubic inches
1 cubic yard	27 cubic feet
1 cubic yard	0.7646 cubic meters
1 fluid ounce	8 fluid drams
1 fluid ounce	.03125 U.S. quarts
1 fluid ounce	.007812 U.S. gallons
1 fluid ounce	.29573 liters
1 U.S. quart	256 fluid drams
1 U.S. quart	32 fluid ounces
1 U.S. quart	.94633 liters
1 U.S. gallon	3.78533 liters
1 U.S. gallon	268.8 cubic inches
1 cubic centimeter	0.061 cubic inches
1 milliliter	0.03381 fluid ounces
1 milliliter	0.001057 U.S. quarts
1 milliliter	0.000264 U.S. gallons
1 cubic decimeter	1 liter
1 liter	61.023 cubic inches
1 liter	0.0353 cubic feet
1 liter	1.0567 U.S. quarts
1 liter	0.2642 U.S. gallons
1 liter	33.8147 fluid ounces

TABLE C.10. CONVERSION FORMULAS

To Convert	into	Multiply by
U.S. pints	liters	0.4732
U.S. quarts	liters	0.9463
U.S. gallons	liters	3.78533
U.S. gallons	cubic meters	0.00455
cubic inches	cubic centimeters	16.387064
cubic feet	liters	28.316847
cubic yards	cubic meters	0.764555
liters	U.S. pints	2.1134
liters	U.S. quarts	1.0567
liters	U.S. gallons	0.2642

TABLE C.11.   DRY MEASURE

1 pint	0.5506	liters
1 quart	1.1012	liters
1 peck	8.8096	liters
1 bushel	35.2383	liters
1 liter	1.8162	pints
1 liter	.9081	quarts
1 liter	.1135	pecks
1 liter	.0284	bushels

TABLE C.12.   LIQUID MEASURE

1 dash	6 drops—1⅓ milliliters
1 teaspoon	⅛ fluid ounces—4 milliliters
1 tablespoon	½ fluid ounces—15 milliliters
1 pony	1 fluid ounce—28⅓ milliliters
1 jigger	1½ fluid ounces—42½ milliliters
1 miniature	3.4 fluid ounces—100 milliliters
1 nip (split)	6.3 fluid ounces—187 milliliters
1 half-bottle	12.7 fluid ounces—375 milliliters
1 pint	16 fluid ounces—378½ milliliters
1 bottle	25.4 fluid ounces—750 milliliters
1 fifth	25.6 fluid ounces—757 milliliters
1 quart	32 fluid ounces—946⅓ milliliters
1 liter	33.8 fluid ounces—1000 milliliters
1½ liters	50.7 fluid ounces—1500 milliliters
3 liters	101.4 fluid ounces—3000 milliliters
1 gallon	128.0 fluid ounces—3785 milliliters

TABLE C.13. U.S. GALLONS INTO LITERS

Gal.	0	1	2	3	4	5	6	7	8	9
0	—	3.785	7.571	11.356	15.141	18.927	22.712	26.497	30.283	34.068
10	37.853	41.639	45.424	49.209	52.995	56.780	50.565	64.351	68.136	71.921
20	75.707	79.492	83.277	87.063	90.848	94.633	98.419	102.204	105.989	109.775
30	113.560	117.345	121.131	124.916	128.701	132.487	136.272	140.057	143.843	147.628
40	151.413	155.199	158.984	162.769	166.555	170.340	174.125	177.911	181.696	185.481
50	189.267	193.052	196.837	200.622	204.408	208.193	211.978	215.764	219.549	223.334
60	227.120	230.905	234.690	238.476	242.261	246.046	249.832	253.617	257.402	261.188
70	264.973	268.758	272.544	276.329	280.114	283.900	287.685	291.470	295.256	299.041
80	302.826	306.612	310.397	314.182	317.968	321.753	325.538	329.324	333.109	336.894
90	340.680	344.465	348.250	352.036	355.821	359.606	363.392	367.177	370.962	374.748
100	378.533	382.318	386.104	389.889	393.674	397.460	401.245	405.030	408.816	412.601

TABLE C.14. LITERS INTO U.S. GALLONS

Liters	0	1	2	3	4	5	6	7	8	9
0	—	.265	.528	.793	1.057	1.321	1.585	1.849	2.114	2.378
10	2.642	2.906	3.170	3.435	3.699	3.963	4.227	4.491	4.756	5.020
20	5.284	5.548	5.812	6.077	6.341	6.605	6.869	7.133	7.398	7.662
30	7.926	8.190	8.454	8.719	8.983	9.247	9.511	9.775	10.040	10.304
40	10.568	10.832	11.096	11.361	11.625	11.889	12.153	12.417	12.682	12.946
50	13.210	13.474	13.738	14.003	14.267	14.531	14.795	15.059	15.324	15.588
60	15.852	16.116	16.380	16.645	16.909	17.173	17.437	17.701	17.966	18.230
70	18.494	18.758	19.022	19.287	19.551	19.815	20.079	20.343	20.608	20.872
80	21.136	21.400	21.664	21.929	22.193	22.457	22.721	22.985	23.250	23.514
90	23.778	24.042	24.306	24.571	24.835	25.099	25.363	25.627	25.892	26.156
100	26.420	26.684	26.948	27.213	27.477	27.741	28.005	28.269	28.534	28.798

TABLE C.15.   CUBIC INCHES INTO CUBIC CENTIMETERS

Cu. In.	0	1	2	3	4	5	6	7	8	9
					Cu. Cm					
0	—	16.39	32.77	49.16	65.55	81.94	98.32	114.71	131.10	147.48
10	163.87	180.26	196.64	213.03	229.42	245.81	262.19	278.58	294.97	311.35
20	327.74	344.13	360.51	376.90	393.29	409.68	426.06	442.45	458.84	475.22
30	491.61	508.00	524.38	540.77	557.16	573.55	589.93	606.32	622.71	639.09
40	655.48	671.87	688.25	704.64	721.03	737.42	753.80	770.19	786.58	802.96
50	819.35	835.74	852.12	868.51	884.90	901.29	917.67	934.06	950.45	966.83
60	983.22	999.61	1016.0	1032.4	1048.8	1065.2	1081.5	1097.9	1114.3	1130.7
70	1147.1	1163.5	1179.9	1196.3	1212.6	1229.0	1245.4	1261.8	1278.2	1294.6
80	1311.0	1327.4	1343.7	1360.1	1376.5	1392.9	1409.3	1425.7	1442.1	1458.5
90	1474.8	1491.2	1507.6	1524.0	1540.4	1556.8	1573.2	1589.6	1605.9	1622.3
100	1638.7	1655.1	1671.5	1687.9	1704.3	1720.7	1737.1	1753.4	1769.8	1786.2

TABLE C.16.   CUBIC CENTIMETERS INTO CUBIC INCHES

Cu. Cm	0	1	2	3	4	5	6	7	8	9
					Cu. In.					
0	—	0.0610	0.1221	0.1831	0.2441	0.3051	0.3661	0.4272	0.4882	0.5492
10	0.6102	0.6713	0.7323	0.7933	0.8543	0.9154	0.9764	1.0374	1.0984	1.1595
20	1.2205	1.2815	1.3425	1.4036	1.4646	1.5256	1.5866	1.6477	1.7087	1.7697
30	1.8307	1.8917	1.9528	2.0138	2.0748	2.1358	2.1969	2.2579	2.3189	2.3799
40	2.4410	2.5020	2.5630	2.6240	2.6850	2.7461	2.8071	2.8681	2.9291	2.9902
50	3.0512	3.1122	3.1732	3.2343	3.2953	3.3563	3.4173	3.4784	3.5394	3.6004
60	3.6614	3.7225	3.7835	3.8445	3.9055	3.9665	4.0276	4.0886	4.1496	4.2106
70	4.2717	4.3327	4.3937	4.4548	4.5158	4.5768	4.6378	4.6988	4.7599	4.8209
80	4.8819	4.9429	5.0040	5.0650	5.1260	5.1870	5.2480	5.3091	5.3701	5.4311
90	5.4921	5.5532	5.6142	5.6752	5.7362	5.7973	5.8583	5.9193	5.9803	6.0414
100	6.1024	6.1634	6.2244	6.2854	6.3465	6.4075	6.4685	6.5295	6.5906	6.6516

TABLE C.17.   CUBIC FEET INTO LITERS (CUBIC DECIMETERS)

Cu. Ft	0	1	2	3	4	5	6	7	8	9
					Liters					
0	—	28.32	56.63	84.95	113.26	141.58	169.90	198.21	226.53	254.84
10	283.16	311.48	339.79	368.11	396.42	424.74	453.06	481.37	509.69	538.01
20	566.32	594.64	622.95	651.27	679.59	707.90	736.22	764.53	792.85	821.17
30	849.48	877.80	906.11	934.43	962.75	991.06	1019.4	1047.7	1076.0	1104.3
40	1132.6	1161.0	1189.3	1217.6	1245.9	1274.2	1302.5	1330.9	1359.2	1387.5
50	1415.8	1444.1	1472.4	1500.8	1529.1	1557.4	1585.7	1614.0	1642.3	1670.7
60	1699.0	1727.3	1755.6	1783.9	1812.2	1840.5	1868.9	1897.2	1925.5	1953.8
70	1982.1	2010.4	2038.8	2067.1	2095.4	2123.7	2152.0	2180.3	2208.7	2237.0
80	2265.3	2293.6	2321.9	2350.2	2378.6	2406.9	2435.2	2463.5	2491.8	2520.1
90	2548.4	2576.8	2605.1	2633.4	2661.7	2690.0	2718.3	2746.7	2775.0	2803.3
100	2831.6	2859.9	2888.2	2916.6	2944.9	2973.2	3001.5	3029.8	3058.1	3086.5

TABLE C.18.   LITERS (CUBIC DECIMETERS) INTO CUBIC FEET

Liters	0	1	2	3	4	5	6	7	8	9
					Cu. Ft					
0	—	0.0353	0.0706	0.1060	0.1413	0.1766	0.2119	0.2472	0.2825	0.3178
10	0.3532	0.3885	0.4238	0.4591	0.4944	0.5297	0.5651	0.6004	0.6357	0.6710
20	0.7063	0.7416	0.7769	0.8123	0.8476	0.8829	0.9182	0.9535	0.9888	1.0242
30	1.0595	1.0948	1.1301	1.1654	1.2007	1.2361	1.2714	1.3067	1.3420	1.3773
40	1.4126	1.4479	1.4833	1.5186	1.5539	1.5892	1.6245	1.6598	1.6952	1.7305
50	1.7658	1.8011	1.8364	1.8717	1.9071	1.9424	1.9777	2.0130	2.0483	2.0836
60	2.1189	2.1543	2.1896	2.2249	2.2602	2.2955	2.3308	2.3662	2.4015	2.4368
70	2.4721	2.5074	2.5427	2.5780	2.6134	2.6487	2.6840	2.7193	2.7546	2.7899
80	2.8253	2.8606	2.8959	2.9312	2.9665	3.0018	3.0372	3.0725	3.1078	3.1431
90	3.1784	3.2137	3.2490	3.2844	3.3197	3.3550	3.3903	3.4256	3.4609	3.4963
100	3.5315	3.5669	3.6021	3.6375	3.6728	3.7081	3.7434	3.7787	3.8140	3.8493

TABLE C.19.   CUBIC FEET INTO CUBIC METERS

Cu. Ft	0	1	2	3	4	5	6	7	8	9
					Cu. M					
0	—	0.0283	0.0566	0.0850	0.1133	0.1416	0.1689	0.1982	0.2265	0.2549
10	0.2832	0.3115	0.3398	0.3681	0.3964	0.4248	0.4531	0.4814	0.5097	0.5380
20	0.5663	0.5947	0.6230	0.6513	0.6796	0.7079	0.7352	0.7646	0.7929	0.8212
30	0.8495	0.8778	0.9061	0.9345	0.9628	0.9911	1.0194	1.0477	1.6760	1.1044
40	1.1327	1.1610	1.1893	1.2176	1.2459	1.2743	1.3026	1.3309	1.3592	1.3875
50	1.4158	1.4442	1.4725	1.5008	1.5291	1.5574	1.5857	1.6141	1.6424	1.6707
60	1.6990	1.7273	1.7556	1.7840	1.8123	1.8406	1.8689	1.8972	1.9256	1.9539
70	1.9822	2.0105	2.0388	2.0671	2.0955	2.1238	2.1521	2.1804	2.2087	2.2370
80	2.2654	2.2937	2.3220	2.3503	2.3786	2.4069	2.4353	2.4636	2.4919	2.5202
90	2.5485	2.5768	2.6052	2.6335	2.6618	2.6901	2.7184	2.7467	2.7751	2.8034
100	2.8317	2.8600	2.8883	2.9166	2.9450	2.9733	3.0016	3.0299	3.0582	3.0865

TABLE C.20.   CUBIC METERS INTO CUBIC FEET

Cu. M	0	1	2	3	4	5	6	7	8	9
					Cu. Ft					
0	—	35.3	70.6	105.9	141.3	176.6	211.9	247.2	282.5	317.8
10	353.1	388.5	423.8	459.1	494.4	529.7	565.0	600.3	635.7	671.0
20	706.3	741.6	776.9	812.2	847.6	882.9	918.2	953.5	988.8	1024.1
30	1059.4	1094.8	1130.1	1165.4	1200.7	1236.0	1271.3	1306.6	1342.0	1377.3
40	1412.6	1447.9	1483.2	1518.5	1553.9	1589.2	1624.5	1659.8	1695.1	1730.4
50	1765.7	1801.1	1836.4	1871.7	1907.0	1942.3	1977.6	2012.9	2048.3	2083.6
60	2118.9	2154.2	2189.5	2224.8	2260.1	2295.5	2330.8	2366.1	2401.4	2436.7
70	2472.0	2507.3	2542.7	2578.0	2613.3	2648.6	2683.9	2719.2	2754.5	2789.9
80	2825.2	2860.5	2895.8	2931.1	2966.4	3001.8	3037.1	3072.4	3107.7	3143.0
90	3178.3	3213.6	3249.0	3284.3	3319.6	3354.9	3390.2	3425.5	3460.8	3496.2
100	3531.5	3566.8	3602.1	3637.4	3672.7	3708.0	3743.4	3778.7	3814.0	3849.3

UNITS OF LENGTH

TABLE C.21.   BASIC CONVERSIONS: UNITS OF LENGTH

1 inch	25.4 millimeters
1 inch	2.54 centimeters
1 inch	0.0254 meters
1 inch	0.08333 feet
1 inch	0.02777 yards
1 foot	304.8 millimeters
1 foot	30.48 centimeters
1 foot	0.3048 meters
1 foot	12 inches
1 foot	0.3333 yards
1 yard	91.44 centimeters
1 yard	0.9144 meters
1 yard	36 inches
1 yard	3 feet
1 mile	1609.34 meters
1 mile	63,360 inches
1 mile	5,280 feet
1 mile	1760 yards
1 rod	16.5 feet
1 furlong	40 rods
8 furlongs	1 mile
1 league	3 miles
1 knot	6,085 feet
1 knot	1.1526 miles
1 millimeter	.03937 inches
1 millimeter	1000 micrometers (μms)
1 centimeter	0.3937 inches
1 centimeter	0.03281 feet
1 centimeter	0.01094 yards
1 centimeter	0.01 meters
1 meter	39.37 inches
1 meter	3.2808 feet
1 meter	1.0936 yards
1 meter	100 centimeters
1 kilometer	0.62137 miles

TABLE C.22.   CONVERSION FORMULAS

To Convert	into	Multiply by
inches	millimeters	25.4
inches	centimeters	2.54
feet	meters	0.304800
yards	meters	0.914400
miles	kilometers	1.609344
centimeters	inches	0.393701
meters	feet	3.280840
meters	yards	1.093613
kilometers	miles	0.6213712

TABLE C.23.   MILLIMETERS INTO INCHES

Mm	In.	Mm	In.
1	.03937	19	.74803
2	.07874	20	.78740
3	.11811	21	.82677
4	.15743	22	.86614
5	.19685	23	.90551
6	.23622	24	.94488
7	.27559	25	.98425
8	.31496	25.4	1.0
9	.35433	38.1	1.5
10	.39370	50.0	1.968
11	.43307	50.8	2.0
12	.47244	75.0	2.953
13	.51181	76.2	3.0
14	.55118	100.0	3.937
15	.59055	101.6	4.0
16	.62992	127.0	5.0
17	.66929	152.4	6.0
18	.70866		

TABLE C.24.   EQUIVALENTS: FRACTIONS OF AN INCH, DECIMAL INCHES AND MILLI-METERS

Fraction In.	Decimal In.	Mm
1/64	.015625	0.397
1/32	.03125	0.794
3/64	.046875	1.191
1/16	.0625	1.588
5/64	.078125	1.984
3/32	.09375	2.381
1/8	.125	3.175
3/16	.1875	4.763
1/4	.25	6.35
5/16	.3125	7.938
3/8	.375	9.525
3/16	.4375	11.113
1/2	.50	12.70
9/16	.5625	14.288
5/8	.625	15.875
11/16	.6875	17.463
3/4	.75	19.05
13/16	.8125	20.638
7/8	.875	22.225
15/16	.9375	23.813
1	1.0	25.40

TABLE C.25.   HUNDREDTHS OF A MILLIMETER INTO INCHES

Mm	In.	Mm	In.	Mm	In.	Mm	In.	Mm	In.
0.01	0.0004	0.21	0.0083	0.41	0.0161	0.61	0.0240	0.81	0.0319
0.02	0.0008	0.22	0.0087	0.42	0.0165	0.62	0.0244	0.82	0.0323
0.03	0.0012	0.23	0.0091	0.43	0.0169	0.63	0.0248	0.83	0.0327
0.04	0.0016	0.24	0.0094	0.44	0.0172	0.64	0.0252	0.84	0.0331
0.05	0.0020	0.25	0.0098	0.45	0.0177	0.65	0.0256	0.85	0.0335
0.06	0.0024	0.26	0.0102	0.46	0.0181	0.66	0.0260	0.86	0.0339
0.07	0.0028	0.27	0.0106	0.47	0.0185	0.67	0.0264	0.87	0.0343
0.08	0.0032	0.28	0.0110	0.48	0.0189	0.68	0.0268	0.88	0.0346
0.09	0.0035	0.29	0.0114	0.49	0.0193	0.69	0.0272	0.89	0.0350
0.10	0.0039	0.30	0.0118	0.50	0.0197	0.70	0.0276	0.90	0.0354
0.11	0.0043	0.31	0.0122	0.51	0.0201	0.71	0.0280	0.91	0.0358
0.12	0.0047	0.32	0.0126	0.52	0.0205	0.72	0.0283	0.92	0.0362
0.13	0.0051	0.33	0.0130	0.53	0.0209	0.73	0.0287	0.93	0.0366
0.14	0.0055	0.34	0.0134	0.54	0.0213	0.74	0.0291	0.94	0.0370
0.15	0.0059	0.35	0.0138	0.55	0.0217	0.75	0.0295	0.95	0.0374
0.16	0.0063	0.36	0.0142	0.56	0.0220	0.76	0.0299	0.96	0.0378
0.17	0.0067	0.37	0.0146	0.57	0.0224	0.77	0.0303	0.97	0.0382
0.18	0.0071	0.38	0.0150	0.58	0.0228	0.78	0.0307	0.98	0.0386
0.19	0.0075	0.39	0.0154	0.59	0.0232	0.79	0.0311	0.99	0.0390
0.20	0.0079	0.40	0.0157	0.60	0.0236	0.80	0.0315	1.00	0.0394

TABLE C.26.   INCHES INTO MILLIMETERS

In.	Mm	In.	Mm	In.	Mm
1/64	0.3969	51/64	20.2406	2-5/32	54.7688
1/32	0.7938	13/16	20.5375	2-3/16	55.5625
3/64	1.1906	53/64	21.0344	2-9/32	56.3563
1/16	1.5875	27/32	21.4313	2-1/4	57.1500
5/64	1.9844	55/64	21.8281	2-9/32	57.9438
3/32	2.3813	7/8	22.2250	2-5/16	58.7375
7/64	2.7781	57/64	22.6219	2-11/32	59.5313
1/8	3.1750	29/32	23.0188	2-3/8	60.3250
9/64	3.5719	59/64	23.4156	2-13/32	61.1188
5/32	3.9688	15/16	23.8125	2-7/16	61.9125
11/64	4.3656	61/64	24.2094	2-15/32	62.7063
3/16	4.7625	31/32	24.6063	2-1/2	63.5000
13/64	5.1594	63/64	25.0031	2-17/32	64.2938
7/32	5.5563	1	25.4000	2-9/16	65.0875
15/64	5.9531	1-1/32	26.1938	2-19/32	65.8813
1/4	6.3500	1-1/16	26.9875	2-5/8	66.6750
17/64	6.7469	1-3/32	27.7813	2-21/32	67.4688
9/32	7.1438	1-1/6	28.5750	2-11/16	68.2625
19/64	7.5406	1-5/32	29.3688	2-23/32	69.0563
5/16	7.9375	1-3/16	30.1625	2-3/4	69.8500
21/64	8.3344	1-7/32	30.9563	2-25/32	70.6438
11/32	8.7313	1-1/4	31.7500	2-13/16	71.4375
23/64	9.1281	1-9/32	32.5438	2-27/32	72.2313
3/8	9.5250	1-5/16	33.3375	2-7/8	73.0250
25/64	9.9219	1-11/32	34.1313	2-29/32	73.8188
13/32	10.3188	1-3/8	34.9250	2-15/16	74.6125
27/64	10.7156	1-13/32	35.7188	2-31/32	75.4063
7/16	11.1125	1-7/16	36.5125	3	76.2000
29/64	11.5094	1-15/32	37.3063	3-1/32	76.9938
15/32	11.9063	1-1/2	38.1000	3-1/16	77.7875
31/64	12.3031	1-17/32	38.8938	3-3/32	78.5813
1/2	12.7000	1-9/16	39.6875	3-1/8	79.3750
33/64	13.0969	1-19/32	40.4813	3-5/32	80.1688
17/32	13.4938	1-5/8	41.2750	3-3/16	80.9625
35/64	13.8906	1-21/32	42.0688	3-7/32	81.7563
9/16	14.2875	1-11/16	42.8625	3-1/4	82.5500
37/64	14.6844	1-23/32	43.6563	3-9/32	83.3438
19/32	15.0813	1-3/4	44.4500	3-5/16	84.1375
39/64	15.4781	1-25/32	45.2438	3-11/32	84.9313
5/8	15.8750	1-13/16	46.0375	3-3/8	85.7250
41/64	16.2719	1-27/32	46.8313	3-13/32	86.5188
21/32	16.6688	1-7/8	47.6250	3-7/16	87.3125
43/64	17.0656	1-29/32	48.4188	3-15/32	88.1063
11/16	17.4625	1-15/16	49.2125	3-1/2	88.9000
45/64	17.8594	1-31/32	50.0063	3-17/32	89.6938
23/32	18.2563	2	50.8000	3-9/16	90.4875
47/64	18.6531	2-1/32	51.5938	3-19/32	91.2813
3/4	19.0550	2-1/16	52.3875	3-5/8	92.0750
49/64	19.4469	2-3/32	53.1813	3-21/32	92.8688
25/32	19.8438	2-1/8	53.9750	3-11/16	93.6625

TABLE C.26.   *(Cont'd)*

In.	Mm	In.	Mm	In.	Mm
3-23/32	94.4563	5-9/32	134.144	7-11/16	195.262
3-3/4	95.2500	5-5/16	134.938	7-3/4	196.850
3-25/32	96.0438	5-11/32	135.731	7-13/16	198.438
3-13/16	96.8375	5-3/8	136.525	7-7/8	200.025
3-27/32	97.6313	5-13/32	137.319	7-15/16	201.612
3-7/8	98.4250	5-7/16	138.112	8	203.200
3-29/32	99.2188	5-15/32	138.906	8-1/16	204.783
3-15/16	100.012	5-1/2	139.700	8-1/8	206.375
3-31/32	100.806	5-17/32	140.494	8-3/16	207.962
4	101.600	5-9/16	141.288	8-1/4	209.550
4-1/32	102.394	5-19/32	142.081	8-5/16	211.138
4-1/16	103.188	5-5/8	142.875	8-3/8	212.725
4-3/32	103.981	5-21/32	143.669	8-7/16	214.312
4-1/8	104.775	5-11/16	144.462	8-1/2	215.900
4-5/32	105.569	5-23/32	145.256	8-9/16	217.488
4-3/16	106.362	5-3/4	146.050	8-6/8	219.075
4-7/32	107.156	5-25/32	146.844	8-11/16	220.662
4-1/4	107.950	5-13/16	147.638	8-3/4	222.250
4-9/32	108.744	5-27/32	148.431	8-13/16	223.838
4-5/16	109.538	5-7/8	149.225	8-7/8	225.425
4-11/32	110.331	5-29/32	150.019	8-15/16	227.012
4-3/8	111.125	5-15/16	150.812	9	228.600
4-13/32	111.919	5-31/32	151.606	9-1/16	230.188
4-7/16	112.712	6	152.400	9-1/8	231.775
4-15/32	113.506	6-1/16	153.988	9-3/16	233.362
4-1/2	114.300	6-1/8	155.575	9-1/4	234.950
4-17/32	115.094	6-3/16	157.162	9-5/16	236.538
4-9/16	115.888	6-1/4	158.750	9-3/8	238.125
4-19/32	116.681	6-5/16	160.338	9-7/16	239.712
4-5/8	117.475	6-3/8	161.925	9-1/2	241.300
4-21/32	118.269	6-7/16	163.512	9-9/16	242.888
4-11/16	119.062	6-1/2	165.100	9-5/8	244.475
4-23/32	119.856	6-9/16	166.688	9-11/16	246.062
4-3/4	120.650	6-5/8	168.275	9-3/4	247.650
4-25/32	121.444	6-11/16	169.862	9-13/16	249.238
4-13/18	122.238	6-3/4	171.450	9-7/8	250.825
4-27/32	123.031	6-13/16	173.038	9-15/16	252.412
4-7/8	123.825	6-7/8	174.625	10	254.000
4-29/32	124.619	6-15/16	176.212	10-1/16	255.588
4-15/16	125.412	7	177.800	10-1/8	257.175
4-31/32	126.206	7-1/16	179.388	10-3/16	258.762
5	127.000	7-1/8	180.975	10-1/4	260.350
5-1/32	127.794	7-3/16	182.562	10-5/16	261.938
5-1/16	128.588	7-1/4	184.150	10-3/8	263.525
5-3/32	129.381	7-5/16	185.738	10-7/16	265.112
5-1/8	130.175	7-3/8	187.325	10-1/2	266.700
5-5/32	130.969	7-7/16	188.912	10-9/16	268.288
5-3/16	131.762	7-1/2	190.500	10-5/8	269.875
5-7/32	132.556	7-9/16	192.088	10-11/16	271.462
5-1/4	133.350	7-5/8	193.675	10-3/4	273.050

TABLE C.26.   *(Cont'd)*

In.	Mm	In.	Mm	In.	Mm
10-13/16	274.638	13	330.200	28	711.200
10-7/8	276.225	14	355.600	29	736.600
10-15/16	277.812	15	381.000	30	762.000
11	279.400	16	406.400	31	787.400
11-1/16	280.988	17	431.800	32	812.800
11-1/8	282.575	18	457.200	33	838.200
11-3/16	284.162	19	482.600	34	863.600
11-1/4	285.750	20	508.000	35	889.000
11-5/16	287.338	21	533.400	36	914.400
11-3/8	288.925	22	558.800	37	939.800
11-7/16	290.512	23	584.200	38	965.200
11-1/2	292.100	24	609.600	39	990.600
11-9/16	293.688	25	635.000	40	1016.00
11-5/8	295.275	26	660.400	41	1041.40
11-11/16	296.862	27	685.800	42	1066.80
11-3/4	298.450				
11-13/16	300.038				
11-7/8	301.625				
11-15/16	303.212				
12	304.800				

TABLE C.27.   MILLIMETERS INTO INCHES

Mm	In.	Mm	In.	Mm	In.	Mm	In.
1	0.0394	51	2.0079	101	3.9764	151	5.9449
2	0.0787	52	2.0472	102	4.0158	152	5.9843
3	0.1181	53	2.0866	103	4.0551	153	6.0236
4	0.1575	54	2.1260	104	4.0945	154	6.0630
5	0.1969	55	2.1654	105	4.1339	155	6.1024
6	0.2362	56	2.2047	106	4.1732	156	6.1417
7	0.2756	57	2.2441	107	4.2126	157	6.1811
8	0.3150	58	2.2835	108	4.2520	158	6.2205
9	0.3543	59	2.3228	109	4.2913	159	6.2599
10	0.3937	60	2.3622	110	4.3307	160	6.2992
11	0.4331	61	2.4016	111	4.3701	161	6.3386
12	0.4724	62	2.4409	112	4.4095	162	6.3780
13	0.5118	63	2.4803	113	4.4488	163	6.4173
14	0.5512	64	2.5197	114	4.4882	164	6.4567
15	0.5906	65	2.5591	115	4.5276	165	6.4961
16	0.6299	66	2.5984	116	4.5669	166	6.5354
17	0.6693	67	2.6378	117	4.6063	167	6.5748
18	0.7087	68	2.6772	118	4.6457	168	6.6142
19	0.7480	69	2.7165	119	4.6850	169	6.6535
20	0.7874	70	2.7559	120	4.7244	170	6.6929
21	0.8268	71	2.7953	121	4.7638	171	6.7323
22	0.8661	72	2.8347	122	4.8032	172	6.7717
23	0.9055	73	2.8740	123	4.8425	173	6.8110
24	0.9449	74	2.9134	124	4.8819	174	6.8504
25	0.9843	75	2.9528	125	4.9213	175	6.8898
26	1.0236	76	2.9921	126	4.9606	176	6.9291
27	1.0630	77	3.0315	127	5.0000	177	6.9685
28	1.1024	78	3.0709	128	5.0394	178	7.0079
29	1.1417	79	3.1102	129	5.0787	179	7.0472
30	1.1811	80	3.1496	130	5.1181	180	7.0866
31	1.2205	81	3.1890	131	5.1575	181	7.1260
32	1.2598	82	3.2284	132	5.1969	182	7.1654
33	1.2992	83	3.2677	133	5.2362	183	7.2047
34	1.3386	84	3.3071	134	5.2756	184	7.2441
35	1.3780	85	3.3465	135	5.3150	185	7.2835
36	1.4173	86	3.3858	136	5.3543	186	7.3228
37	1.4567	87	3.4252	137	5.3937	187	7.3622
38	1.4961	88	3.4646	138	5.4331	188	7.4016
39	1.5354	89	3.5039	139	5.4724	189	7.4409
40	1.5748	90	3.5433	140	5.5118	190	7.4803
41	1.6142	91	3.5827	141	5.5512	191	7.5197
42	1.6535	92	3.6221	142	5.5906	192	7.5591
43	1.6929	93	3.6614	143	5.6299	193	7.5984
44	1.7323	94	3.7008	144	5.6693	194	7.6378
45	1.7717	95	3.7402	145	5.7087	195	7.6772
46	1.8110	96	3.7795	146	5.7480	196	7.7165
47	1.8504	97	3.8189	147	5.7874	197	7.7559
48	1.8898	98	3.8583	148	5.8268	198	7.7953
49	1.9291	99	3.8976	149	5.8661	199	7.8347
50	1.9685	100	3.9370	150	5.9055	200	7.8740

TABLE C.27.   *(Cont'd)*

Mm	In.	Mm	In.	Mm	In.	Mm	In.
201	7.9134	251	9.8819	301	11.8504	351	13.8189
202	7.9528	252	9.9213	302	11.8898	352	13.8583
203	7.9921	253	9.9606	303	11.9291	353	13.8976
204	8.0315	254	10.0000	304	11.9686	354	13.9370
205	8.0709	255	10.0393	305	12.0079	355	13.9764
206	8.1102	256	10.0787	306	12.0472	356	14.0157
207	8.1496	257	10.1181	307	12.0866	357	14.0551
208	8.1890	258	10.1575	308	12.1260	358	14.0945
209	8.2284	259	10.1969	309	12.1654	359	14.1339
210	8.2677	260	10.2362	310	12.2047	360	14.1732
211	8.3071	261	10.2756	311	12.2441	361	14.2126
212	8.3465	262	10.3150	312	12.2835	362	14.2520
213	8.3858	263	10.3543	313	12.3228	363	14.2913
214	8.4252	264	10.3937	314	12.3622	364	14.3307
215	8.4646	265	10.4331	315	12.4016	365	14.3701
216	8.5039	266	10.4724	316	12.4409	366	14.4094
217	8.5433	267	10.5118	317	12.4803	367	14.4488
218	8.5827	268	10.5512	318	12.5197	368	14.4882
219	8.6221	269	10.5906	319	12.5591	369	14.5276
220	8.6614	270	10.6299	320	12.5984	370	14.5669
221	8.7008	271	10.6693	321	12.6378	371	14.6063
222	8.7402	272	10.7087	322	12.6772	372	14.6457
223	8.7795	273	10.7480	323	12.7165	373	14.6850
224	8.8189	274	10.7874	324	12.7559	374	14.7244
225	8.8583	275	10.8268	325	12.7953	375	14.7638
226	8.8976	276	10.8661	326	12.8346	376	14.8031
227	8.9370	277	10.9055	327	12.8740	377	14.8425
228	8.9764	278	10.9449	328	12.9134	378	14.8819
229	9.0158	279	10.9843	329	12.9528	379	14.9213
230	9.0551	280	11.0236	330	12.9921	380	14.9606
231	9.0945	281	11.0630	331	13.0315	381	15.0000
232	9.1339	282	11.1024	332	13.0709	382	15.0394
233	9.1732	283	11.1417	333	13.1102	383	15.0787
234	9.2126	284	11.1811	334	13.1496	384	15.1181
235	9.2520	285	11.2205	335	13.1890	385	15.1575
236	9.2913	286	11.2598	336	13.2283	386	15.1969
237	9.3307	287	11.2992	337	13.2677	387	15.2362
238	9.3701	288	11.3386	338	13.3071	388	15.2756
239	9.4095	289	11.3780	339	13.3465	389	15.3150
240	9.4488	290	11.4173	340	13.3858	390	15.3543
241	9.4882	291	11.4567	341	13.4252	391	15.3937
242	9.5276	292	11.4961	342	13.4646	392	15.4331
243	9.5669	293	11.5354	343	13.5039	393	15.4724
244	9.6063	294	11.5748	344	13.5433	394	15.5118
245	9.6457	295	11.6142	345	13.5827	395	15.5512
246	9.6850	296	11.6535	346	13.6220	396	15.5906
247	9.7244	297	11.6929	347	13.6614	397	15.6299
248	9.7638	298	11.7323	348	13.7008	398	15.6693
249	9.8031	299	11.7717	349	13.7402	399	15.7087
250	9.8425	300	11.8110	350	13.7795	400	15.7480

TABLE C.27. *(Cont'd)*

Mm	In.	Mm	In.	Mm	In.	Mm	In.
401	15.7874	451	17.7559	501	19.7244	551	21.6929
402	15.8268	452	17.7963	502	19.7638	552	21.7323
403	15.8661	453	17.8346	503	19.8031	553	21.7717
404	15.9055	454	17.8740	504	19.8425	554	21.8110
405	15.9449	455	17.9134	505	19.8819	555	21.8504
406	15.9832	456	17.9528	506	19.9213	556	21.8898
407	16.0236	457	17.9921	507	19.9606	557	21.9291
408	16.0630	458	18.0315	508	20.0000	558	21.9685
409	16.1024	459	18.0709	509	20.0394	559	22.0079
410	16.1417	460	18.1102	510	20.0787	560	22.0472
411	16.1811	461	18.1496	511	20.1181	561	22.0866
412	16.2205	462	18.1890	512	20.1575	562	22.1260
413	16.2598	463	18.2283	513	20.1969	563	22.1654
414	16.2992	464	18.2677	514	20.2362	564	22.2047
415	16.3386	465	18.3071	515	20.2756	565	22.2441
416	16.3780	466	18.3465	516	20.3150	566	22.2835
417	16.4173	467	18.3358	517	20.3543	567	22.3228
418	16.4567	468	18.4252	518	20.3937	568	22.3622
419	16.4961	469	18.4646	519	20.4331	569	22.4016
420	16.5354	470	18.5039	520	20.4724	570	22.4409
421	16.5748	471	18.5433	521	20.5118	571	22.4803
422	16.6142	472	18.5827	522	20.5512	572	22.5197
423	16.6535	473	18.6220	523	20.5906	573	22.5591
424	16.6929	474	18.6614	524	20.6299	574	22.5984
425	16.7323	475	18.7008	525	20.6693	575	22.6378
426	16.7716	476	18.7402	526	20.7087	576	22.6772
427	16.8110	477	18.7795	527	20.7480	577	22.7165
428	16.8504	478	18.8189	528	20.7874	578	22.7559
429	16.8898	479	18.8583	529	20.8268	579	22.7953
430	16.9291	480	18.8976	530	20.8661	580	22.8346
431	16.9685	481	18.9370	531	20.9055	581	22.8740
432	17.0079	482	18.9764	532	20.9449	582	22.9134
433	17.0472	483	19.0157	533	20.9843	583	22.9528
434	17.0866	484	19.0551	534	21.0236	584	22.9921
435	17.1260	485	19.0945	535	21.0630	585	23.0315
436	17.1654	486	19.1339	536	21.1024	586	23.0709
437	17.2047	487	19.1732	537	21.1417	587	23.1102
438	17.2441	488	19.2126	538	21.1811	588	23.1496
439	17.2835	489	19.2520	539	21.2205	589	23.1890
440	17.3228	490	19.2813	540	21.2598	590	23.2283
441	17.3622	491	19.3307	541	21.2992	591	23.2677
442	17.4016	492	19.3701	542	21.3386	592	23.3071
443	17.4409	493	19.4094	543	21.3780	593	23.3465
444	17.4803	494	19.4488	544	21.4173	594	23.3858
445	17.5197	495	19.4882	545	21.4567	595	23.4252
446	17.5591	496	19.5276	546	21.4961	596	23.4646
447	17.5984	497	19.5669	547	21.5354	597	23.5039
448	17.6378	498	19.6063	548	21.5743	598	23.5433
449	17.6772	499	19.6457	549	21.6142	599	23.5827
450	17.7165	500	19.6850	550	21.6535	600	23.6220

TABLE C.27.   *(Cont'd)*

Mm	In.	Mm	In.	Mm	In.	Mm	In.
601	23.6614	651	25.6299	701	27.5984	751	29.5669
602	23.7008	652	25.6693	702	27.6378	752	29.6063
603	23.7402	653	25.7087	703	27.6772	753	29.6457
604	23.7795	654	25.7480	704	27.7165	754	29.6850
605	23.8189	655	25.7874	705	27.7559	755	29.7244
606	23.8583	656	25.8268	706	27.7953	756	29.7638
607	23.8976	657	25.8661	707	27.8346	757	29.8031
608	23.9370	658	25.9055	708	27.8740	758	29.8425
609	23.9764	659	25.9449	709	27.9134	759	29.8819
610	24.0157	660	25.9843	710	27.9528	760	29.9213
611	24.0551	661	26.0236	711	27.9921	761	29.9606
612	24.0945	662	26.0630	712	28.0315	762	30.0000
613	24.1339	663	26.1024	713	28.0709	763	30.0394
614	24.1732	664	26.1417	714	28.1102	764	30.0787
615	24.2126	665	26.1811	715	28.1496	765	30.1181
616	24.2520	666	26.2205	716	28.1890	766	30.1575
617	24.2913	667	26.2598	717	28.2283	767	30.1969
618	24.3307	668	26.2992	718	28.2677	768	30.2362
619	24.3701	669	26.3386	719	28.3071	769	30.2756
620	24.4094	670	26.3780	720	28.3465	770	30.3150
621	24.4488	671	26.4173	721	28.3858	771	30.3543
622	24.4882	672	26.4567	722	28.4252	772	30.3937
623	24.5276	673	26.4961	723	28.4646	773	30.4331
624	24.5669	674	26.5354	724	28.5039	774	30.4724
625	24.6063	675	26.5748	725	28.5433	775	30.5118
626	24.6457	676	26.6142	726	28.5827	776	30.5512
627	24.6850	677	26.6535	727	28.6220	777	30.5906
628	24.7244	678	26.6929	728	28.6614	778	30.6299
629	24.7638	679	26.7323	729	28.7008	779	30.6693
630	24.8031	680	26.7717	730	28.7042	780	30.7087
631	24.8425	681	26.8110	731	28.7795	781	30.7480
632	24.8819	682	26.8504	732	28.8189	782	30.7874
633	24.9213	683	26.8898	733	28.8583	783	30.8268
634	24.9606	684	26.9291	734	28.8976	784	30.8661
635	25.0000	685	26.9685	735	28.9370	785	30.9055
636	25.0394	686	27.0079	736	28.9764	786	30.9449
637	25.0787	687	27.0472	737	29.0157	787	30.9843
638	25.1181	688	27.0866	738	29.0551	788	31.0236
639	25.1575	689	27.1260	739	29.0945	789	31.0630
640	25.1969	690	27.1654	740	29.1339	790	31.1024
641	25.2362	691	27.2047	741	29.1732	791	31.1417
642	25.2756	692	27.2441	742	29.2126	792	31.1811
643	25.3150	693	27.2835	743	29.2520	793	31.2205
644	25.3543	694	27.3228	744	29.2913	794	31.2598
645	25.3937	695	27.3622	745	29.3307	795	31.2992
646	25.4331	696	27.4016	746	29.3701	796	31.3386
647	25.4724	697	27.4409	747	29.4094	797	31.3780
648	25.5118	698	27.4803	748	29.4488	798	31.4173
649	25.5512	699	27.5197	749	29.4882	799	31.4567
650	25.5906	700	27.5591	750	29.5276	800	31.4961

TABLE C.27.   *(Cont'd)*

Mm	In.	Mm	In.	Mm	In.	Mm	In.
801	31.5354	851	33.5039	901	35.4724	951	37.4409
802	31.5748	852	33.5433	902	35.5118	952	37.4803
803	31.6142	853	33.5827	903	35.5512	953	37.5197
804	31.6535	854	33.6220	904	35.5906	954	37.5591
805	31.6929	855	33.6614	905	35.6299	955	37.5984
806	31.7323	856	33.7008	906	35.6693	956	37.6378
807	31.7717	857	33.7402	907	35.7087	957	37.6772
808	31.8110	858	33.7795	908	35.7480	958	37.7165
809	31.8504	859	33.8189	909	35.7874	959	37.7559
810	31.8898	860	33.8583	910	35.8268	960	37.7953
811	31.9291	861	33.8976	911	35.8661	961	37.8346
812	31.9685	862	33.9370	912	35.9055	962	37.8740
813	32.0079	863	33.9764	913	35.9449	963	37.9134
814	32.0472	864	34.0157	914	35.9843	964	37.9528
815	32.0866	865	34.0551	915	36.0236	965	37.9921
816	32.1260	866	34.0945	916	36.0630	966	38.0315
817	32.1654	867	34.1339	917	36.1024	967	38.0709
818	32.2047	868	34.1732	918	36.1417	968	38.1102
819	32.2441	869	34.2126	919	36.1811	969	38.1496
820	32.2835	870	34.2520	920	36.2205	970	38.1890
821	32.3228	871	34.2913	921	36.2598	971	38.2283
822	32.3622	872	34.3307	922	36.2992	972	38.2677
823	32.4016	873	34.3701	923	36.3386	973	38.3071
824	32.4409	874	34.4094	924	36.3780	974	38.3465
825	32.4803	875	34.4488	925	36.4173	975	38.3858
826	32.5197	876	34.4882	926	36.4567	976	38.4252
827	32.5591	877	34.5276	927	36.4961	977	38.4646
828	32.5984	878	34.5670	928	36.5354	978	38.5039
829	32.6378	879	34.6063	929	36.5748	979	38.5433
830	32.6772	880	34.6457	930	36.6142	980	38.5827
831	32.7165	881	34.6850	931	36.6535	981	38.6220
832	32.7559	882	34.7244	932	36.6929	982	38.6614
833	32.7953	883	34.7638	933	36.7323	983	38.7008
834	32.8346	884	34.8031	934	36.7717	984	38.7402
835	32.8740	885	34.8425	935	36.8110	985	38.7795
836	32.9134	886	34.8819	936	36.8504	986	38.8189
837	32.9528	887	34.9213	937	36.8898	987	38.8583
838	32.9921	888	34.9606	938	36.9291	988	38.8976
839	33.0315	889	35.0000	939	36.9685	989	38.9370
840	33.0709	890	35.0394	940	37.0079	990	38.9764
841	33.1102	891	35.0787	941	37.0472	991	39.0157
842	33.1496	892	35.1181	942	37.0866	992	39.0551
843	33.1890	893	35.1575	943	37.1260	993	39.0945
844	33.2283	894	35.1969	944	37.1654	994	39.1339
845	33.2677	895	35.2362	945	37.2047	995	39.1732
846	33.3071	896	35.2756	946	37.2441	996	39.2126
847	33.3465	897	35.3150	947	37.2835	997	39.2520
848	33.3858	898	35.3543	948	37.3228	998	39.2913
849	33.4252	899	35.3937	949	37.3622	999	39.3307
850	33.4646	900	35.4331	950	37.4016	1000	39.3701

TABLE C.28.   DECIMALS OF AN INCH INTO MILLIMETERS

In.	Mm	In.	Mm	In.	Mm	In.	Mm
0.001	0.025	0.200	5.08	0.470	11.94	0.740	18.80
0.002	0.051	0.210	5.33	0.480	12.19	0.750	19.05
0.003	0.076	0.220	5.59	0.490	12.45	0.760	19.30
0.004	0.102	0.230	5.84	0.500	12.70	0.770	19.56
0.005	0.127	0.240	6.10	0.510	12.95	0.780	19.81
0.006	0.152	0.250	6.35	0.520	13.21	0.790	20.07
0.007	0.178	0.260	6.60	0.530	13.46	0.800	20.32
0.008	0.203	0.270	6.86	0.540	13.72	0.810	20.57
0.009	0.229	0.280	7.11	0.550	13.97	0.820	20.83
0.010	0.254	0.290	7.37	0.560	14.22	0.830	21.08
0.020	0.508	0.300	7.62	0.570	14.48	0.840	21.34
0.030	0.762	0.310	7.87	0.580	14.73	0.850	21.59
0.040	1.016	0.320	8.13	0.590	14.99	0.860	21.84
0.050	1.270	0.330	8.38	0.600	15.24	0.870	22.10
0.060	1.524	0.340	8.65	0.610	15.49	0.880	22.35
0.070	1.778	0.350	8.89	0.620	15.75	0.890	22.61
0.080	2.032	0.360	9.14	0.630	16.00	0.900	22.86
0.090	2.286	0.370	9.40	0.640	16.26	0.910	23.11
0.100	2.540	0.380	9.65	0.650	16.51	0.920	23.37
0.110	2.794	0.390	9.91	0.660	16.76	0.930	23.62
0.120	3.048	0.400	10.16	0.670	17.02	0.940	23.88
0.130	3.302	0.410	10.41	0.680	17.27	0.950	24.13
0.140	3.56	0.420	10.67	0.690	17.53	0.960	24.38
0.150	3.81	0.430	10.92	0.700	17.78	0.970	24.64
0.160	4.06	0.440	11.18	0.710	18.03	0.980	24.89
0.170	4.32	0.450	11.43	0.720	18.29	0.990	25.15
0.180	4.57	0.460	11.68	0.730	18.54	1.000	25.40
0.190	4.83						

TABLE C.29.   INCHES INTO CENTIMETERS

In.	0	1	2	3	4	5	6	7	8	9
	Cm	Cm	Cm	Cm	Cm	Cm	Cm	Cm	Cm	Cm
0	—	2.54	5.08	7.62	10.16	12.70	15.24	17.78	20.32	22.86
10	25.40	27.94	30.48	33.02	35.56	38.10	40.64	43.18	45.72	48.26
20	50.80	53.34	55.88	58.42	60.96	63.50	66.04	68.58	71.12	73.66
30	76.20	78.74	81.28	83.82	86.36	88.90	91.44	93.98	96.52	99.06
40	101.60	104.14	106.68	109.22	111.76	114.30	116.84	119.38	121.92	124.46
50	127.00	129.54	132.08	134.62	137.16	139.70	142.24	144.78	147.32	149.86
60	152.40	154.94	157.48	160.02	162.56	165.10	167.64	170.18	172.72	175.26
70	177.80	180.34	182.88	185.42	187.96	190.50	193.04	195.58	198.12	200.66
80	203.20	205.74	208.28	210.82	213.36	215.90	218.44	220.98	223.52	226.06
90	228.60	231.14	233.68	236.22	238.76	241.30	243.84	246.38	248.92	251.46
100	254.00	256.54	259.08	261.62	264.16	266.70	269.24	271.78	274.32	276.86

TABLE C.30.   CENTIMETERS INTO INCHES

Cm	0	1	2	3	4	5	6	7	8	9
	In.	In.	In.	In.	In.	In.	In.	In.	In.	In.
0	—	0.394	0.787	1.181	1.575	1.969	2.362	2.756	3.150	3.543
10	3.937	4.331	4.724	5.118	5.512	5.906	6.299	6.693	7.087	7.480
20	7.874	8.268	8.661	9.055	9.449	9.843	10.236	10.630	11.024	11.417
30	11.811	12.205	12.598	12.992	13.386	13.780	14.173	14.567	14.961	15.354
40	15.748	16.142	16.535	16.929	17.323	17.717	18.110	18.504	18.898	19.291
50	19.685	20.079	20.472	20.866	21.260	21.654	22.047	22.441	22.835	23.228
60	23.622	24.016	24.409	24.803	25.197	25.591	25.984	26.378	26.772	27.164
70	27.559	27.953	28.346	28.740	29.134	29.528	29.921	30.315	30.709	31.102
80	31.496	31.890	32.283	32.677	33.071	33.465	33.858	34.252	34.646	35.039
90	35.433	35.827	36.220	36.614	37.008	37.402	37.795	38.189	38.583	38.976
100	39.370	39.764	40.157	40.551	40.945	41.339	41.732	42.126	42.520	42.913

TABLE C.31. FEET INTO METERS

Ft	0	1	2	3	4	5	6	7	8	9
	M	M	M	M	M	M	M	M	M	M
0	—	0.305	0.610	0.914	1.219	1.524	1.829	2.134	2.438	2.743
10	3.048	3.353	3.658	3.962	4.267	4.572	4.877	5.182	5.486	5.791
20	6.096	6.401	6.706	7.010	7.315	7.620	7.925	8.230	8.534	8.839
30	9.144	9.449	9.754	10.058	10.363	10.668	10.973	11.278	11.582	11.887
40	12.192	12.497	12.802	13.106	13.411	13.716	14.021	14.326	14.630	14.935
50	15.240	15.545	15.850	16.154	16.459	16.764	17.069	17.374	17.678	17.983
60	18.288	18.593	18.898	19.202	19.507	19.812	20.177	20.422	20.726	21.031
70	21.336	21.641	21.946	22.250	22.555	22.860	23.165	23.470	23.774	24.079
80	24.384	24.689	24.994	25.298	25.603	25.908	26.213	26.518	26.822	27.127
90	27.432	27.737	28.042	28.346	28.651	28.956	29.261	29.566	29.870	30.175
100	30.480	30.785	31.090	31.394	31.699	32.004	32.309	32.614	32.918	33.223

TABLE C.32. METERS INTO FEET

M	0	1	2	3	4	5	6	7	8	9
	Ft	Ft	Ft	Ft	Ft	Ft	Ft	Ft	Ft	Ft
0	—	3.281	6.562	9.842	13.123	16.404	19.685	22.966	26.247	29.528
10	32.808	36.089	39.370	42.661	45.932	49.212	52.493	55.774	59.055	62.336
20	65.617	68.897	72.178	75.459	78.740	82.021	85.302	88.582	91.863	95.144
30	98.425	101.71	104.99	108.27	111.55	114.83	118.11	121.39	124.67	127.95
40	131.23	134.51	137.79	141.08	144.36	147.64	150.92	154.20	157.48	160.76
50	164.04	167.32	170.60	173.88	177.16	180.45	183.73	187.01	190.29	193.57
60	196.85	200.13	203.41	206.69	209.97	213.25	216.53	219.82	223.10	226.38
70	229.66	232.94	236.22	239.50	242.78	246.06	249.34	252.62	255.90	259.19
80	262.47	265.75	269.03	272.31	275.59	278.87	282.15	285.43	288.71	291.99
90	295.27	298.56	301.84	305.12	308.40	311.68	314.96	318.24	321.52	324.80
100	328.08	331.36	334.64	337.93	341.21	344.49	347.77	351.05	354.33	357.61

TABLE C.33. MILES INTO KILOMETERS

Miles	0	1	2	3	4	5	6	7	8	9
	Km	Km	Km	Km	Km	Km	Km	Km	Km	Km
0	—	1.609	3.219	4.828	6.437	8.047	9.656	11.265	12.875	14.484
10	16.093	17.703	19.312	20.922	22.531	24.140	25.750	27.359	28.968	30.578
20	32.187	33.796	35.406	37.015	38.624	40.234	41.843	43.452	45.062	46.671
30	48.280	49.890	51.499	53.108	54.718	56.327	57.936	59.546	61.155	62.764
40	64.374	65.983	67.592	69.202	70.811	72.421	74.030	75.639	77.249	78.858
50	80.467	82.077	83.686	85.295	86.905	88.514	90.123	91.733	93.342	94.951
60	96.561	98.170	99.779	101.39	103.00	104.61	106.22	107.83	109.44	111.05
70	112.65	114.26	115.87	117.48	119.09	120.70	122.31	123.92	125.53	127.14
80	128.75	130.36	131.97	133.58	135.19	136.79	138.40	140.01	141.62	143.23
90	144.84	146.45	148.06	149.67	151.28	152.89	154.50	156.11	157.72	159.33
100	160.93	162.54	164.15	165.76	167.37	168.98	170.59	172.20	173.81	175.42

TABLE C.34. KILOMETERS INTO MILES

Km	0	1	2	3	4	5	6	7	8	9
	Miles	Miles	Miles	Miles	Miles	Miles	Miles	Miles	Miles	Miles
0	—	0.621	1.243	1.864	2.486	3.107	3.728	4.350	4.971	5.592
10	6.214	6.835	7.457	8.078	8.699	9.321	9.942	10.563	11.185	11.806
20	12.427	13.049	13.670	14.292	14.913	15.534	16.156	16.777	17.398	18.020
30	18.641	19.263	19.884	20.505	21.127	21.748	22.369	22.991	23.612	24.234
40	24.855	25.476	26.098	26.719	27.340	27.962	28.583	29.204	29.826	30.447
50	31.069	31.690	32.311	32.933	33.554	34.175	34.797	35.418	36.040	36.661
60	37.282	37.904	38.525	39.146	39.768	40.389	41.011	41.632	42.253	42.875
70	43.496	44.117	44.739	45.370	45.982	46.603	47.224	47.846	48.467	49.088
80	49.710	50.331	50.952	51.574	52.195	52.817	53.438	54.059	54.681	55.302
90	55.923	56.545	57.166	57.788	58.409	59.030	59.652	60.273	60.894	61.516
100	62.137	62.759	63.380	64.001	64.623	65.244	65.865	66.487	67.108	67.730

UNITS OF POWER AND PRESSURE

TABLE C.35.   UNITS OF POWER

1 watt	0.73756 foot pound per second
1 foot pound per second	1.35582 watts
1 watt	0.056884 BTU per minute
1 BTU per minute	17.580 watts
1 watt	0.001341 U.S. horsepower
1 U.S. horsepower	745.7 watts
1 watt	0.01433 kilogram-calorie per minute
1 kilogram-calorie per minute	69.767 watts
1 watt	1×10^7 ergs per second
1 lumen	0.001496 watt

TABLE C.36.   POUNDS PER SQUARE INCH INTO KILOGRAMS PER SQUARE CENTIMETER

Lb per Sq. In.	0	1	2	3	4	5	6	7	8	9
					Kg per Sq. Cm					
0	—	0.0703	0.1406	0.2109	0.2812	0.3515	0.4218	0.4922	0.5625	0.6328
10	0.7031	0.7734	0.8437	0.9140	0.9843	1.0546	1.1249	1.1952	1.2655	1.3358
20	1.4061	1.4765	1.5468	1.6171	1.6874	1.7577	1.8280	1.8983	1.9686	2.0389
30	2.1092	2.1795	2.2498	2.3201	2.3904	2.4607	2.5311	2.6014	2.6717	2.7420
40	2.8123	2.8826	2.9529	3.0232	3.0935	3.1638	3.2341	3.3044	3.3747	3.4450
50	3.5154	3.5857	3.6560	3.7263	3.7966	3.8669	3.9372	4.0075	4.0778	4.1481
60	4.2184	4.2887	4.3590	4.4293	4.4997	4.5700	4.6403	4.7106	4.7809	4.8512
70	4.9215	4.9918	5.0621	5.1324	5.2027	5.2730	5.3433	5.4136	5.4839	5.5543
80	5.6246	5.6949	5.7652	5.8355	5.9058	5.9761	6.0464	6.1167	6.1870	6.2573
90	6.3276	6.3980	6.4682	6.5386	6.6089	6.6792	6.7495	6.8198	6.8901	6.9604
100	7.0307	7.1010	7.1713	7.2416	7.3120	7.3822	7.4525	7.5228	7.5932	7.6635

TABLE C.37.   KILOGRAMS PER SQUARE CENTIMETER INTO POUNDS PER SQUARE INCH

Kg per Sq. Cm	0	1	2	3	4	5	6	7	8	9
					Lb per Sq. In.					
0	—	14.22	28.45	42.67	56.89	71.12	85.34	99.56	113.79	128.01
10	142.23	156.46	170.68	184.90	199.13	213.35	227.57	241.80	256.02	270.24
20	284.47	298.69	312.91	327.14	341.36	355.58	369.81	384.03	398.25	412.48
30	426.70	440.92	455.15	469.37	483.59	497.82	512.04	526.26	540.49	554.71
40	568.93	583.16	597.38	611.60	625.83	640.05	654.27	668.50	682.72	696.94
50	711.17	725.39	739.61	753.84	768.06	782.28	796.51	810.73	824.95	839.18
60	853.40	867.62	881.85	896.07	910.29	924.52	938.74	952.96	967.19	981.41
70	995.63	1009.9	1024.1	1038.3	1052.5	1066.8	1081.0	1095.2	1109.4	1123.6
80	1137.9	1152.1	1166.3	1180.5	1194.8	1209.0	1223.2	1237.4	1251.7	1265.9
90	1280.1	1294.3	1308.6	1322.8	1337.0	1351.2	1365.4	1379.7	1393.9	1408.1
100	1422.3	1436.6	1450.8	1465.0	1479.2	1493.4	1507.7	1521.9	1536.1	1550.3

TABLE C.38. POUNDS PER SQUARE FOOT INTO KILOGRAMS PER SQUARE METER

Lb per Sq. Ft	0	1	2	3	4	5	6	7	8	9
					Kg per Sq. M					
0	....	4.88	9.77	14.65	19.53	24.41	29.30	34.18	39.06	43.94
10	48.82	53.70	58.59	63.47	68.35	73.23	78.12	83.00	87.88	92.76
20	97.65	102.53	107.42	112.30	117.18	122.06	126.95	131.83	136.71	141.59
30	146.47	151.35	156.24	161.12	166.00	170.88	175.77	180.65	185.53	190.41
40	195.30	200.18	205.07	209.95	214.83	219.71	224.60	229.48	234.36	239.24
50	244.12	249.00	253.89	258.77	263.65	268.53	273.42	278.30	283.18	288.06
60	292.95	297.83	302.72	307.60	312.48	317.36	322.25	327.13	332.01	336.89
70	341.77	346.65	351.54	356.42	361.30	366.18	371.07	375.95	380.83	385.71
80	390.59	395.47	400.36	405.24	410.12	415.00	419.89	424.77	429.65	434.53
90	439.43	444.30	449.19	454.07	458.95	463.83	468.72	473.60	478.48	483.36
100	488.24	493.12	498.01	502.89	507.77	512.65	517.54	522.42	527.30	532.18

TABLE C.39. KILOGRAMS PER SQUARE METER INTO POUNDS PER SQUARE FOOT

Kg per Sq. M	0	1	2	3	4	5	6	7	8	9
					Lb per Sq.Ft					
0	....	0.2048	0.4096	0.6144	0.8193	1.0241	1.2289	1.4337	1.6385	1.8433
10	2.0481	2.2530	2.4578	2.6626	2.8674	3.0722	3.2771	3.4819	3.6867	3.8915
20	4.0963	4.3011	4.5060	4.7108	4.9156	5.1204	5.3252	5.5300	5.7349	5.9397
30	6.1445	6.3493	6.5541	6.7589	6.9638	7.1686	7.3734	7.5782	7.7830	7.9878
40	8.1927	8.3975	8.6023	8.8071	9.0119	9.2167	9.4215	9.6262	9.8310	10.036
50	10.241	10.446	10.650	10.855	11.060	11.265	11.470	11.675	11.879	12.084
60	12.289	12.494	12.698	12.903	13.108	13.313	13.518	13.723	13.927	14.132
70	14.337	14.542	14.747	14.952	15.156	15.361	15.566	15.771	15.976	16.181
80	16.385	16.590	16.795	17.000	17.205	17.409	17.614	17.819	18.024	18.229
90	18.434	18.638	18.843	19.048	19.253	19.458	19.662	19.867	20.072	20.277
100	20.482	20.686	20.891	21.096	21.301	21.506	21.711	21.915	22.120	22.325

TABLE C.40.   WATER HEADS AND EQUIVALENT PRESSURES

Head in Feet	Pressure in Lb Sq. In.	Pressure in Lb Sq. Ft
	(weight of water at 62.4 lb per cubic ft)	
5	2.17	312
10	4.33	624
15	6.50	936
20	8.66	1,248
30	12.99	1,872
40	17.32	2,496
50	21.65	3,120
60	25.98	3,744
70	30.31	4,368
80	34.64	4,992
90	38.97	5,616
100	43.30	6,240
125	54.13	7,800
150	64.95	9,360
175	75.78	10,920
200	86.60	12,480

TABLE C.41.   INCHES VACUUM INTO FEET SUCTION

In. Vacuum	Ft Suction
.5	.56
1.0	1.13
1.5	1.70
2.0	2.27
2.5	2.84
3.0	3.41
4.0	4.54
5.0	5.67
6.0	6.80
7.0	7.94
8.0	9.07
9.0	10.21
10.0	11.34
15.0	17.01
20.0	22.68
25.0	28.35
30.0	34.02

TABLE C.42.   FRICTION OF WATER IN PIPES

(approximate loss of heat in ft due to friction per 100 ft of pipeline)				
U.S. Gal. per Min.	1 In. I.D.	1½ In. I.D.	2 In. I.D.	4 In. I.D.
10	11.7	1.4	0.5	
20	42.0	5.2	1.8	
30	89.0	11.0	3.8	
40	152.0	18.8	6.6	0.2
50		28.4	9.9	0.3
70		53.0	18.4	0.6
100		102.0	35.8	1.2
125			54.0	1.9
150			76.0	2.6
175			102.0	3.4
200			129.0	4.4
250				6.7

TABLE C.43.   FRICTION IN FITTINGS

	(reduced to equivalent feet of pipeline loss)			
Type of Fitting	1 In. I.D.	1½ In. I.D.	2 In. I.D.	4 In. I.D.
90° Elbow	2.8	4.3	5.5	11.0
45° Elbow	1.3	2.0	2.6	5.0
Tee Side Outlet	5.6	9.1	12.0	22.0
Close Return Bend	6.3	10.2	13.0	24.0
Gate Valve	.6	.9	1.2	2.3
Check Valve	10.5	15.8	21.1	42.3
Ball Valve (ported)	27.0	43.0	55.0	115.0

Examples:

A 2-in. inside diameter pipeline loses 29.7 ft of head pumping at 50 gal. per min for 300 ft of straight length. (9.9 per hundred ft × 3 hundreds)

If the pipeline in the above example also contained, say, three 90° elbows, three 45° elbows, two close return bends, two check valves and one ported ball valve, we could add to the loss of 29.7 ft of head as follows:

Three 90° elbows	= 3 × 5.5	=	16.5 additional pipeline feet of friction
Three 45° elbows	= 3 × 2.6	=	7.8 additional pipeline feet of friction
Two close return ends	= 2 × 13.0	=	26.0 additional pipeline feet of friction
Two check valves	= 2 × 21.1	=	42.2 additional pipeline feet of friction
One ported ball valve	= 1 × 55.0	=	55.0 additional pipeline feet of friction
Total fittings		=	147.5 additional pipeline feet of friction

Then 1.475 (hundreds of feet) × 9.9 (per 100 at 2 in. I.D.) = 14.6 additional ft of head loss

Grand total pipeline (300 straight ft plus 11 fittings) = 44.3 ft of head loss

44.3 ft of head = approximately 19.2 lb pressure per sq. in.

UNITS OF TEMPERATURE

TABLE C.44.   TEMPERATURE CONVERSIONS

°F	°C	°F	°C	°F	°C	°F	°C
−40.	−40.	2.	−16.67	44.60	7.	87.	30.56
−39.	−39.44	3.	−16.11	45.	7.22	87.80	31
−38.20	−39.	3.20	−16.	46.	7.78	88.	31.11
−38.	−38.89	4.	−15.56	46.40	8.	89.	31.67
−37.	−38.33	5.	−15.	47.	8.33	89.60	32.
−36.40	−38.	6.	−14.44	48.	8.89	90.	32.22
−36.	−37.78	6.80	−14.	48.20	9.	91.	32.78
−35.	−37.22	7.	−13.89	49.	9.44	91.40	33.
−34.60	−37.	8.	−13.33	50.	10.	92.	33.33
−34.	−36.67	8.60	−13.	51.	10.56	93.	33.89
−33.	−36.11	9.	−12.78	51.80	11.	93.20	34.
−32.80	−36.	10.	−12.22	52.	11.11	94.	34.44
−32.	−35.56	10.40	−12.	53.	11.67	95.	35.
−31.	−35.	11.	−11.67	53.60	12.	96.	35.56
−30.	−34.44	12.	−11.11	54.	12.22	96.80	36.
−29.20	−34.	12.20	−11.	55.	12.78	97.	36.11
−29.	−33.89	13.	−10.56	55.40	13.	98.	36.67
−28.	−33.33	14.	−10.	56.	13.33	98.60	37.
−27.40	−33.	15.	− 9.44	57.	13.89	99.	37.22
−27.	−32.78	15.80	− 9	57.20	14.	100.	37.38
−26.	−32.22	16.	− 8.89	58.	14.44	100.40	38.
−25.60	−32	17.	− 8.33	59.	15.	101.	38.33
−25.	−31.67	17.60	− 8.	60.	15.56	102.	38.89
−24.	−31.11	18.	− 7.78	60.80	16.	102.20	39.
−23.80	−31.	19.	− 7.22	61.	16.11	103.	39.44
−23.	−30.56	19.40	− 7.	62.	16.67	104.	40.
−22.	−30.	20.	− 6.67	62.60	17.	105.	40.56
−21.	−29.44	21.	− 6.11	63.	17.22	105.80	41.
−20.20	−29.	21.20	− 6.	64.	17.78	106.	41.11
−20.	−28.89	22.	− 5.56	64.40	18.	107.	41.67
−19.	−28.33	23.	− 5.	65.	18.33	107.60	42.
−18.40	−28.	24.	− 4.44	66.	18.89	108.	42.22
−18.	−27.78	24.80	− 4.	66.20	19.	109.	42.78
−17.	−27.22	25.	− 3.89	67.	19.44	109.40	43.
−16.60	−27.	26.	− 3.33	68.	20.	110.	43.33
−16.	−26.67	26.60	− 3.	69.	20.56	111.	43.89
−15.	−26.11	27.	− 2.78	69.80	21.	111.20	44.
−14.80	−26.	28.	− 2.22	70.	21.11	112.	44.44
−14.	−25.56	28.40	− 2.	71.	21.67	113.	45.
−13.	−25.	29.	− 1.67	71.60	22.	114.	45.56
−12.	−24.44	30.	− 1.11	72.	22.22	114.80	46.
−11.20	−24.	30.20	− 1.	73.	22.78	115.	46.11
−11.	−23.89	31.	− 0.56	73.40	23.	116.	46.67
−10.	−23.33	32.	0.	74.	23.33	116.60	47.
− 9.40	−23.	33.	0.56	75.	23.89	117.	47.22
− 9.	−22.78	33.80	1.	75.20	24.	118.	47.78
− 8.	−22.22	34.	1.11	76.	24.44	118.40	48.
− 7.60	−22.	35.	1.67	77.	25.	119.	48.33
− 7.	−21.67	35.60	2.	78.	25.56	120.	48.89
− 6.	−21.11	36.	2.22	78.80	26.	120.20	49.
− 5.80	−21.	37.	2.78	79.	26.11	121.	49.44
− 5.	−20.56	37.40	3.	80.	26.67	122.	50.
− 4.	−20.	38.	3.33	80.60	27.	123.	50.56
− 3.	−19.44	39.	3.89	81.	27.22	123.80	51.
− 2.20	−19	39.20	4.	82.	27.78	124.	51.11
− 2.	−18.89	40.	4.44	82.40	28.	125.	51.67
− 1.	−18.33	41.	5.	83.	28.33	125.60	52.
− 0.40	−18.	42.	5.56	84.	28.89	126.	52.22
0.	−17.78	42.80	6.00	84.20	29.	127.	52.78
1.	−17.22	43.	6.11	85.	29.44	127.40	53.
1.40	−17.	44.	6.67	86.	30.	128.	53.33

TABLE C.44.   *(Cont'd)*

°F	°C	°F	°C	°F	°C	°F	°C
129.	53.89	159.	70.56	188.60	87.	217.40	103.
129.20	54.	159.80	71.	189.	87.22	218.	103.33
130.	54.44	160.	71.11	190.	87.78	219.	103.89
131.	55.	161.	71.67	190.40	88.	219.20	104.
132.	55.56	161.60	72.	191.	88.33	220.	104.44
132.80	56.	162.	72.22	192.	88.89	221.	105.
133.	56.11	163.	72.78	192.20	89.	222.	105.56
134.	56.67	163.40	73.	193.	89.44	222.80	106.
134.60	57.	164.	73.33	194.	90.	223.	106.11
135.	57.22	165.	73.89	195.	90.56	224.	106.67
136.	57.78	165.20	74.	195.80	91.	224.60	107.
136.40	58.	166.	74.44	196.	91.11	225.	107.22
137.	58.33	167.	75.	197.	91.67	226.	107.78
138.	58.89	168.	75.56	197.60	92.	226.40	108.
138.20	59.	168.80	76.	198.	92.22	227.	108.33
139.	59.44	169.	76.11	199.	92.78	228.	108.89
140.	60.	170.	76.67	199.40	93.	228.20	109.
141.	60.56	170.60	77.	200.	93.33	229.	109.44
141.80	61.	171.	77.22	201.	93.89	230.	110.
142.	61.11	172.	77.78	201.20	94.	231.	110.56
143.	61.67	172.40	78.	202.	94.44	231.80	111.
143.60	62.	173.	78.33	203.	95	232.	111.11
144.	62.22	174.	78.89	204.	95.56	233.	111.67
145.	62.78	174.20	79.	204.80	96.	233.60	112.
145.40	63.	175.	79.44	205.	96.11	234.	112.22
146.	63.33	176.	80.	206.	96.67	235.	112.78
147.	63.89	177.	80.56	206.60	97.	235.40	113.
147.20	64.	177.80	81.	207.	97.22	236.	113.33
148.	64.44	178.	81.11	208.	97.78	237.	113.89
149.	65.	179.	81.67	208.40	98.	237.20	114.
150.	65.56	179.60	82.	209.	98.33	238.	114.44
150.80	66.	180.	82.22	210.	98.89	239.	114
151.	66.11	181.	82.78	210.20	99.	240.	115.56
152.	66.67	181.40	83.	211.	99.44	240.80	116.
152.60	67.	182.	83.33	212.	100.	241.	116.11
153.	67.22	183.	83.89	213.	100.56	242.	116.67
154.	67.78	183.20	84.	213.80	101.	242.60	117.
154.40	68.	184.	84.44	214.	101.11	243.	117.32
155.	68.33	185.	85.	215.	101.67	244.	117.78
156.	68.89	186.	85.56	215.60	102.	244.40	118.
156.20	69.	186.80	86.	216.	102.22	245.	118.33
157.	69.44	187.	86.11	217.	102.78	246.	118.89
158.	70.	188.	86.67				

UNITS OF WEIGHT

TABLE C.45.   BASIC CONVERSIONS: UNITS OF WEIGHT

1 grain	.0029 ounces
1 grain	.000143 pounds
1 grain	.064799 grams
1 dram	60 grains
1 dram	.1371 ounces
1 dram	.008571 pounds
1 dram	3.88794 grams
1 ounce	437.5 grains
1 ounce	28.3495 grams
1 ounce	.0625 pounds
1 pound	453.592 grams
1 pound	16 ounces
1 ton	2000 pounds
1 ton	907.18581 kilograms
1 gram	.03527 ounces
1 gram	.002205 pounds
1 gram	15.432 grains
1 gram	.001 milligram
1 kilogram	1000 grams
1 kilogram	35.2739 ounces
1 kilogram	2.20462 pounds

TABLE C.46.   CONVERSION FORMULAS

To Convert	Into	Multiply By
grains	grams	0.0648
drams	grams	1.7718
ounces	grams	28.3495
pounds	grams	435.5924
tons	kilograms	907.18581
grams	grains	15.4324
grams	drams	0.5644
grams	ounces	0.0353
kilograms	pounds	2.2046

TABLE C.47.   OUNCES INTO FRACTIONS OF A POUND, DECIMALS OF A POUND, AND GRAMS

Oz	Fraction of a Lb	Decimal of a Lb	G	Oz	Fraction of a Lb	Decimal of a Lb	G
1/4	1/64	.0156	7.09	8-1/4	33/65	.5156	233.88
1/2	1/32	.0313	14.17	8-1/2	17/32	.5313	240.97
3/4	3/64	.0469	21.26	8-3/4	35/64	.5469	248.06
1	1/16	.0625	28.35	9	9/16	.5625	255.15
1-1/4	5/64	.0781	35.44	9-1/4	37/64	.5781	262.23
1-1/2	3/32	.0938	42.52	9-1/2	19/32	.5938	269.30
1-3/4	7/64	.1094	49.61	9-3/4	39/64	.6094	276.41
2	1/8	.125	56.70	10	5/8	.625	283.50
2-1/4	9/64	.1406	63.79	10-1/4	41/64	.6406	290.58
2-1/2	5/32	.1563	70.87	10-1/2	21/32	.6563	297.67
2-3/4	11/64	.1719	77.96	10-3/4	43/64	.6719	304.76
3	3/16	.1875	85.05	11	11/16	.6875	311.84
3-1/4	13/64	.2031	92.14	11-1/4	45/64	.7031	318.93
3-1/2	7/32	.2188	99.22	11-1/2	23/32	.7188	326.02
3-3/4	15/64	.2344	106.31	11-3/4	47/64	.7344	333.11
4	1/4	.250	113.40	12	3/4	.750	340.19
4-1/4	17/64	.2656	120.49	12-1/4	49/64	.7656	347.28
4-1/2	9/32	.2813	127.57	12-1/2	25/32	.7813	354.37
4-3/4	19/64	.2969	134.66	12-3/4	51/64	.7969	361.46
5	5/16	.3125	141.75	13	13/16	.8125	368.54
5-1/4	21/64	.3281	148.84	13-1/4	53/64	.8281	375.63
5-1/2	11/32	.3438	155.92	13-1/2	27/32	.8438	382.72
5-3/4	23/64	.3594	163.01	13-3/4	55/64	.8594	389.81
6	3/8	.375	170.10	14	7/8	.875	396.89
6-1/4	25/64	.3906	177.18	14-1/4	57/64	.8906	403.93
6-1/2	13/32	.4063	184.27	14-1/2	29/32	.9063	411.07
6-3/4	27/64	.4219	191.36	14-3/4	59/64	.9219	418.16
7	7/16	.4375	198.45	15	15/16	.9375	425.24
7-1/4	29/64	.4531	205.53	15-1/4	61/64	.9531	432.33
7-1/2	15/32	.4688	212.62	15-1/2	31/32	.9688	439.42
7-3/4	31/64	.4844	219.71	15-3/4	63/64	.9844	446.50
8	1/2	.500	226.80	16	1	1.000	453.59

TABLE C.48.   OUNCES INTO GRAMS

Oz	0	1	2	3	4	5	6	7	8	9
					G					
0	—	28.350	56.699	85.049	113.40	141.75	170.10	198.45	226.80	255.15
10	283.50	311.84	340.19	368.54	396.89	425.24	453.59	481.94	510.29	538.64
20	566.99	595.34	623.69	652.04	680.39	708.74	737.09	765.44	793.79	822.14
30	850.49	878.84	907.19	935.53	963.88	992.23	1020.6	1048.9	1077.3	1105.6
40	1134.0	1162.3	1190.7	1219.0	1247.4	1275.7	1304.1	1332.4	1360.8	1389.1
50	1417.5	1445.8	1474.2	1502.5	1530.9	1559.2	1587.6	1615.9	1644.3	1672.6
60	1701.0	1729.3	1757.7	1786.0	1814.4	1842.7	1871.1	1899.4	1927.8	1956.1
70	1984.5	2012.8	2041.2	2069.5	2097.9	2126.2	2154.6	2182.9	2211.3	2239.6
80	2268.0	2296.3	2324.7	2353.0	2381.4	2409.7	2438.1	2466.4	2494.8	2523.1
90	2551.5	2579.8	2608.2	2636.5	2664.9	2693.2	2721.6	2749.9	2778.3	2806.6
100	2835.0	2863.3	2891.6	2920.0	2948.3	2976.7	3005.0	3033.4	3061.7	3090.1

TABLE C.49.   GRAMS INTO OUNCES

G	0	1	2	3	4	5	6	7	8	9
					Oz					
0	—	0.035274	0.070548	0.10582	0.14110	0.17637	0.21164	0.24692	0.28219	0.31747
10	0.35274	0.38801	0.42329	0.45856	0.49384	0.52911	0.56438	0.59966	0.63493	0.67021
20	0.70548	0.74075	0.77603	0.81130	0.84658	0.88185	0.91712	0.95240	0.98767	1.0229
30	1.0582	1.0935	1.1288	1.1640	1.1993	1.2346	1.2699	1.3051	1.3404	1.3757
40	1.4110	1.4462	1.4815	1.5168	1.5521	1.5873	1.6226	1.6579	1.6932	1.7284
50	1.7637	1.7990	1.8348	1.8695	1.9048	1.9401	1.9753	2.0106	2.0459	2.0812
60	2.1164	2.1517	2.1870	2.2223	2.2575	2.2928	2.3281	2.3634	2.3986	2.4339
70	2.4692	2.5045	2.5397	2.5750	2.6103	2.6456	2.6808	2.7161	2.7514	2.7866
80	2.8219	2.8572	2.8925	2.9277	2.9630	2.9983	3.0336	3.0688	3.1041	3.1394
90	3.1747	3.2099	3.2452	3.2805	3.3158	3.3510	3.3863	3.4216	3.4569	3.4921
100	3.5274	3.5627	3.5979	3.6332	3.6685	3.7038	3.7390	3.7743	3.8096	3.8449

TABLE C.50.   POUNDS INTO KILOGRAMS

Lb	0	1	2	3	4	5	6	7	8	9
					Kg					
0	—	0.454	0.907	1.361	1.814	2.268	2.722	3.175	3.629	4.082
10	4.536	4.990	5.443	5.897	6.350	6.804	7.257	7.711	8.165	8.618
20	9.072	9.525	9.979	10.433	10.886	11.340	11.793	12.247	12.701	13.154
30	13.608	14.061	14.515	14.969	15.422	15.876	16.329	16.783	17.237	17.690
40	18.144	18.597	19.051	19.504	19.958	20.412	20.865	21.319	21.772	22.226
50	22.680	23.133	23.587	24.040	24.494	24.948	25.401	25.855	26.308	26.762
60	27.216	27.669	28.123	28.576	29.030	29.484	29.937	30.391	30.844	31.298
70	31.752	32.205	32.659	33.112	33.566	34.019	34.473	34.927	35.380	35.834
80	36.287	36.741	37.195	37.648	38.102	38.555	39.009	39.463	39.916	40.370
90	40.823	41.277	41.731	42.184	42.638	43.091	43.545	43.999	44.452	44.906
100	45.359	45.813	46.266	46.720	47.174	47.627	48.081	48.534	48.988	49.442

TABLE C.51.   KILOGRAMS INTO POUNDS

Kg	0	1	2	3	4	5	6	7	8	9
					Lb					
0	—	2.205	4.409	6.614	8.819	11.023	13.228	15.432	17.637	19.842
10	22.046	24.251	26.456	28.660	30.865	33.069	35.274	37.479	39.683	41.888
20	44.093	46.297	48.502	50.706	52.911	55.116	57.320	59.525	61.729	63.934
30	66.139	68.343	70.548	72.753	74.957	77.162	79.366	81.571	83.776	85.980
40	88.185	90.390	92.594	94.799	97.003	99.208	101.41	103.62	105.82	108.03
50	110.23	112.44	114.64	116.85	119.05	121.25	123.46	125.66	127.87	130.07
60	132.28	134.48	136.69	138.89	141.10	143.30	145.51	147.71	149.91	152.12
70	154.32	156.53	158.73	160.94	163.14	165.35	167.55	169.76	171.96	174.17
80	176.37	178.57	180.78	182.98	185.19	187.39	189.60	191.80	194.01	196.21
90	198.42	200.62	202.83	205.03	207.24	209.44	211.64	213.85	216.05	218.26
100	220.46	222.67	224.87	227.08	229.28	231.49	233.69	235.90	238.10	240.30

Glossary

Abscission Breakdown of cells causing the drop of fruit or leaves.

Absolute zero Represents the temperature at which all thermal motion ceases in the kinetic theory of heat.

Absorption Intake of chemicals by root system or through foliage cuticle and stomata; surface adhesion phenomenon exhibited by molecules of gases, liquids or dissolved substances in contact with solids.

Acetic acid Colorless, pungent substance commonly known as vinegar, CH_3CO_2H.

Acetification Turning into acetic acid; the oxidation of ethanol into acetic acid by *Acetobacter*.

Acid Will neutralize alkaline substances; also the pH scale from 0 to 6.99; and the fruit profile of grapes and wines.

Acid soil Soil with a pH of 6.99 or less.

Acidity Used to indicate tartness or sharpness on the palate. Does not indicate astringency or dryness.

Active ingredient A chemical agent that produces an effect in a formulation.

Adsorption Concentration of ions or molecules on the surface of colloidal or solid materials.

Adjuvants Materials used with sprays to achieve penetration, sticking, spreading or wetting.

Adventitious bud Bud development other than in the axil of a leaf: on the margin of a leaf blade, or from a root.

Adventitious roots Root formation from leaf or stem tissue.

Adventitious shoots Shoots resulting from adventitious buds.

Aerobic Microorganisms that require free atmospheric oxygen in order to grow.

After-ripening A time of cooling before germination will commence from buds and seeds.

Aftertaste Sensory evaluation of wine after swallowing—an olfactory function produced by receptors in the mouth and nasal passages.

Age   The length of time a wine has existed. Often confused by an association with quality, i.e., old wines are not always good wines.

Aged   Wines having been kept in storage (either in bulk or in bottle) under conditions designed to improve organoleptic qualities.

Aging   The process by which wines become aged.

Alcohol   Hydroxides of organic radicals; in wines, the common reference to ethyl alcohol or ethanol.

Alkali soil   Often a reference to ion-exchangeable sodium in soils that restrict vine growth.

Alkaline soil   Soil with a pH of 7.1 or more.

Amelioration   The addition of water and/or sugar to juice or wine.

Amontillado   A popular type of Spanish sherry: may be dry or noticeably sweet.

Amoroso   A Spanish sherry that is darker and sweeter than Amontillado.

Ampullae   Sealed containers, usually made of glass, in which sterilized liquids intended for hypodermic injection are stored.

Ampuls   Same as **ampullae.**

Anaerobic   Microorganisms that do not require free atmospheric oxygen in order to grow.

Angstrom   Unit used in expressing the length of light waves: 1/10,000 of a micron; 1/100,000,000 of a centimeter.

Anhydrous   Without water; absence of water in crystallization.

Anion   Negatively charged ion; the particle in an electrolytic solution which moves toward the anode under electrical influence.

Anode   Negative pole of a battery; an electrode upon which oxidation takes place; positive pole in an electrolytic solution under electrical influence.

Anther   The pollen-producing portion of the stamen in a flower.

Anthesis   Flower in full bloom after the calyptra is removed.

Anthocyanin   Red and purple color pigment in leaves and fruit.

Aperitif   Wines that have added essences and flavors of spices, herbs, roots, etc. Vermouths are an example.

Apex   Extreme tip of shoots, leaf lobes and roots.

Apical dominance   Apical meristem restriction of bud growth.

Apical meristem   Leaf, root or stem tissue at the apex where cell division develops.

Apices   Plural of **apex.**

Appearance   The visual portion of organoleptic evaluation, usually concerned with color and clarity.

Appellation Contrôlée   The authorization to use the name of a controlled winegrowing region in France.

Appellation d'Origine   The geographic origin of a wine, much the same as the **appellation contrôlée,** but less apt to be an indication of quality.

Appetizer wine Wines preferred for consumption before meals or for cocktail use: dry sherry, cocktail sherry and vermouths are common types.

Arid Very low rainfall climates: hot, dry land where water evaporates quickly.

Aroma The fragrance in a juice, must or wine that is contributed by the fruit. The aroma is part of the bouquet.

Aromatized wines The same as **aperitif** wines.

Asexual propagation Reproduction by budding, cuttings, grafting and layering.

Astringency Sensory response on the palate usually due to tannins. Similar to the values of taste from alum or aspirin.

Astringent A term used to describe the palate's reaction to tannins in organoleptic evaluation.

Atom The smallest whole portion of an element in atomic theory; the smallest whole portion of an element that can stand alone or contribute to a chemical bond.

Atomic theory The hypothesis which states that all matter is composed of atoms, and that all atoms of any one element are virtually the same size and weight.

Atomic weight The weight of an element atom in relation to the standard weight of 12 adopted for carbon.

Auslese A German wine made from specially selected ripe grapes with all unripe grape berries removed and discarded.

Auxins Stimulators for cell elongation; plant growth hormones.

Available water Soil moisture that can be readily absorbed.

Avogadro's number Avagadro's number is equal to 6.02×10^{23} and is also called the mole. It is the number of atoms or molecules contained in a gram-atomic weight or gram molecular weight of any given substance. For instance, there are 6.02×10^{23} atoms in 16 g of atomic oxygen, O; and the same number of molecules in 32 g of molecular oxygen, O_2.

Bacteria One-celled organisms that do not contain chlorophyll.

Balance A term used in organoleptic evaluation describing the proportions of dryness or sweetness and acidity.

Balling A graduated specific gravity scale relating to dissolved solids and alcohol content on a Brix hydrometer.

Band application The restricted spreading of chemicals in strips, as opposed to **broadcast application**.

Bar Pressure unit equal to about 750.1 mm of mercury, or one megadyne per sq. cm.

Bark The outermost layer of a woody stem.

Basal bud The small bud on a cane or spur innermost to the source; the portion of a whorl that may not develop unless outer buds fail.

Beerenauslese A German wine made from individually selected ripened grapes that have been exposed to and affected by *Botrytis cinerea*, the "noble mold".

Bereich German for "area", which generally denotes wine having originated in the area of, or in a sub-region of, a larger, more noteworthy region.

Binning Storage of bottled wines in bins, usually for aging.

Biological control Fungus and insect control by the use of parasites, predators, and organisms.

Bilateral cordon A method of vineyard training, such as the Geneva Double Curtain, in which the trunk is horizontally extended in two opposite directions, either on one or two wires.

Blade The broad photosynthetic portion of a leaf which is sessile or attached to the stem by a petiole.

Blanc de blancs French for "white of whites", denoting a white wine having been made entirely from white grapes.

Blanc de noirs French for "white of blacks", denoting a white wine having been made entirely from black grapes.

Bleeding The emission of sap dripping from canes cut after the dormant season has ended.

Blending The combining of two or more wines together.

Bloom The powdery substance on the skin of grapes that is comprised of natural yeasts.

Body The fullness of a wine cause by dissolved solids, generally discounting dissolved sugar solids; the viscosity or "mouth-feel" of a wine as light bodied and thin, or heavy bodied and thick.

Bond The status of an alcoholic beverage when excise taxes have not yet been paid; also a guarantee required by the Bureau of Alcohol, Tobacco and Firearms before a basic winery permit will be issued.

Bottle aging A program designed to improve wine quality through holding periods in bottles.

Bottle fermentation The secondary fermentation of wine in bottles to capture carbon dioxide gas in the production of sparkling wines.

Bottle sickness A term in wine production usually referring to wines just having been bottled which exhibit a temporary loss of bouquet and flavor value, usually due to the addition of preservatives and cask-to-bottle cellar treatment.

Bottoms The sediment in wine tanks after fermentation, racking, clarification, etc.; more often called **lees**.

Bouquet The fragrance of a mature wine, comprised of the fruit aroma and the volatile constituents resulting from cellaring techniques and materials.

Boyle's law The volume of a gas will vary inversely to the amount of pressure applied to it, given that the temperature remains constant.

Brandy   Distilled wine.

Brilliant   A clear wine totally free of any suspended solids.

British Thermal Unit   The amount of heat required to increase the temperature of one pound of pure water one degree Fahrenheit at the point of maximum density—39.1°F; equivalent to .252 kilogram-calorie.

Brix   A hydrometer scale used to measure dissolved solids in grape juice; not to be confused with Balling, which includes the effects of alcohol in solution along with dissolved solids.

Broadcast application   The spreading of a spray or solid application over an entire area, rather than a **band application**.

Brut   French for natural, unsweetened (usually champagne) wines; the very driest.

Bud   The embryonic shoot on a cane comprised of an apical meristem and leaf primordia, protected during dormancy by a scale cap; commonly used in numbers as a crude measure of crop potential.

Bud sport   A mutation of cell genes which causes a particular portion of a vine to evolve differently from a comparative portion.

Buffer   A water-dissolved substance producing a solution which resists change in pH values.

Bulk process   The production of sparkling wine in tanks rather than bottle fermentation; also known as the Charmat process.

Bung   The stopper used in a keg or barrel.

Bunghole   The opening in the top of a keg or barrel.

Butt   An English term relating to Spanish cooperage: approximately 126 Imperial gal. or about 151 U.S. gal.

Calcareous soil   Soil that contains calcium carbonate or magnesium carbonate. Both are basic substances which will effervesce and give off carbon dioxide when neutralized with hydrochloric acid.

Callus   The formation of parenchyma tissue around a graft union.

Calyx   The external sepal portion of a flower.

Calyptra   Grape flower petals that are fused together and drop at anthesis.

Cambium   The thin layer between bark and wood consisting of meristematic tissue.

Candling   The process of judging clarity in a bottle of wine by holding it in front of a filament bulb or candle.

Cap stem   Individual stems of fruit or flowers in bunches.

Capillarity   The ascent or descent of a liquid in contact with a solid. It is dependent upon the relative attraction of the molecules in both the liquid and the solid. A capillary tube is a good example.

Capillary water   Water retained in porous soil materials after surface drainage.

Capsule   The seal over the closure of a wine bottle used to protect the neck and closure, and to improve bottle package appearance.

Carbohydrate   Carbon, hydrogen and oxygen chemically bonded as an energy compound.

Carbonated wines   Wines injected with carbon dioxide gas, rendering the wine effervescent.

Carbon dioxide CO_2—the gas produced by fermentation (see **Gay-Lussac equation**).

Carbon-to-nitrogen ratio   Proportion of carbon to nitrogen in plant tissue and soil.

Carboy   A glass bottle, usually with a capacity of 5 gal.

Case   A container in which bottled wines are held, usually made of heavy paperboard, wood or plastic.

Cask   A wooden container for wine, usually made of oak and containing more than 200 U.S. gal. Not to be confused with butts, hogsheads, pipes and puncheons which are smaller, or with tuns which are larger.

Caskiness   A flavor in wine generally attributed to casks being used without being properly cleaned.

Casse   A haze that develops in wines usually as a result of excessive metal content.

Catch wire   Any wire that serves for vine tendril attachment.

Cation   Positively charged ion; the particle in an electrolytic solution which moves toward the cathode under electrical influence.

Cave   French for wine "cellar".

Cavitation   A localized gaseous condition within a liquid stream which occurs when the pressure is reduced to the vapor pressure of the liquid.

Cellar   Any building, either above or below ground level, used in a particular phase (or group of phases) in the winemaking process.

Cellar treatment   The materials and methods used in a particular phase (or group of phases) in the winemaking process.

Cépage   French for "grape cultivar".

Certified planting stock   California Department of Agriculture certified virus-free vine propagating material.

Champagne   A white wine producing district located northeast of Paris in France; also sparkling white wines in general, but particularly those produced exclusively in the French Champagne district.

Champagne rouge   A misnomer; such a wine does not exist. Refers to red sparkling wines labeled as sparkling burgundy.

Chaptalisation   The addition of sugar to juice or must in order to increase the alcoholic strength resulting from fermentation (compare to **amelioration**, the addition of water and/or sugar).

Character   The whole of organoleptic attributes that distinguish a wine: the color, taste and bouquet values normally associated with a particular wine type.

Charles law   The volume of a gas will vary directly with the temperature, at constant pressure. The volume will vary 1/273 of total volume at 0°C for each degree Centigrade.

Charmat   The bulk, or tank, method of producing sparkling wines; usually a much faster process (and generally less respected) than the traditional French "methode champenoise" bottle-fermentation technique.

Château   French for "castle"; the homestead estate of vineyards and winery in France, usually in Bordeaux.

Chimera   Tissues of varying genetic constituency in the same plant.

Chisel   An implement designed to loosen subsoils with one or more weighted metal points forced beneath the soil surface.

Chloroplast   Chlorophyll-containing plastid in leaf cells.

Chlorophyll   The green pigment in chloroplasts that absorbs light energy and converts it to chemical energy in the function of photosynthesis.

Chlorosis   The effect of disease, nutritional deficiencies, chemical damage, and other factors that cause a blanching or yellowing effect on green portions of a plant.

Clarify   To clear a wine; the addition of agents to a wine in order to precipitate solids.

Classified growth   A particular vineyard estate or chateau in Bordeaux that was classified for the Paris Exposition of 1855. There are only five first-growths, but many second, third, fourth and fifth-growth chateaux.

Claypan   A dense subsoil type that is firm and difficult to penetrate when dry, and kneadable when moist or wet.

Clean   An organoleptic term to describe a wine which has not been subjected to unsanitary cellar conditions; the full varietal effect on the palate without interference from other varieties having been blended; the absence of overprocessing.

Clear   A wine which has been clarified and/or filtered successfully; free from visible solids but not brilliant.

Clone   An individual plant group resulting from the asexual reproduction of germplasm from a single parent.

Clos   French for "walled vineyard"; generally found in Burgundy.

Cloudy   A wine with a large content of suspended solids.

Coarse   A wine that has harsh or overpowering organoleptic properties; may refer to young wines not having had the mellowing effect of further cellar treatment and aging.

Colloidal suspension   Hazy or slightly cloudy suspensions of semi-solid particles in wine (not to be confused with metal casse).

Color   The hue of a wine in organoleptic evaluation. Terms such as "light straw" or "ruby" or "tawny" etc. relate to the influence of grape variety, cellar treatment, aging, etc.

Compatible chemicals   Chemicals that can be mixed without introducing toxic effects in a wine.

Compatibility   The genetic and physical ability of scion and rootstock to unite in a graft union. Also, female and male nuclei that can unite to form a viable egg.

Complete flower   An entire flower comprised of petals, pistil(s), sepals and stamens.

Complex   A wine in which bouquet and/or flavors are composed of many different constituents which may make the wine difficult to describe.

Concentrate   Dehydrated grape juice (either red or white) with or without volatile essences returned. Used for sweetening grape juice and wines that are deficient in natural grape sugar.

Contact herbicide   A chemical used for weed control that is toxic to the part of the weed to which it is applied.

Cooperage   A term that traditionally refers to wine containers that are made from wood (the derivation of the word is from "cooper"—a British sailing term used to designate the ship's barrelmaker). In modern times cooperage refers to any container used for wine, whether of wood, steel, glass, etc.

Cooperative   A winery owned by more than one grape-grower. Generally formed so that the capital required for winery establishment and operation can be greatly reduced on a per-member basis.

Cork   The bark of the cork oak, grown extensively in Mediterranean countries, harvested and processed into stoppers for both still and sparkling wine bottles.

Corkscrew   A spiral metal device with a sharp tip used for removal of wine bottle corks.

Corkiness   An unpleasant flavor and bouquet in a wine that was bottled with a defective cork, usually because there was not a complete seal and outside air was allowed to enter the bottle.

Corolla   Flower petals.

Cradle   A service device in which a bottle of wine is placed at an angle; usually made of wicker in a basket weave.

Cream of tartar   The colorless crystalline deposit of potassium bitartrate that precipitates from unstable wines under refrigeration.

Crown suckering   The removal of shoots and suckers from the trunk of head-pruned vines.

Cru   French for a specific vineyard "growth"; classified for a specific echelon of quality such as "premier cru classe" or "grand cru classe".

Crust   Sediment of unstable solids from wine which have collected and solidified on the surface in the bottle. Common in red wines; most often associated with very old wines, especially ports.

Cutting   One-year wood severed at two or more internode lengths used for propagation.

Cuvée   French for "tank", or "tubful"; a blend of wines, either of the same or different vintages, prepared for secondary fermentation into sparkling wine.

Decant   The operation of delicately transferring wine from a bottle to a decanter to separate any sediment that may have formed in the bottle while aging.

Definition   The visual clarity or resolution achieved by the microscope in magnifying a specimen.

Dégorgement   French for "disgorging"; the removal of sediment from a bottle of sparkling wine, after the remuage and prior to the addition of dosage.

Degrees Balling   Divisions for dissolved solids on the Balling hydrometer scale when used in solutions containing ethanol (such as wine). Often used as a term for degrees Brix, which is the same scale expressed without ethanol influence.

Degrees Brix   Divisions for dissolved solids on the Brix hydrometer scale.

Delicate   An organoleptic term used to describe wines with subtle bouquet and flavor values.

Demijohn   Small glass containers for wine that are usually wicker-covered; a misnomer often used for carboys.

Demi-sec   French for "nearly-dry" or "semi-dry"; most often used in the description of sparkling wines.

Dentate   With teeth; usually meaning the outward direction of teeth forms such as in leaves.

Density   Mass per unit volume at a specific temperature; quantity of electric charge in space per unit volume; quantity of energy per unit volume.

Depth of focus   The depth or thickness of a specimen in focus in the microscope.

Dessert wine   In the U.S., a wine that has had the addition of brandy, or has been "fortified"; usually sweetened rather heavily so as to result in port and sherry-type wines; may also refer to any wine that is served with dessert courses.

Dew point   The condensation point of a vapor; the point where liquefaction of a vapor takes place.

Dextro   A turning motion to the right, or clockwise rotation; often used in the application of a plane of polarized light.

Dioecious   A state in which the male and female organs of flowers occur on different plants.

Dinner wines   Table wines.

Disgorging   An English form of the French term for sparkling wine, *dégorgement*, meaning to expel the frozen plug of sediment in the neck of a sparkling wine bottle with the carbon dioxide gas pressure which has developed during secondary fermentation.

Dissociation   The division of a compound or element into more simple atoms, ions or molecules: usually applied to the effects of energy on gases, or of solvents on substances in solution.

Dissociation, degree of   The quantitative ratio of the amount of a compound which has dissociated, relative to the total amount of the compound originally present.

Distillation   Heating a volatile liquid in a still pot and condensing the vapors that form so that the liquid condensate is collected in a separate container.

Domaine   French for "vineyard estate" or "wine estate" (most commonly used in Burgundy).

Dormant   Not actively growing.

Dosage   The addition of a high-alcohol, very sweet syrup to sparkling wines directly after *dégorgement* in order to slightly sweeten the wine. The dosage usually contains a preservative to prevent a third fermentation.

Doux   French for "sweet". Most often used in the description of sparkling wines.

Dregs   The precipitated sediment in wine, more commonly referred to as **lees**.

Dry   The absence of fermentable sugar; opposite of sweet.

Dry inches   The space devoid of wine in the top of a wine container which may be measured by the Bureau of Alcohol, Tobacco and Firearms inspectors in order to calculate an exact volume to deduct from the total capacity.

Drying ratio   The amount of fresh grapes required to yield one pound (16 oz) of raisins.

Earthy   The contribution of the soil on which the grapes were grown to the bouquet and/or flavor of a wine: the chalk of Champagne and flint of Chablis are good examples.

Electrode   The part of an electrolytic cell through which a current enters or exits.

Electron   The smallest negatively charged particle in matter; the opposite electrical entity of a proton; equal to 4.77×10^{-10} cgs; constitutes beta and cathode rays.

Emasculation   The severing of stamens in a flower, often done by plant breeders for cross-pollination.

Enations   Outgrowths of leaf tissue from lower surfaces near the principal leaf veins.

Enology   The art, science and study of making wine.

Entire   The absence of divisions, indentations or lobes in leaves.

Enzymes   Complex organic compounds that are produced by living cells and catalyze or control chemical reactions.

Epinasty   The downward stress of leaves, usually caused by sprays, rain, etc.

Épluchage   A French term for the removal of defective berries from bunches of grapes. Similar to the German **auslese**.

Equilibrium constant   A numerical constant at a specific temperature used in the law of mass action for any reaction.

Equivalent weight   The weight of a substance which is required to completely react with a given amount of another substance.

Erosion   The effect of water, wind and other sources wearing away land surfaces.

Essential oils   Organic oils generated in flowers and fruit that are distinctive in odor and taste; the aroma and flavor profile of a particular variety of grape or other fruit.

Estate-bottled   Generally signifies that the same authority who vinified a wine also grew the grapes. This often misused term may not be an indication of superior wine.

Esters   Volatile organic flavor constituents.

Ethylene   A gas produced by plant tissues, often described as a hormone that triggers ripening; a compound with the $CH_2 = CH_2$ structure.

Extract   The expression of total dissolved solids in wine, including sugar, color pigments, glycerols, etc.

Extra dry   Bone-dry; totally devoid of sugar. Often used incorrectly in labeling sparkling wines that are slightly sweet, but not as dry as Brut or Sec.

Eye   The whole bud, i.e., primary, secondary, and tertiary buds compounded under one cap; also refers to multibud formations.

Facultative   Microorganisms that can grow with or without free atmospheric oxygen.

False wines   Wines made from sources other than grapes.

Fasciation   An oval, rather than round, profile assumed by the cane, usually because of multi-bud growth from the same eye.

Feeder roots   Shallow-growth roots with expansive surface area for the intake of moisture and minerals necessary for proper plant metabolism.

Fermentation   Generally referred to as the breakdown of sugar(s) to produce ethyl alcohol, carbon dioxide and energy by the action of yeasts, although bacterial fermentations also occur.

Fermentation lock   A one-way device on fermenters that allows gases to escape, but keeps outside elements from entering.

Fermenter   A container in which fermentation takes place.

Fertility   A measure of soil quality associated with the ability to provide nutrients and other essential requirements for plant growth.

Fertilization   The application of materials to increase fertility; also the formation of a zygote from gametes in sexual reproduction.

Field capacity   Soil retention capability measured as the quantity of water held against the force of gravity; the water-holding capacity of a soil.

Fifth   Equal to 757 ml, one-fifth of a U.S. gallon or four-fifths of a quart; a common wine bottle size which was replaced by the 750 ml, or three-fourths liter size under Bureau of Alcohol, Tobacco and Firearms regulation.

Filament   The stalk that supports a stamen in a flower.

Filter element   The porous device which performs the actual process of filtration.

Filtering   The act of passing a wine through a filter medium in order to render clarification.

Filtration   The forcing of a wine through a porous medium to remove suspended solids, often to such low porosities as to remove yeast and bacteria cells.

Fine   The same as to **clarify**.

Fine wine   A term usually applied to wines of very good quality—but not great.

Fining   The application of specific agents to clarify and stabilize wines. Tannins, gelatin, Sparkolloid and bentonite are good examples of fining agents.

Fino   The lightest and driest of the Spanish sherries.

Flat   A wine generally described as devoid of interesting qualities, as lacking finesse or polish; also, sparkling wine that has lost effervescence.

Flavored wines   Wines that have qualities of taste and aroma that were added from sources other than the grape used to make the wine, such as vermouth and some pop wines.

Flavorous   An organoleptic term used to describe full or extra-full flavor value.

Flavors   An organoleptic term used to describe distinct taste values perceived by the human palate and olfactory receptors.

Flor   Spanish for "flower"; generally used in reference to the surface-growing yeasts that synthesize acetaldehyde, the "nutty" flavor common to sherry-type wines.

Flow rate   The volume, mass, or weight of a fluid passing through any conductor per unit of time.

Flower buds   Plant buds that produce flowers.

Flowers of wine   A white film on the surface of wine that denotes the growth of *Acetobacter*; the acetification process.

Flowery   The aroma and bouquet of wine that arouses the nasal sensory organs in a manner similar to flowers in blossom, such as the effect of white wine made properly from the variety Johannisberg Riesling.

Foliage wire   The same as **catch wire**.

Foliar feeding   The application of dust or spray fertilizers on the leaves of a plant.

Formative effects   The abnormal arrangement of veins in a leaf.

Formula weight   The sum total of the atomic weights calculated for an entire chemical formula.

Fortified wine   Wines in which the alcohol content has been increased by the addition of brandy. Examples are sherry, port, and most other dessert wines.

Fortify   The act of adding brandy to wine in order to increase alcohol percentage.

Foxiness   The aroma and flavor value generally attributed to grapes from the *V. labrusca* species, among others, but more properly attributed to varieties of *V. rotundifolia*, which includes Great Fox.

Free-run   The juice or wine that flows from the press without benefit of pressure.

Frizzante   Italian for "petillant" or "slightly sparkling".

Fruit   The mature berry which has grown from a fertilized pistil.

Fruit set   The development of fruit and seeds from pistils and fertilized ovules.

Fruit wines   Wines made from fruits, fruit essences or fruit concentrates other than grapes. (See **true fruit wines**)

Fruity   An organoleptic term applied to wines having high values of bouquet and flavor captured from the grape or other fruit from which it was made. Often used in evaluating wines made from *V. labrusca* and *V. rotundifolia*.

Full   An organoleptic term applied to wines that are heavy bodied or strong in values of bouquet and/or flavor.

Fungicide   A chemical material used to control the growth of fungi on plants.

Fungus   A simple plant lacking chlorophyll which lives on other life forms.

Gamete   A sex cell, either female or male, which can fuse with opposite sex cells to form a zygote in sexual reproduction.

Gas constant   The term R in the ideal gas law equation: $PV = nRT$. An energy term equal to 0.06205 liter-atmospheres per degree per mole.

Gay-Lussac equation   One molecule of sugar will ferment into two molecules each of ethyl alcohol and carbon dioxide.

Germination   The initiation of growth by the embryo of a seed.

Generic   A term applied to popular wine types which are usually common to or famous for a particular viticultural region, such as Burgundy or Champagne, but not necessarily produced in the region.

Geneva double curtain   A cordon system of training grape vines on two top wires spread several feet apart at equal distances from the soil surface in order to gain additional exposure of the foliage to the sun.

Genus   Closely related species comprising a specific group of plants.

Gibberellin   A plant hormone that promotes cell elongation in the growth of fruit and shoots.

Girdling   The circumcision of the bark and phloem on a cane, shoot or trunk so as to inhibit the circulatory system from functioning.

Glabrous   A smooth even surface as opposed to rough, scale-type or other uneven types.

Glucose   A simple monosaccharide sugar.

Grafting   The introduction of a bud or a scion to a rootstock or other portion of a vine so that a callous forms in the process of uniting two separate tissues.

Gram-atomic weight   The numerical weight in grams equal to the atomic weight of a given element.

Gram-molecular-weight   The numerical weight in grams equal to the molecular weight of an element or compound.

Grand cru   French for "grand growth" or "great growth". Used primarily to classify vineyards in France, *grand cru* has a slightly different interpretation from one wine region to another.

Green   An organoleptic term used to describe undeveloped wines with bouquet and flavor of a "grassy" nature; also, grapes that are not yet ripe.

Green-manure crop   A cover crop grown to be mulched back into the soil to improve the content of organic matter.

Guard cells   Special cells that grow on leaf epidermis which direct stomata to open and close.

Guttation   The release of moisture from the margins or tips of leaves in uninjured plants.

Hard   An organoleptic term used to describe wines that exhibit coarse and harsh sensations on the human palate; the opposite of delicate and soft.

Hardpan   A subsoil that is of a dense texture, usually composed of a clay-type consistency.

Harshness   An organoleptic term usually referring to a wine that is excessively high in fixed or total acidity; not to be confused with astringent.

Haut   A French term meaning "high", but more often used in wine nomenclature as a reference to "higher than" or "further away from" or "better than", etc.

Head   The upper portion of a grapevine trunk from which canes or cordons extrude. Also, the height of a column or body of fluid above a given point expressed in linear units (i.e., feet): pressure is equal to the height times the density of the fluid. For water, the pressure (in pounds per square inch) is equal to the head (in feet) times 0.433.

Head, static   The height of a column or body of fluid above a given point.

Head velocity   The equivalent head through which the liquid would have to fall to obtain a given velocity. Mathematically it is equal to the square of the velocity (in feet per second) divided by 64.4 feet per second squared: $h = V^2/2g$.

Headland   Land space left untilled at the ends of vineyard rows for the manipulation of tractors and equipment entering from between the rows.

Heady   An organoleptic term usually referring to wines with excessive alcohol content, roughly equivalent to "strong" wines. Also, loosely used as a term referring to persistent foam on the surface of sparkling wine in a glass.

Heartwood   The older wood in the center of a trunk or stem which no longer functions in the circulatory system of the plant.

Heavy   Heady or strong wines without corresponding excessive values of bouquet and flavor. Also used in describing the viscosity of wines, especially sweet dessert wines.

Henry's law   Gas solubility in a liquid is directly proportional to gas pressure on the liquid at a specific temperature.

Herbicide   A chemical used to control or kill plants, usually weeds, that interfere with crop-production.

Hermaphrodite   Flower with both female and male parts.

Heterofermentative   A type of lactic acid bacteria from which the final fermentation products are lactic acid and carbon dioxide gas.

Hirsute   An uneven, hairy or rough surface; the opposite of **glabrous**.

Hogshead   A small wine cask, usually found in Bordeaux, containing approximately 225 liters, or 59.445 gallons.

Homofermentative   A type of lactic acid bacteria from which the final fermentation product is lactic acid; no carbon dioxide gas is given off.

Hormone   A chemical regulator that influences plant processes. Natural hormones are generally utilized in an area of the plant which is different from the site of production.

Host   Plant that harbors parasites or predators.

Hybrid   The crossing of two different cultivars of vines (in viticulture), resulting in seedlings of new cultivars which may exhibit various properties contributed by each parent.

Hydathode   The leaf structure that releases plant liquid during guttation.

Hydrocarbon   A chemical compound that is comprised of only carbon and hydrogen.

Hydrometer   A floating instrument used to measure the density or specific gravity of liquids.

Hydroponics   Plant culture in nutrient aqueous solutions.

Hygroscopic   A state during which moisture can be retained in quantities less than sufficient to constitute a liquid; materials which take up moisture from the environment.

Hypha   Fungal thread or filament.

Illumination   The light provided by some source, either reflected or built into the microscope, which is used to observe the specimen.

Imperfect flowers   Flowers that have normal female or male parts, but not both.

Incompatibility   Failure of rootstock and scion to callous and properly unite in the formation of a viable graft; the inability of chemicals to be mixed or used together because of toxicity or other negative factors; also, the inability of male and female gametes to properly form a viable zygote.

Incomplete flower   A flower that is missing at least one of the four vital organs: sepals, petals, stamens or pistils.

Indexing   The determination of the presence of virus or other disease by the growth of plant tissues in or on an independent medium.

Indicator   A substance that will change color, or react in some other visible manner, so that a change in the nature of a solution (usually during titrations with reagents) is readily apparent at a distinguishable end point.

Indigenous   Vines native to a particular area, district or region.

Indoleacetic acid   A naturally-produced plant hormone which regulates some growth reactions.

Infiltration rate   The percolation rate of water absorbed into the soil from the surface.

Inflorescence   The flower cluster of a grapevine.

Internode   The cane or shoot between two contiguous buds or nodes.

Ion   A particle bearing either a negative or positive charge, formed when a neutral atom or molecule gains or loses one or more electrons. The valence of the ion is indicated by the number of charges gained or lost by the atom or molecule.

Irrigation   An artificial application of water to the soil.

Jereboam   A large wine bottle, generally four times the size of a common champagne bottle and five times the size of a typical Bordeaux bottle: i.e., from about three to four liters in capacity.

K-electron   An electron from the innermost ring of electrons in an atom.

Kabinett   A German wine made without the use of added sugar, generally considered the driest and lowest grade of *Qualitatswain Mit Pradikat*.

Keg   A very small wooden container used to store wine. Usually less than 30 U.S. gallons in capacity.

Kellar   German for "cellar". Often used as short for "wine cellar".

Kelter   German for "press". Often used as short for "grape press" or "wine press".

Kniffin   A training method for vine trellising. Usually used with 4 or 6 cane pruned vines upon 2 or 3 catch wires, the canes being tied to the vine in a strict horizontal position.

Kosher wine   A wine made under Rabbinical law and supervision which is used for sacramental purposes during Jewish holidays and religious rites.

Labrusca   Short for *Vitis labrusca*, the native grapes in the northeastern part of the U.S.

Lage   German for "locale". Generally used in the description of a specific vineyard site.

Land leveling   The transfer of soil and/or subsoil from higher to lower topography so as to create a level surface.

Latent bud   A bud in dormancy for more than one year.

Lateral   A cane or shoot developed from a main cane or shoot; also, shoulder(s) on bunches or a bunch of flowers.

Leach   The removal of moisture and nutrients from the soil by plant roots; also, the removal of soluble materials by water passage through soil.

Leaf bud   A bud that grows into a stem that bears leaves and perhaps tendrils, but no flowers.

Leaf scar   A lesion remaining on a cane or stem after petiole separation.

Lees   The sediment that precipitates from young wines during and after fermentation which is composed primarily of grape pulp, yeasts, color pigments, acid salts, etc.; also used in combination with clarification procedures as "fining lees", denoting the precipitation of the fining agents.

Lenticel   A pore on stems, pedicels and grape berries for the exchange of gases.

Levo   A turning motion to the left, or counterclockwise rotation, often used in the application of plane polarized light.

Light   An organoleptic term used to describe low values of viscosity or mouth feel in wines, usually dry white wines; also, may be properly used to describe a wine lacking in value of bouquet or flavor.

Light saturation   The point at which an increase in the intensity of light will not increase the rate of photosynthesis.

Luminescence   A light emission taking place at a low temperature not caused by incandescence.

Madeirization   The oxidation of ethyl alcohol and acetic acid into aldehydes, as in the making of sherry.

Magnification   The virtual image of the specimen observed in the microscope as compared to the same specimen viewed with the naked eye from a distance of 10 in. Usually expressed as a numerical factor such as $100\times$ or $400\times$, etc.

Magnum   A wine bottle usually twice the size of a normal bottle.

Malo-lactic fermentation   The transformation of malic acid to lactic acid, carbon dioxide and energy by the action of specific bacteria.

Marque   French for "mark"—usually used in connection with a trade mark, or "marque déposée".

Mass number   The whole number nearest the isotopic weight of an element compared with the value of 12.0000 as the mass of carbon; the number of neutrons and protons in the nucleus of an atom.

Mature   A term used for a wine that has been properly aged so as to have reached full development of all organoleptic qualities.

Maturity   Full ripeness. A state in which the fruit has developed properly for the use intended.

May wine   Traditionally a light, rather sweet German white wine infused with woodruff.

Mellow   A general term usually referring to a wine that is not soft or harsh, and/or has benefited from positive cellaring techniques.

Meristem   A plant area in which cell division takes place rapidly.

Metabolism   The chemical processes in living cells by which nutrients are utilized for energy, cell division and repair.

Methuselah   A large wine bottle, most often found in Champagne, generally about 8 times the normal bottle size.

Micron   A millionth of a meter (about 0.00004 inch).

Mildew   A fungal disease of a vine which cripples both green tissue and the fruit. Two types are noteworthy which are commonly known as "downy" mildew and "powdery" mildew.

Millerandage   An extraordinary looseness of grape berry clusters caused by poor fruit set.

Millesime   French for "vintage", as used in determining the year the grapes were grown to make a specific wine.

Mis en bouteilles au château   French for "bottled at the winery estate". Used as an indication of authenticity of origin and quality.

Mis en bouteilles au domaine   French for "bottled at the vineyard estate". Used as an indication of guaranteed origin and quality.

Molality   A solution strength described in gram-moles of solute per kilogram of solvent.

Molarity   A solution strength described in gram-moles of solute per liter of solvent.

Moldy   An organoleptic term referring to "off" values of bouquet and flavor. This usually occurs among wines that have been made from grapes that have been infected with mold, or wines that have been stored or aged in cooperage that has harbored mold, or both.

Mole fraction   A quotient resulting from the division of the number of moles of a particular substance in a system by the entire number of moles of all substances in the system.

Molecular weight   The sum total weight of all atomic weights of all the elements comprising a compound.

Molecule   The smallest combination of atoms chemically united which still retains the properties of the substance mass.

Monoecious   A plant having both female and male flowers.

Mousseux   French for "sparkling".

Mulch   Artificial or natural material applied to or in the soil to increase temperature control, weed control or water retention, or to supply nutrients after a mulch commences to decompose.

Must   Crushed grapes that have been destemmed.

Musty   An organoleptic term often used interchangeably with "moldy" or "mousey", but may be somewhat different in that musty wines can result from aging in cooperage that has decayed or become waterlogged.

Mutage   The process of adding brandy to fermenting juice or must in order to arrest fermentation, retaining some of the natural grape sugar.

Mutation   A genetic change which alters the hereditary material in living organisms.

Mycelia   A concentration of fungus filaments.

Mycorrhiza   The hyphae or mycelium in symbiosis with plant roots.

Natural fermentation   A fermentation taking place with natural, rather than cultured, yeast cells. Also called a normal fermentation procedure.

Natural wines   Wines resulting from a natural fermentation; also, wines produced without the addition of sugar; also, wines which have not been fortified.

Nature   French for "natural". Generally refers to wines made without added sugar.

Négociant   French for "shipper". Generally used in reference to wine buyers in France who negotiate for the purchase of bulk wines from individual winegrowers, resulting in the *negociant* aging and bottling the wines, and eventually shipping the goods to other distribution outlets.

Nematode   A soil-borne small parasitic worm that can live in or on grape roots.

Neutron   A particle with no electrical charge that is approximately 1800 times heavier than an electron, and of about the same mass as a hydrogen atom. Along with protons, neutrons comprise atomic nuclei.

Noble rot   A fungal penetration of grape skins, usually from the strain *Botrytis cinerea*, that results in dehydration of the grape berry and consequently an increased percentage of sugar in the fruit.

Node   The bud site on a cane. The region on the cane from which buds develop into shoots.

Nonsaline-alkali soil   Soil with sufficiently high exchangeable sodium content as to inhibit vine growth, yet which does not contain extraordinary quantities of soluble salts.

Normal solution   A solution concentration containing one gram-equivalent of a substance per liter of the solution.

Nose   The reaction of the human senses to wine odor. Often used in describing the bouquet of a wine or the aroma of grapes, grape juice and grape must.

Numerical aperture   The determination of the angle from which the maximum cone of light enters the objective of the microscope.

Nutty   A descriptive term in wine evaluation meaning that the wine has a nut-like bouquet and/or flavor, usually derived from one or another processes of madeirization, such as in the making of marsala or sherry.

Odor   Usually used by the wine tasters in a negative manner during description of aroma and bouquet.

Oenology   British spelling of enology.

Oidium   See **mildew**.

Oloroso   The darkest and sweetest of Spanish sherries.

Optical activity   The dextro and levo rotation of polarized light planes.

Organoleptic   The evaluation of wine by the use of human sensory organs for sight, smell and taste.

Originalabfullung   German for "bottled by the grower". A more strictly enforced form of "estate-bottled", but recently discontinued in favor of new German wine labeling laws.

Osmosis   The passage of materials, usually liquid, through a semipermeable membrane from areas of higher to lower concentrations.

Ovary   The large basal part of the female pistil which contains ovules.

Overcropping   Higher crop yield than a vine can bring to maturity, or than a vine can yield and remain healthy. Usually results from disease, insects, light pruning or water stress.

Oxidation   The reaction of wine constituents with oxygen, such as the "browning" of color and the formation of acetaldehyde from ethanol. Wine aging is a form of controlled oxidation.

Palisade parenchyma   Cylindrical cells containing chloroplasts that are located in leaves just below the upper epidermis.

Parasite   An organism that lives on, or in, and obtains its food from the body of another organism.

Parenchyma   Plant tissue composed of thin-walled cells whose intercellular spaces fit together loosely, usually making up the soft parts of plants.

Parthenocarpy   The development of fruit without the production of seeds.

Pasteurization   The process of killing harmful organisms by the application of heat, in wine from 140° to about 180°F. May also be termed "flash" pasteurization when the process is such that a higher temperature is held for only a short time, perhaps less than 30 seconds. The wine is then rapidly cooled to room temperature. Either is generally looked upon as a process that reduces wine quality.

Pedicel   The stem of one berry or flower in a grape cluster.

Peduncle   The cluster stem from the point of shoot attachment to the first lateral or shoulder branch on the cluster.

Perfect flower   A grape flower having viable pistil and stamens.

Perfume   An organoleptic term referring usually to very flowery values of aroma and bouquet in wine.

Periodic table   A table in which the elements are arranged in horizontal rows according to increasing atomic number, so that elements with similar chemical properties are aligned in vertical columns or families. This forms a progressive change in chemical properties from one end of the table to the other.

Permanent wilting percent   The moisture percentage in a soil at which a plant will wilt and not recover when exposed to a relative humidity atmosphere of 100 percent.

Petiolar sinus   An opening in the margin of a leaf at the petiole junction.

Petiole   A stalk that attaches a leaf blade to a stem.

Petillant   French for "slightly sparkling"; *crémant* is another term used in France to describe such wines.

pH   The entire scale from the strongest acidity at pH 1 to neutrality at pH 7, to the strongest alkalinity at pH 14.

Phloem   The tissue in a plant comprised of parenchyma and sieve tubes that translocate nutritional materials produced by the leaves.

Photosynthesis   The natural plant process of producing carbohydrates from carbon dioxide and water, by the use of ultraviolet light energy.

Phylloxera   A root louse (some airborne, leaf-galling types also exist), more precisely known as *Phylloxera vastatrix*, which is parasitic to grape vines. The great Phylloxera epidemic of the mid-1800s killed most European vineyards which today exist grafted upon American rootstocks.

Phytotoxic   To cause death or injury to plants.

Pipe   A wine container equal to 2 hogsheads, about 105 Imperial gallons, or 81 U.S. gallons.

Piquant   A French term usually referring to an appropriate level of acidity in a wine.

Pistil   The female portion of a flower, comprised of an ovary, a stigma and a style.

Pistillate flower   An imperfect flower lacking stamens.

Plant propagation   The reproduction of plants by seeds, cuttings, etc.

Plant regulator   Non-nutritional substances that modify plant processes such as auxins, cytokinins, ethylene, gibberellins, etc.

Plaque A visible group of mycelium mold strands on a plane, generally shaped in a fan.

Polarization (light) Polarization orients light rays, which are normally scattered in random directions, so that they all travel along a plane in the same direction. Many asymmetric organic compounds rotate the plane of polarized light either clockwise or counterclockwise to some characteristic angle when viewed in a polarimeter.

Pollen Male gametophytes and microspores produced by the anther of a flower stamen.

Pollination Pollen transferred from the anther to the pistil stigma.

Pomace The pressed seeds and skins that remain in the press after the juice or wine has been removed.

Pop wine Usually refers to light wines that have been infused with fruit flavors to produce a "soda pop" type of wine.

Porosity In soils, the ratio of the volume of space filled with water and air to the total volume of the soil, water and air; in filtration, the size limitation of materials that may pass through a medium.

Port A sweet red dessert wine made famous by the Portuguese.

Pourriture noble French for "noble mold", or *Botrytis cinerea.*

Ppm Parts per million—an expression of the number of units existing per a total of one million units; the same as one milligram per liter.

Press wine The wine that is extracted from a fermented must after the "free-run" has been collected; generally considered somewhat inferior to the "free-run".

Pressure Force per unit area, usually expressed in pounds per square inch.

Pressure, back The pressure encountered on the return side of a system.

Pressure, working The pressure which overcomes the resistance of the working device.

Primordia Rudimentary leaves, vegetative buds or floral buds.

Proof The measure of alcoholic strength usually used for distilled spirits, seldom referenced for wine; one degree of proof is approximately equal to ½ of one percent by volume.

Prophylls Bud scales or small scaled leaves that emerge from primary and secondary buds prior to foliage leaves.

Proton A particle with a positive charge that is of about the same mass as a hydrogen atom and approximately 1800 times heavier than an electron. Protons, along with neutrons, comprise atomic nuclei.

Pubescent Hairy, with hair-like growth.

Pump A device which converts mechanical force and motion into hydraulic or fluid force.

Pump, non-positive displacement A pump which does not necessarily discharge the same amount of fluid per cycle, such as a centrifugal pump.

Pump, positive displacement A pump which discharges the same quantity of fluid per cycle, such as a piston pump.

Puncheon A British wine container measure equal to 70 Imperial gallons or 56 U.S. gallons.

Punt The indentation in the bottom of a wine bottle. It was originally intended to provide added strength to the container, although in modern times it is being dispensed with as a result of technological advancement and economics of production.

Qualitätswein German for "quality wine" and short for *Qualitätswein bestimmter Anbaugebiete* (often designated as "Q. b. A."). Signifies that the wine was grown and produced in one of eleven specific German wine districts, as disclosed on the label.

Qualitätswein mit prädikat German for "Quality Wine with Special Properties". The same as "Qualitätswein" except that no sugar is allowed in the production processes. The Prädikat categories, in ascending order of regard are *Kabinett, Spätlese, Auslese, Beerenauslese,* and *Trockenbeerenauslese.*

Quantum A radiant energy bundle emitted as a unit by a resonator, vibrating with a frequency v. The "quantum" energy is hv, where h is Planck's constant.

Quarantine The legal action taken, usually to prevent the infestation or spread of disease or pests, by limiting or prohibiting the distribution, sale or shipment of plant propagation materials.

Quiescence A time of slow growth progress due to unfavorable environmental conditions.

Quinta Portugese for "vineyard estate". Generally used in reference to the origins of port wines.

Racking The transferring of wines from one vessel to another, usually to separate the lees.

Raya A low grade of Spanish sherry that is usually distilled into brandy or sold for very modest prices.

Récolte French for "vintage season" or "harvest".

Red table wine Wines, usually made by including the skins during fermentation in order to leach out red color pigments (although some are made by heating the must before pressing and fermentation). They are predominantly dry, and are typified by the French wines of Medoc and Pomerol in Bordeaux and the wines of Burgundy made from the cultivars Gamay and Pinot Noir.

Reduction A decrease in oxygen content or increase in the hydrogen content of a substance; the decrease in positive valence or the increase in negative valence of an element.

Refraction   The deflection of a light ray when passed obliquely from one medium to another which changes the velocity of the light ray.

Regional   Wines originating from a general area rather than from a specific village or vineyard, generally denoting a lesser quality.

Rehoboam   A large wine bottle generally found in Champagne, and the equivalent of about six normal sized bottles.

Remuage   French for "riddling". The process of shaking sparkling wine bottles being stored in an inverted postion so that the yeast sediment from tirage precipitates to the neck of the bottle. This allows degorgement which follows the *remuage.*

Reserva   Spanish for "reserve". Wines selected for special aging and/or usage.

Resolution   The quality or extent to which small details of a specimen are visible in the microscope.

Respiration   The "breathing" of plants during which nutritional elements are metabolized, oxygen is absorbed and carbon dioxide is released.

Rest   A dormant period. The same as **quiescence** except that growth will not resume when positive environmental conditions for growth are applied.

Retsina   A Greek wine that has been infused with pine resin, producing a rather turpentine-like bouquet.

Rhine wine   The wines of Germany produced along the Rhine river; also, loosely refers to any white wine made in the German style or from grapes indigenous to Germany.

Rich   An organoleptic term used to describe wines that are heavy-bodied, or with bouquet and flavor values that are abundant and robust.

Riddling   The French **remuage**, or the process of working the sediment in bottle-fermented sparkling wine into the neck of the bottle, usually performed on a rack or table that holds the bottles inverted. The "riddling" method is generally to raise each bottle slightly and, with a quarter-turn, firmly reinsert the bottle back to its resting place so as to jar the sediment loose from the sides of the bottle. After several weeks of thrice-daily "riddling" the sediment spirals downward into the bottle neck.

Ringing   The same as **girdling**.

Ripe   An organoleptic term usually used in reference to wines that have reached a full term of aging or have achieved a proper state of bouquet and flavor development.

Risers   The pipes that support sprinkler heads in sprinkler irrigation systems.

Riserva   Italian for "reserve". See **Reserva**.

Rootstock   The plant propagation material that becomes the lower portion, or root section, of a benchgraft. Usually has properties which reduce susceptibility to disease and pests in the soil than if the scion variety was propagated as a self-rooted cutting.

Rosato   Italian for **rosé**.

Rosé   A generic name for pink table wines that can be dry, near-dry or rather sweet.

Rosette   Leaves appearing to be "bunched" due to short internodes of canes on a vine.

Rosso   Italian for "red".

Rounded   An organoleptic term usually referring to wines with good balance of all characteristics, or a wine that has been blended harmoniously.

Rugose   Uneven, furrowed, or wrinkled.

Sack   Usually refers to a dry, or nearly dry, light amber Spanish sherry, the favorite wine of Shakespeare.

Saline soil   A soil that contains sufficient soluble salts (not necessarily alkaline) to injure a crop.

Salmanazar   The largest commercially available wine bottle. Generally found only in Champagne and the equivalent of twelve normal-sized bottles.

Sapwood   The younger, physiologically active wood, between the vascular cambium and the center of the stem, which translocates water and minerals.

Scion   The plant portion that is budded or grafted on a rootstock or a vine.

Scud   A mold that may develop in wines that are low in alcohol content.

Sec   French for "dry", although the term may be used in Champagne to denote "near-dry" wines.

Secco   Italian for "dry", although the term may be used to describe "near-dry" wines.

Sediment   The precipitated solids from fermentation, clarification, or some other cellar process, that are more commonly referred to as "lees". (However, in bottled wines this precipitation is referred to only as sediment, not lees.)

Seed   A fertilized ovule that contains an embryo yet to germinate.

Sekt   German for **sparkling wine**.

Selective herbicide   A compound that has properties toxic to weeds, but is slightly or non-toxic to crop-producing plants.

Self-pollinated   Pollen transferred to a pistil stigma from stamen anthers of the same, or identical, plant.

Sepal   The outer flower organ; a modified leaf of the calyx.

Serrations   The teeth-like, jagged indentations about the edge or margin of a leaf.

Set   The swollen appearance of a fertilized pistil in the early development of a grape berry.

Sexual reproduction   The fusion of gametes into a zygote that results in the generation of offspring.

Sherry   An English word for the fortified wine of the Jerez region of southern Spain that is characterized by a "nut-like" bouquet and flavor; more loosely used in the description of any wine that has a high aldehyde content purposely generated in order to make a wine of similar type.

Shoot   A stem or cane grown during the course of the current season.

Shouldered cluster   Lateral clusters of grapes extending from the peduncle base of the main cluster.

Siphon   A method of decanting by which a hose is used to transfer wine from a higher elevation to a lower one.

Sod culture   A vineyard management technique by which ground cover is maintained and often mowed during the course of a growing season.

Soft   An organoleptic term usually referring to wines that are low in both acidity and astringency, generally white or rosé table wines; much the same as smooth.

Soil structure   The arrangement of soil particles in profile.

Soil texture   The proportion of particles of clay, sand and silt.

Solera   The Spanish fractional blending system used to age sherry wines. Older wines are only gradually removed from aging casks and transferred in progressive stages, the last stage prior to bottling being the *Solera*.

Solubility   The quantity of solute which can be dissolved in a given amount of solvent at a certain temperature.

Solute   The substance which is dissolved to form a solution.

Solution   A homogeneous mixture resulting from one substance being dissolved in another.

Sour   An organoleptic term often used incorrectly in describing the acidity or astringency of a wine.

Soutirage   French for **racking**.

Sparkling burgundy   An American term describing sparkling red wine.

Sparkling wines   Wines in which natural carbon dioxide gas is captured during secondary fermentation in a closed container. The traditional French method of bottle-fermentation is called the "Methode Champenoise". The bulk, or tank, method is called the "Charmat" process. The most famous example of a sparkling wine is champagne.

Spätlese   German for "late-picked", which refers to grapes left on the vine for several days or weeks after ripening in order to dehydrate some of the natural water from the grape berries. This yields high sugar concentrations in the fruit.

Specific gravity   The ratio of the weight of a given volume of liquid as compared to an equal volume of water at the same temperature.

Specific heat   The quantity of heat necessary to increase a unit weight of substance one degree in temperature at either a constant pressure or a constant volume.

Spectrum   A series of radiant energies in order of wavelength.

Sperm   A male gamete.

Spicy   An organoleptic term referring to bouquet and flavor values that are spice-like on the nasal and palate receptors; the best example would be wine properly made from the cultivar Gewurztraminer which is indigenous to Alsace in France.

Sulfur   A chemical element used in the basal form to combat mildew in the vineyard.

Sulfur dioxide   A compound of sulfur and oxygen (SO_2) used in winemaking as an antioxidant, and to inhibit bacteria and yeast; also used in cleaning procedures as a disinfectant and preservative.

Surface tension   The molecular force common to all liquids which contracts volume into the physical shape having the least surface area.

Sweetness   A human sensation which is not bitter, salty or sour. In wine it is associated with taste sensations resulting from added or residual sugar content. Wines with sweetness levels that are barely detectable may be properly called "near-dry"; sweet wines are obviously so, such as port and sauternes.

Systemic   Affecting the whole system, as when a chemical invades the entire system of a plant causing severe injury or death.

Table wine   In the U.S, any wine below 14% alcohol by volume so as to qualify for the $.17 per gallon tax category, but generally refers to wines consumed at the table with food.

Tannic   An organoleptic term generally used in reference to wines that are astringent or "stemmy" (as grape stems contain astringent flavor values).

Tannic acid   An astringent acid with a "leather" value of taste. Normally added in order to increase the life expectancy of a wine by slowing down the aging process.

Tannins   Special phenolic compounds found in grape stems, seeds and skins which contribute to astringency, particularly in young red wines. Tannins are also introduced in wines from the wood of aging vessels and may lengthen the life of wines due to a slowing of oxidation reactions.

Tart   An organoleptic term used in reference to wines that are high in fixed or total acidity.

Tartaric acid   A natural acid of grapes which is unstable in wine stored at cold temperatures, causing the acid salt potassium bitartrate (cream of tartar) to precipitate.

Taste   The sensations on the human palate that detect sourness, saltiness, bitterness, and sweetness, as well as flavor and viscosity values of wine.

Tastevin   A small tasting cup, the best made of silver, used for tasting wines. Used predominantly in Burgundy.

Tawny   The amber character of red wines that have been exposed to high temperature and/or long aging terms. Tawny port is a good example.

Teinturier   A very heavily pigmented black grape usually used for blending to enhance color intensity.

Temperature inversion   A phenomenon of atmospheric layering whereupon temperatures increase with increased altitude rather than the normal condition of reduced temperature with increased altitude.

Tendril   An appendage of a shoot that will grasp, in a coil-like fashion, any adjacent object to help support the weight of the shoot.

Tensiometer   An instrument used for soil water tension measurements.

Tenuta   Italian for "estate". Usually denotes a wine and/or vineyard estate.

Tirage   The laying of bottles on their sides in piles for secondary fermentation into sparkling wines.

Titer   The concentration, determined by titration, of a dissolved substance in a solution; also, the least amount required to provide a given titration result.

Tolerance   The lawful limit of toxic residue in, or on, an edible substance.

Tomentum   Pubescence, with epidermal hair.

Translocation   Moved from one location to another, as in plants transferring chemicals, nutrients, water, etc.

Transpiration   The loss of water from leaves by evaporation.

Trockenbeerenauslese   German for "dry berry special selection". The separation of only those berries on each bunch of grapes that have been "raisinized" by the "noble mold", *Botrytis cinerea*. The juice is highly concentrated in sugar content and may have a rather caramel-like flavor. Slowly fermented with a very sweet finish, *Trockenbeerenauslese* is one of the most expensive wines made in the world.

True fruit wines   Wines made entirely from fresh fruit or fresh fruit concentrate other than grapes.

Trunk   The main body of a grape vine between the root system and branch appendages.

Turgidity   The distension caused by internal pressure from liquids in a cell.

Ullage   The air space in a wooden aging vessel generated by seepage, evaporation and assimilation of the wine in the pores. This space is usually refilled weekly in order to remove the oxygen.

Valance   The combining power of an element or radical expressed by a number equal to unity, or some multiple of unity, for hydrogen.

Vapor density   At a stated pressure and temperature, the weight of a vapor per unit volume.

Varietal wine   A wine labeled for the variety of grape from which it was predominantly, or entirely, made. Good examples are California Cabernet Sauvignon and New York State Seyval Blanc.

Variety A strain or a cultivar within a group of closely related plants which have different characteristics, yet not such as to require a separate species.

Vat A tub, or some other wine container which is positioned vertically, as opposed to a cask which is horizontal.

Vegetative propagation Plant reproduction utilizing buds and cuttings as opposed to seedlings.

Venation The abnormal formation of veins in a leaf.

Vendange French for "grape harvest" as used in the picking of the grapes and making of new wines.

Vendemnia Italian for "grape harvest". Used in the same way as the French, **vendange**.

Véraison A French term used to describe a point at which grapes start to ripen, becoming soft and starting to change color.

Vermouth A wine that contains aromatic essences derived from herbs, roots, spices, etc. Dry vermouth is generally known as the French type and is white. Sweet vermouth is normally very dark from the addition of caramel coloring and is often referred to as the Italian type.

Vigne French for "vine".

Vignoble French for "vineyard".

Vigorous vines Vines that produce large volumes of foliage.

Vin French for "wine".

Vina Spanish for "vineyard".

Vin de pays French for "country wine", as applied in describing or discussing the local wines of a given village or geographical area.

Vinifera Short for *Vitis vinifera,* the most prolific species of wine grape grown. Also known as the "Old World" vine. Generally considered to be the finest wine-grape species cultivated, the cultivars Chardonnay, Pinot Noir, Cabernet Sauvignon and Johannisberg Riesling are classic examples of *Vitis vinifera.*

Vinification The process of making grapes into wine.

Vino Italian and Spanish for "wine".

Vinosity An organoleptic term that refers to the "vinous", as opposed to "fruity" values of bouquet and flavor.

Vintage A crop of grapes being harvested; the wine from a crop of grapes during a particular year. A "vintage" year describes a good or great growing season resulting in a distinguished wine.

Vintner A winemaker, usually in reference to the ownership and/or management of a winemaking facility.

Virtual image The apparent size of a specimen as it appears in the microscope.

Viscidity   The lack of tannic acid having rendered a wine prematurely aged.

Viscosity   The resistance of a gas or liquid to flow; a measure of shearing stress in relation to velocity.

Viticulture   The art, science and study of grape growing.

Volatile acid   An acid that evaporates.

Volatile oil   An oil that evaporates.

Water berries   Grape berries having suffered a disorder that results in unripened individual berries that seem to contain an extraordinary amount of liquid.

Water-holding capacity   Same as **field capacity**.

Waterlogged   In viticulture, a poor draining soil that is deficient in oxygen for proper root functions; in enology, a wooden wine container that is "punky" or rotted to the extent that water will soak very deeply into the pores of the vessel.

Water sprouts   Shoots that grow rapidly from latent buds.

Watt   A unit of electric power.

Wein   German for "wine".

Weingut   German for "wine estate" or "vineyard and winery estate".

Weinkeller   German for "wine cellar".

Wettable powder   A chemical powder that can be suspended in water for spray applications.

White table wines   Wines made dry, or nearly dry, from the juice of either white or red grapes, usually with an alcohol content of less than 14% by volume. The French wines from Chablis typify "vinous", or "flinty" dry whites. "Flowery" white wines are common to Germany, and "fruity" white wines are found in New York State made chiefly from the Niagara grape cultivar. Sauternes from Bordeaux in France is legally taxed as white table wine but is so sweet as to be used primarily for dessert courses.

Wild vine   An uncultivated vine existing in the wild.

Wine   A beverage produced from the fermentation of the juice or must derived from grapes.

Winegrowing   The art, science and study of growing grapes and making wine.

Winery   The building, cave, room, vault, etc. in which grapes are made into wine.

Wing   A large shoulder cluster attached to the main cluster of grapes.

Winzer   German for "vintner" or "winegrower".

Winzergenossenschaft   German for "winegrowers cooperative association".

Winzerverein   German for "winegrowers cooperative".

Woody   An organoleptic term that refers to wines that have a bouquet and flavor value characteristic of wet wood, usually as a result of overaging.

Working distance   The distance between the top of the cover slip and the front of the microscope objective.

Xylem   Woody tissue that conducts minerals, nutrients and water in plants.

Yeasts   One-celled, principally asexually reproducing fungi, about 4 millimicrons in diameter. They secrete the enzyme *zymase* which converts sugar(s) into ethyl alcohol, carbon dioxide gas and energy.

Yeasty   An organoleptic term used in a negative sense when describing the characteristic bouquet and flavor values of yeast in wines, usually due to excessive contact with the lees; used in a positive sense when evaluating champagne or other bottle-fermented sparkling wines due to a good dissolution of yeast protoplasm during tirage aging.

Zygote   The cell resulting from the fusion of female and male gametes.

Zymase   Enzyme produced by wine yeasts. See **yeasts**.

Bibliography

ADAMS, L.D. 1978. The Wines of America, 2nd Edition. McGraw-Hill, New York.

ALLEN, H.W. 1963. The Wines of Portugal. McGraw-Hill, New York.

AMERICAN OPTICAL CORPORATION. 1977. Series One-Ten MICRO-STAR Advanced Laboratory Microscopes, Reference Manual. Scientific Instrument Division, Buffalo, NY.

AMERINE, M.A. *et al.* 1980. The Technology of Wine Making, 4th Edition. AVI Publishing Co., Westport, Conn.

AMERINE, M.A. and OUGH, C.S. 1974. Wine and Must Analysis. Wiley-Interscience, New York.

AMERINE, M.A. and ROESSLER, E.B. 1976. Wines: Their Sensory Evaluation. Freeman, San Francisco.

AMERINE, M.A. and SINGLETON, V.L. 1965. Wine. University of California Press, Berkeley and Los Angeles.

AUSTIN, C. 1968. The Science of Wine. American Elsevier Publishing, New York.

BLANC, G.H. *et al.* 1970. The Great Book of Wine. Edita Lausanne, Switzerland.

BROWN, L.R. and TISCHER, R.G. 1967. Elementary Microbiology. Department of Microbiology, Mississippi State University, Mississippi State.

CHABANON, R.L. and PUISAIS, J. 1974. Initiation into the Art of Wine Tasting. Interpublish, Madison, Wisconsin.

CHAMBERLAIN, B.P. 1931. The Making of Palatable Table Wines. Private Printing.

CHROMAN, N. 1973. The Treasury of American Wines. Rutledge-Crown, New York.

COLES PUBLISHING COMPANY LIMITED. 1975. Handy Metric Conversion Tables. Toronto.

CONN, H.J., DARROW, M.A. and EMMEL, V.M. 1960. Staining Procedures. Williams and Wilkins, Baltimore.

COOK, A.H.   1964.   The Chemistry and Biology of Yeasts.   Academic Press, New York.

EMERSON, E.R.   1908.   Beverages, Past and Present.   New York and London.

FADIMAN, C. and AARON, S.   1975.   The Joys of Wine.   Henry N. Abrams, New York.

FISHER, M.F.K.   1962.   The Story of Wine in California.   University of California Press, Berkeley and Los Angeles.

GROSSMAN, H.J.   1964.   Grossman's Guide to Wines, Spirits, and Beers, 4th Edition.   Scribner's, New York.

GURR, E.   1971.   Synthetic Dyes.   Academic Press, New York.

HEREFORD, K.T.   1979.   Sensory Analysis and Evaluation: Wines, An Advanced Course Of Study.   Virginia Polytechnic Institute and State University, Blacksburg.

HOYNAK, P.X. and BOLLENBACK, G.N.   1966.   This Is Liquid Sugar, 2nd Edition.   Corn Products Company, Yonkers, NY.

HYAMS, E.   1965.   Dionysus.   Macmillan, New York.

JACQUELIN, L. and POULAIN, R.   1966.   The Wines and Vineyards of France, 3rd Edition.   Paul Hamlyn, London.

JOHNSON, H.   1969.   Wine, 4th Edition.   Simon and Schuster, New York.

JOHNSON, H.   1971.   The World Atlas of Wine.   Simon and Schuster, New York.

JOSLYN, M.A. and AMERINE, M.A.   1964.   Dessert, Appetizer and Related Flavored Wines, University of California, Division of Agricultural Sciences.

LAYTON, T.A.   1961.   Wines of Italy.   Harper, London.

LICHINE, A.   1969.   Wines of France, 5th Edition.   Alfred A. Knopf, New York.

LOUBÉRE, L.A.   1978.   The Red and the White.   State University of New York Press, Albany.

LUCIA, S.P.   1971.   Wine and Your Well-Being.   Popular Library, New York.

MASSEL, A.   1969.   Applied Wine Chemistry and Technology.   Heidelberg Publishers, London.

MILLIGAN, D.   1974.   All Color Book of Wine.   Octopus, London.

RAY, C.   1966.   The Wines of Italy.   McGraw-Hill, New York.

READ, J.   1973.   The Wines of Spain and Portugal.   Faber and Faber, London.

SCHOENMAN, T.   1979.   The Father of California Wine: Agostin Haraszthy.   Capra Press, Santa Barbara.

SCHOONMAKER, F. and MARVEL, T.   1941.   American Wines.   Duell, Sloan and Pearce, New York.

SCHULTZ, H.W.   1967.   The Chemistry and Physiology of Flavors.   AVI Publishing Co., Westport, Conn.

SIMON, A.L.   1962.   Champagne.   McGraw-Hill, New York.

SIMON, A.L.   1962.   Wines of the World.   McGraw-Hill, New York.

SIMON, A.L.   1968.   The Noble Grapes and the Great Wines of France.   Mc-Graw-Hill, New York.

SIMON, A.L. and HALLGARTEN, S.F.   1963.   The Great Wines of Germany. McGraw-Hill, New York.

TAYLOR, W.S. and VINE, R.P.   1968.   Home Winemaker's Handbook. Harper and Row, New York.

TOPOLOS, M., DOPSON, B. and CALDEWEY, J.   1977.   Napa Valley.   Vintage Image, St. Helena, Calif.

WAGNER, P.   1963.   American Wines and Wine-making, 5th Edition.   Alfred A. Knopf, New York.

WAGNER, P.   1972.   A Wine-Grower's Guide, 2nd Edition.   Alfred A. Knopf, New York.

WARNER, C.K.   1960.   The Winegrowers of France and the Government since 1875.   Columbia University Press, New York.

WASSERMAN, S.   1977.   The Wines of the Cotes du Rhone.   Stein and Day, New York.

WAUGH, A.   1959.   In Praise of Wine.   William Sloane, New York.

WEAVER, R.J.   1976.   Grape Growing.   Wiley-Interscience, New York.

WEBB, A.D.   1974.   Chemistry of Winemaking.   American Chemical Society, Washington.

WINKLER, A.J., COOK, J.A, KLIEWER, W.M. and LIDER, L.A.   1974. General Viticulture.   University of California Press, Berkeley and Los Angeles.

YOUNGER, W.   1966.   Gods, Men and Wine.   World Publishing, Cleveland.

Index

Other AVI Books

TECHNOLOGY OF WINE MAKING
   Fourth Edition   *Amerine, Berg, Kunkee, Ough, Singleton
   and Webb*
TROPICAL & SUBTROPICAL FRUITS
   Nagy and Shaw
YEAST TECHNOLOGY
   Reed and Peppler

NOTES

NOTES